T0331900

LIFE IN SPACE

LIFE IN SPACE

NASA Life Sciences Research during the Late Twentieth Century

MAURA PHILLIPS MACKOWSKI

University of Florida Press

Gainesville

27 26 25 24 23 22   6 5 4 3 2 1

Library of Congress Cataloging-in-Publication Data
Names: Mackowski, Maura Phillips, author.
Title: Life in space : NASA life sciences research during the late
  twentieth century / Maura Phillips Mackowski.
Description: 1. | Gainesville : University of Florida Press, 2022. |
  Includes bibliographical references and index.
Identifiers: LCCN 2021047909 (print) | LCCN 2021047910 (ebook) | ISBN
  9781683402602 (hardback) | ISBN 9781683403241 (pdf) | ISBN 9781683403128
  (ebook)
Subjects: LCSH: United States. National Aeronautics and Space
  Administration. | Life sciences—Research. | Manned space
  flight—History. | Astronautics—United States. | Space
  biology—Research.
Classification: LCC QH327 .M33 2022 (print) | LCC QH327 (ebook) | DDC
  570.72—dc23/eng/20211015
LC record available at https://lccn.loc.gov/2021047909
LC ebook record available at https://lccn.loc.gov/2021047910

University of Florida Press
2046 NE Waldo Road
Suite 2100
Gainesville, FL 32609
http://upress.ufl.edu

UF PRESS

UNIVERSITY
OF FLORIDA

To my parents

Contents

Abbreviations and Acronyms

AA	Associate Administrator
ACFUSSP/	
AC-FUSSP	Advisory Committee on the Future of the US Space Program
ACRV	assured crew return vehicle
AEM	Animal Enclosure Module
AIBS	American Institute of Biological Sciences
AMAC	Aerospace Medicine Advisory Committee
APL	Applied Physics Laboratory
APO	Advanced Programs Office
ARC	Ames Research Center
ASGSB	American Society for Gravitational and Space Biology
ASI	Agenzia Spaziale Italiana
ASTP	Apollo-Soyuz Test Project
AURA	Association of Universities for Research in Astronomy
BCPR	Bioastronautics Critical Path Roadmap
BNL	Brookhaven National Laboratory
BPR	biological and physical research; also baseline performance review
BPRAC	Biological and Physical Research Advisory Committee
BRYNTRN	BaRYoN TRaNsport
CAM	Computerized Anatomical Man; also Centrifuge Accommodation Module
CAMP	Cooperative Agreement Management Plan

CAN	Cooperative Agreement Notice
CAPS	Crew Altitude Protection System
CELSS	Controlled Ecological Life Support System
CERV	Crew Emergency Rescue (or Return) Vehicle
CFES	Controlled Flow Electrophoresis in Space
CFP	call for proposals; also Centrifuge Facility Project
CNES	Centre National d'Études Spatiales (National Center for Space Studies)
CPCG	Commercial Protein Crystal Growth
CR	centrifuge rotor; also Congressional Record
CRS	Congressional Research Service
CRV	Contingency (or Crew) Return Vehicle
CSA	Canadian Space Agency; also cross-sectional area; also Commercial Space Act
D-1, D-2	Deutschland-1 and -2
DARA	Deutsche Agentur für Raumfahrtangelegenheiten
DASA	Deutsche Aerospace Aktiengesellschaft
DLR	Deutsches Zentrum für Luft- und Raumfahrt e.V.
DFRC	Dryden Flight Research Center
DSLS	Division of Space Life Sciences
DSO	detailed supplementary objective
EDOMP	Extended Duration Orbiter Medical Project
ELV	expendable launch vehicle
EOS	Electrophoresis Operations in Space
ESA	European Space Agency
ET	external tank
EVA	extravehicular activity
FASEB	Federation of American Societies for Experimental Biology
FCO	Flight Crew Operations
FEL	first element launch
FFRDC	Federally Funded Research and Development Center
FOIA	Freedom of Information Act
FY	fiscal year

GAO	General Accounting Office / Government Accountability Office
GAS	Get Away Special
GBX	glovebox
GCR	galactic cosmic ray
GFE	government-furnished equipment
GPS	global positioning system
GSFC	Goddard Space Flight Center
HEDS	Human Exploration and Development of Space
HZETRN	HZE TRaNsport
IBMP	Institute of Biomedical Problems
IG	Inspector General
IGA	Intergovernmental Agreement
IMBP	Institute of Medical and Biological Problems
IML	International Microgravity Laboratory
IRB	institutional review board
ISF	Industrial Space Facility
ISS	International Space Station; also Institute for Space Studies
ISSA	International Space Station Alpha
ISSRI	ISS Research Institute
IVA	intravehicular activity
JAXA	Japan Aerospace Exploration Agency
JEA	joint endeavor agreement
JEM	Japanese Experiment Module
JPL	Jet Propulsion Laboratory
JSC	Johnson Space Center
JWG	joint working group
KSC	Kennedy Space Center
LaRC	Langley Research Center
LBNL	Lawrence Berkeley National Laboratory
LBNP	Lower Body Negative Pressure
LDEF	Long Duration Exposure Facility
LEO	low-Earth orbit

LeRC	Lewis Research Center
LES	Launch and Entry Suit (or Launch/Entry Suit and Launch Entry Suit)
LET	linear energy transfer
LLC	limited liability company
LLNL	Lawrence Livermore National Laboratory
LMEPO	Lunar and Mars Exploration Program Office
LSAC	Life Sciences Advisory Committee
LSAH	Longitudinal Study of Astronaut Health
LSDA	Life Sciences Data Archive
LSG	Life Sciences Glovebox
MDAC	McDonnell Douglas Astronautics Company
MMU	Manned Maneuvering Unit
MOU	memorandum of understanding
MS	Mission Specialist
MSAD	Microgravity Science and Applications Division
MSFC	Marshall Space Flight Center
NACA	National Advisory Committee for Aeronautics
NAS	National Academy of Sciences
NASDA	National Space Development Agency of Japan
NASP	National Aero Space Plane
NGO	nongovernmental organization
NIA	National Institute on Aging
NIH	National Institutes of Health
NOAA	National Oceanographic and Atmospheric Administration
NPO	Nauchnaya Produktsiya Organizatsiya
NRA	NASA research agreement
NRC	National Research Council; also Nuclear Regulatory Commission; also National Research Council of Canada
NRL	National Research Laboratory
NSBRI	National Space Biomedical Research Institute
NSC	National Security Council; National Space Council
NSCORT	NASA Specialized Center of Research and Training

NSF	National Science Foundation
OA	Office of Applications
OAST	Office of Advanced Science and Technology
OBPR	Office of Biological and Physical Research
OCP	Office of Commercial Programs
OI	orthostatic intolerance; also operational instrumentation
OLMSA	Office of Life and Microgravity Sciences and Applications
OMB	Office of Management and Budget
ORINS	Oak Ridge Institute of Nuclear Studies
OSS	Office of Space Science
OSSA	Office of Space Science Applications
OSTP	Office of Science and Technology Policy
OTA	Office of Technology Assessment
PAO	Public Affairs Office/officer
PCG	Protein Crystal Growth
PETA	People for the Ethical Treatment of Animals
PGU	Plant Growth Unit
PI	principal investigator
PS	Payload Specialist
RAHF	Research Animal Holding Facility
R&D	research and development
ReMAP	Research Maximization and Prioritization
RFP	request for proposal
RSA	Russian Space Agency
RTF	Return to Flight
SAA	South Atlantic Anomaly
SAIC	Science Applications International Corporation
SAM	School of Aerospace Medicine; also shuttle activation monitor
SAS	Space Adaptation Syndrome
SBIR	Small Business Innovation Research (Program)
SDI	Space Defense Initiative

SEI	Space Exploration Initiative
SETI	search for extraterrestrial intelligence
SLS	space life sciences
SMS	Space Motion Sickness
SOMS	shuttle orbiter medical system
SOW	statement of work
SPE	solar proton event
SPO	Shuttle Program Office
SR&QA	Safety, Reliability and Quality Assurance Office
SRB	solid rocket booster
SSB	Space Studies Board
SSIP	Space Shuttle Student Involvement Program
SSPO	Space Station Project Office
STS	space transportation system
STScI	Space Telescope Science Institute
SWG	science working group
TBD	to be determined
TDRSS	Tracking and Data Relay Satellite System
TM	transport, modified; also Technical Memorandum
UAB	University of Alabama Birmingham
USDA	US Department of Agriculture
USRA	Universities Space Research Association
UTMB	University of Texas Medical Branch
WETF	weightless environment training facility
WHO	World Health Organization
WSTF	White Sands Test Facility
ZBR	zero-base review

Acknowledgments

This book took much longer to complete than anyone expected and consequently the number of people whose assistance must be acknowledged is enormous.

To begin with, I must thank Steven Dick, then NASA's Chief Historian, for awarding me the contract that started this project and for Historian Stephen Garber for shepherding it for many years. Others at the History Division's Headquarters in Washington, D.C. who were unfailingly helpful include some who have since retired or moved on—Chief Archivist Jane Odom and Senior Archivist Colin Fries, Chief Historian Bill Barry, Editor Yvette Smith—and Program Support Specialist Nadine Andreassen, who continues to hold down the fort. The history staff at Ames Research Center and Johnson Space Center also responded graciously and quickly to my requests and the JSC Oral History Project was an invaluable resource. Thanks also to NASA's Office of the Inspector General for its quick and helpful response to my Freedom of Information Act requests.

Dozens of NASA employees and former employees at Headquarters and various Centers consented to meet with me and in many cases let me tape interviews in which they shared their experiences, thoughts, and feelings about the projects they worked on and about NASA as a whole. In particular I must thank Guy Fogleman, the late John Charles, Emily Morey-Holton, Larry Toups, Nathan Moore, Ronald J. White, David Liskowsky, James Logan, Kenneth Souza, Lynn Harper, William E. Berry, Louis Ostrach, David Tomko, Keith Cowing, Kathie Olsen, Ann Carlson, Charles Sawin, Terri Lomax and Arnauld Nicogossian. A special thanks goes to Lawrence Chambers who left an impressive and very detailed collection of records from his years as a manager on space station, centrifuge, and Cosmos/Bion.

Archivists are a historian's staunchest ally when it comes to maintaining and organizing key records and making them accessible to researchers. My very deep gratitude goes to the staff at University of Houston–Clear Lake's Alfred R. Neumann Library, repository of the JSC History Collection within the UHCL Archives & Special Collections; to Regis University's Dayton Memorial Library, which holds the Richard H. Truly U.S. Space Program Collection; the Library of Congress, which houses the Thomas O. Paine Papers; the National Archives and Records Administration (NARA) in College Park, MD, home to the Records of NASA Administrator Daniel S. Goldin; to the National Air and Space Museum for allowing me access to the David M. Brown Collection and to their videohistory collection on Soviet space medicine. The Presidential Libraries, part of NARA, are wonderful places to research but I was grateful that many holdings at the Jimmy Carter, Ronald Reagan, and George H. W. Bush had been digitized by the time I began this project.

I must certainly thank the half-dozen blind reviewers who scrutinized the very large first draft of this manuscript for the NASA History Department and made so many helpful suggestions, and those who did the same with this shorter version for the University of Florida Press. There were many people involved with this book at UFP, but I wish to especially thank editor Sian Hunter, who was its foremost advocate, and Mary Puckett, who kept it organized.

On a personal level, I want to thank aerospace historians Andrew Butrica and Joseph Tatarewicz who urged me to commence this project. My brother John L. Phillips, his wife Laura Phillips and their children Tim and Alli put me up many times at their home in Houston and John gave me a guided tour of the astronaut training areas at JSC and introduced me to colleagues there. I am beyond grateful for their hospitality and support. My family at home—husband Michael and children Katie, Sarah, and Ben Mackowski—served as research assistants, technology experts, transcribers, typists, photocopiers, scanners, chart makers, and more. Thank you so much.

Introduction

The language of the Space Act of 1958 signed by President Dwight Eisenhower on July 29, declaring the National Aeronautics and Space Administration's purpose, intent, and responsibilities, made no explicit reference to life sciences. Other than broad mention of expanding scientific knowledge, the only allusion to it was the call for "improvement of the . . . safety . . . of aeronautical and space vehicles" and the "development and operation of vehicles capable of carrying instruments, equipment, supplies, and living organisms through space." Machinery was given priority, and the connection with life sciences focused on crew safety. In March, the President's Science Advisory Committee (PSAC) had noted "four factors which give importance, urgency, and inevitability to the advancement of space technology," beginning with exploration, defense, and national prestige. Fourth was "new opportunities for scientific observation and experiment which will add to our knowledge of the earth, the solar system, and the universe." The PSAC-listed scientific goals were chronological—"early," "later," and "still later." In the "early" category, "space physiology" took last place on a list of six science and technology objectives. Among the "later" aims, "biology" was third out of six, and "human flight in earth orbit" was at the end of the list. In the "still later" classification, the very last goal was "Human Lunar Exploration and Return." In its conclusion, a fourth era—"and much later still"—listed a single objective: "Human Planetary Exploration."[1]

Space life sciences, as carried out by (or at the behest of) the National Aeronautics and Space Administration (NASA), came to encompass research, planning, and operational activities in medicine, human safety and habitation, plant and animal biology, commercial biomaterials processing, and exobiology (the study of life in space or on other planetary bodies).[2] Motives and expectations were of three different natures: philosophical/intellectual (extraplanetary origins of life on Earth), commercial (refining pharmaceuticals), and practical (crew health and safety). With the exception of crew issues they were future-oriented, with little or no payoff in the near term—but the bill was due immediately. They were also altruistic, meaning that they were beneficial to someone but not necessarily for the person or agency taking the risk or spending the money. It was a tricky prospect to sell voters, partners abroad, and politicians on hoped-for (but not promised) social and economic benefits in the future that they must pay for today—and every day until the benefit was realized or the effort failed. Even with crew safety, it takes much more than a flag on the Moon or science-fiction visions to motivate a nation to pay (and pay) for the same hopes when some goals have been only partially realized in their lifetime and others are just as distant as when the National Aeronautics and Space Act of 1958 was passed.

Space life sciences had to struggle for an acknowledged and appreciated place at the Agency's table, principally because NASA was formed purposely as an evolution of a predecessor engineering research agency, the National Advisory Committee on Aeronautics (NACA). That in turn had been a creation of early twentieth-century US progressivism and "airmindedness," corralling the brightest engineers to solve the hardest and most specialized problems that were then blocking aeronautical progress in the nation that was the birthplace of the airplane.[3] NACA had been extremely successful, helping the United States to recapture the lead in aircraft design, production, and flight. However, it remained a research and advisory group.

In an essay for the NASA history series *Exploring the Unknown*, former Director of NASA's Life Sciences Division (1993–2000) Joan Vernikos wrote that the decision in 1961 to split life sciences into human versus nonhuman areas of emphasis solidified the dichotomy of life sciences versus engineering/physical science. It furthermore caused an ongoing (and enduring) lack of support from scientists *outside* the space agency.

Thus began the long-running saga of separating various components of the life sciences [among] different organizations, demonstrating the lack of understanding between engineers and medical practitioners on the one hand and of space scientists on the other that has haunted life sciences in NASA to this day. Not least of the

problems was the resulting inability to consolidate life sciences efforts long enough to grow and maintain an adequate life sciences
program budget. It was not in the interest of other factions within
NASA, who had been trying to get a foothold for their own efforts, to see this new entity get established, grow, and prosper,
competing for program dollars. Space life sciences also received
no help from the broader life sciences community . . . [which] was
skeptical of the value of going into space to advance knowledge
or develop new health applications. Funding, predominantly from
the [National Institutes of Health] and [National Science Founda-
tion], was plentiful or, at least, adequate, whereas the paucity and
instability of its budget resulted in the deserved image that NASA
was unreliable as a source of life sciences research funding.[4]

NASA and the Life Sciences, 1980–2004

This book is roughly chronological in organization, though larger themes
may result in some things being slightly out of sequence. *The Human
Factor*, a mid-1980s publication in the NASA History Series, covered life
sciences with an emphasis on medicine and human factors, primarily
within the Agency, from NASA's inception to anticipation of the new
space shuttle, 1958–1980.[5] NASA-sponsored histories of the Agency's
exobiology work, primarily by Chief Historian Steven Dick, allowed
this book, which began life in 2005 as NASA Contract #NNH05CC40C,
to keep that very different field out of the mix. Medical privacy laws
put much of that topic beyond reach. This is not a history of the space
shuttle program or of any space station program. The intent here is to
document and explain the accomplishments and ever-changing status
of NASA's traditional life sciences endeavors from 1980 to the end of
2004, acknowledging specific events, people, challenges, and outcomes.
At times the focus is more on outside agents than NASA's internal workings because the space shuttle and space station programs reframed human space exploration and broadened its scope immensely as corporations, academia, other government agencies, and foreign space agencies
became an integral part of US activities.[6] The time span may seem
awkward, as institutional histories typically examine a major program,
war, event, or administration. In this instance, 20 years (1980–2000),
although ending with the millennium, was insufficient. It ignored the
Columbia disaster, situated the International Space Station (ISS) in orbit

but without so much as a robot arm or Destiny lab, and disregarded all politics after the Clinton-Gore-Goldin years and governmental changes after 9/11. The reorganization of the Office of Life and Microgravity Sciences and Applications (OLMSA, also known as Code U) into its own enterprise, the Office of Biological and Physical Research (OBPR), took place in 2000.[7] However, President George W. Bush's January 2004 announcement of a new Vision for Space Exploration (VSE) essentially put the brakes on most of the existing life sciences research, shifted organizational boundaries significantly, and closed down some lines of inquiry. Thus 2004 became the final calendar year covered in this volume, but with glimpses of 2005 and beyond.

Chapter 1 begins with preparations and goals for the first shuttle launch and goes through the first serious setback, the *Challenger* accident in January 1986. Hopes had soared because the maiden flight marked the permanent return of US astronauts to space, but also because NASA had grossly overpromised what the new vehicle could do. Per Space Act mandate, but also as an attempt at sparking grassroots backing, NASA solicited the participation of high school and college students in designing science experiments to fly on board the shuttle. It courted industry and subsequently the first corporate astronaut flew three times, seeking to develop a marketable drug in orbit.[8] The Agency promised to put the first teacher and the first journalist in space to increase understanding of and support for its work among the public. Originally intended to transport humans, cargos, and experiments between Earth and a space station, the shuttle instead wound up being itself the destination when tight budgets delayed construction of a station until 1998, about halfway through the shuttle program's expected lifetime.

Chapter 2 focuses primarily on astronaut diversity, then discusses environmental issues in space as engineers and scientists worked to keep astronauts healthy and productive on the job, which meant close scrutiny given to clothing, habitats, and the work environment, including life-support systems. Human physiology had always been part of the experiment of space flight, and with the shuttle program the NASA astronaut corps began to include individuals who were not middle-aged, white, male US military pilots. Women, minorities, the relatively untrained, and the elderly drastically altered the sample population available and changed the questions asked, the methodology, the results, the goals, and the Earth applications of both in-flight and ground-based biomedical research forever.[9]

The causes and outcomes of the *Challenger* accident in 1986 have been written about extensively elsewhere and so are not the focus of Chapter 3. However, it and the increasing emphasis in the late 1980s on the anticipated space station Freedom underscored the need for safety and life support systems to be designed into vehicles and habitats, including escape mechanisms and rescue vehicles. This also included planning for medical operations aboard the station and new thinking about ways to put experiments in orbit when no shuttle was available.

Chapter 4 discusses how outside agencies questioned NASA's underlying peer review system in the 1990s, charging that the system was inadequate for selecting the best experiments to fund and that data obtained was insufficient for analysis and inadequately disseminated. The respect and support of outside scientists would be crucial if the planned station centrifuge, on which fundamental life scientists were pinning great hopes, were to achieve credible results. This chapter also focuses on some specific research projects in low-Earth orbit: Spacelab and Extended Duration Orbiter missions. These examined physiological changes due to weightlessness, such as muscle degradation and orthostatic intolerance, and countermeasures to the space environment, and included Neurolab, a Spacelab mission dedicated to brain research.

Work on Earth is the focus of Chapter 5: ground-based research at the Agency and in academia, educational programs, and collaborations. The Clinton administration focused its efforts on the domestic economy, giving rise to the "faster-better-cheaper" years of Administrator Daniel Goldin.[10] Creating consortia of foreign and domestic university and industry members to identify and find solutions for specific medical stumbling blocks was one strategy. Johnson Space Center (JSC) in Houston was lead Center in that effort, supported by the National Space Biomedical Research Institute (NSBRI), Universities Space Research Association (USRA), and the Bioastronautics Critical Path Roadmap strategy for related issues, specifically risk assessment and safety.[11]

Chapter 6 describes the approach NASA took to devising strategies to protect space inhabitants. With the goal of long-term human habitation in mind, be it the planetary surface habitats of the Space Exploration Initiative (SEI) in the early 1990s or the new ISS at the end of the decade, radiation received greater scrutiny. Radiation research was a textbook example of interagency collaboration, and in the 1990s and early 2000s the tools of study became more sophisticated even as knowledge about the real risks increased.

International and domestic politics continually redirected the space station's mission and design but terminated the 1990s studies of lunar and Martian habitats, as discussed in Chapter 7. Early conjecture had been for a station housing a hundred personnel with different backgrounds and experience. This changed to more practical models that still drew on NASA's engineering expertise to design and build something safe, habitable, and with technology integrated sufficiently to minimize the number of humans needed. Keeping astronauts healthy also meant reconsidering old ideas of artificial gravity, based now on four decades of learning about the medical effects of weightlessness.

The end of the Cold War is the topic of Chapter 8. It was pivotal for the US space program in terms of international relations, and not just with Russia. Sometimes the tool of Washington diplomats, NASA was often a ready participant in, even an instigator of, international collaboration, including a 1987 pact that put US biological experiments on board a second series of Soviet Bion satellites. In 1994 Russian cosmonauts flew aboard the shuttle, and beginning the following year NASA astronauts undertook missions to the Russian space station, Mir. This exchange necessitated the sharing of medical data, cross-training, and learning to work with the only real peer NASA had.

International cooperation loomed even larger in the late 1990s with the integration of foreign partners—and their astronauts, equipment, and modules—into the ISS. Chapter 9 introduces some of NASA's newer partners in space. It also touches on the ideas for moving NASA life science functions to nongovernmental organizations and describes how NASA gave up on the centrifuge. Diminished by both budget curtailment and turf wars, a spartan ISS became a reality—something in between the early grandiosity and the free-flyers the Agency had been concerned it might have to settle for.

Chapter 10 reveals that the period from 1980 to 2004 had actually been a "bubble" for life sciences at NASA. It burst with the Vision for Space Exploration (VSE) in early 2004, when domestic and international events had siphoned away money and NASA was redirected into something more like an Apollo mode, focusing nearly all its resources and energy on vehicles that could return humans to the Moon and replace the shuttle. Administrators voiced hope that NASA had already learned enough about survival in space to safely carry out future missions with little, if any, further life sciences preparation. Should additional research be needed, maybe NASA could reassemble its earlier teams and pick up where it had left off. These tensions and the perennial-stepchild status of life sciences research would shape the future of the US space agency.

Doing life sciences research in space during the period under study here always presented both a formidable engineering challenge for the Agency and an unrivaled scientific opportunity for investigators. Joan Vernikos stressed the differences between how the "operations" and the "pure research" mindsets functioned within the engineering organization that was and is NASA.

> Life sciences has usually been subordinate to engineering agency priorities [and] has constantly had to justify to engineers the need for an integrated ground and flight research program whose sole index of success was survival of the specimen, or to the medical operations community who could not see how research-derived knowledge helped them to do their job. Notwithstanding this environment, or maybe because of its difficulties, life sciences leaders have creatively sought out methods and partnerships to promote their goals. Over the years, excellent strategic plans and research strategies have been developed for life sciences. They have not been fully implemented because requests for additional research funding were always hard to defend, and the organizational instability of program elements made it even harder to sustain a stable strategy.[12]

Writing in 2004, before the full impact of the VSE was realized, Vernikos was upbeat about the creativity of the Agency's scientific community in meeting its existential challenges but presciently cautious about financing their work within NASA in an enduring way.

1

Everyone's a Scientist

Students, Industry, and Partners in Space

Enthusiasm literally ran high between April 12, 1981, the date the first space shuttle, *Columbia,* roared skyward and January 28, 1986, when *Challenger* broke up just after liftoff, killing all seven crewmembers. This marked NASA's return to human spaceflight six years after the Apollo-Soyuz Test Project (ASTP), and with much greater frequency, at lower cost, and with room for much larger crews and payloads. The Agency solicited ideas from students to develop life sciences experiments to go aboard the shuttle. An invitation to corporate America resulted in the first industry astronaut. Hopes were soaring, but in a partly NASA-created bubble of optimism and grand plans that were outsize in relation to political and scientific realities. To paraphrase the bumper sticker: NASA flew faster than its angels could fly.

Shuttle Student Involvement Program

The Shuttle Student Involvement Program (SSIP), a joint endeavor with the National Science Teachers Association, was the Agency's vehicle for taking space to future scientists, engineers, and astronauts. Young people and their teachers would design and fly experiments with help from NASA Center scientists, industry mentors, and the Crew Station Integration Section of the Spacecraft Design Division at JSC, which ensured experiments fit into the space available. JSC assigned experiments to specific flights as the missions were confirmed and room on board became available.[1] SSIP created logistical problems and public relations headaches for the Agency, however, and underscored how difficult it was to keep animals alive in space or to launch and land on a biospecimen's schedule.

The first SSIP competition in 1981 brought in about fifteen hundred entries. Ten were selected to fly. Among them was the experiment of Daniel Weber, a junior at Hunter College High School in New York City, that would eventually fly aboard *Challenger* on STS-41B in February 1984.[2] Weber was an undergraduate at Cornell University by that time and received an educational experience achieved only by the handful of other SSIP students over the next five years.

Weber proposed sending six rats into orbit to test a hypothesis that weightlessness would relieve the inflammation of arthritis. One paw of three rats would be injected with Freund's adjuvant, a solution that induces rheumatoid arthritis. Three other rats would fly untreated. Kennedy Space Center (KSC) would maintain two like groups as ground controls. The species selected, the Lewis strain, was particularly sensitive to Freund's compound and astronauts could measure their impairment by observing the amount of movement the rats demonstrated and measuring the swelling of the paws. Researchers had a narrow window of seven to nine days after inoculation to launch the rats, which would then spend a week in orbit.[3]

Because of her experience working with rats, Emily Morey-Holton, an Ames Research Center (ARC) pharmacologist specializing in bone demineralization, became a mentor for SSIP entrants working with rodents and agreed to oversee the Weber project. Ames researchers had developed an animal enclosure module (AEM) to keep animals in good condition but with food, litter, moisture, and waste contained to avoid contaminating the shuttle environment. After a test flight aboard STS-8 validated the AEM, NASA offered it to Weber.[4]

Pfizer Pharmaceutical Company had invited Weber to their Connecticut laboratory on a biweekly basis between January and July 1982. There he did background research and preliminary tests, learned about experiment design and statistics, and worked hands-on with lab animals. NASA invited him to crew briefings and debriefings, pertinent teleconferences, and team meetings in Houston and at KSC on scheduling, ethics, and technical issues in 1982 and 1983. He visited General Dynamics' Convair Division (contractor on the AEM) in July 1982, then spent three weeks at Ames working with Morey-Holton on rat studies, and then returned to Pfizer to test air filter efficiency of the newly delivered AEM. At launch, Weber had already been on hand for six days as part of the preflight team preparing both flight rats and ground controls. If the landing had been diverted to Dryden Flight Research Center (since 2014 the Neil A. Armstrong Flight Research Center), he was on the list to fly there from Ames with the recovery team aboard a NASA aircraft.[5]

Advisors suggested testing Weber's theory on the ground first using a proven technique for simulating weightlessness in rats that Morey-Holton had developed in the early 1970s. Called hindlimb unloading, it was accomplished by fastening rats into a harness that suspended their back legs in midair.[6] That also induced fluid shift toward the head, another phenomenon under study. Preflight testing let Weber perform tasks that would be done by NASA scientists and technicians during the real experiment, such as inoculating rats and measuring paw thickness. It pointed out where a procedure might give false results. For example, the team learned that calipers depressed the swollen paw, so a special tape measure was substituted.[7]

The rats in simulated weightlessness did not develop arthritis. At first the team thought it might be a delayed reaction to the adjuvant, but a second test suggested that being in a gravity-free environment immediately after contracting the disease prevented it from ever manifesting fully. Since mammals and pathogens both evolved in a one-gravity environment, perhaps gravity itself might be a necessary trigger for either the disease agent or the animal's pain response. "These findings are new and presently unpublished," the team reported. "The data were not anticipated and would not have been gathered without the impetus of the SSIP."[8]

The team anticipated some challenges, such as needing an animal-watering system that would work in weightlessness for more than a week. STS-8 had test-flown the first AEM without one, and NASA had sent along potatoes for moisture.[9] The rats consumed all the potatoes by the day prior to landing, and, even though irradiated, the tubers proved to be a stubborn source of bacteria. Five of the seven varieties for which the potatoes were tested still flourished postflight even though random spots inside the cage, feces, the food, and glue all had either no contamination or the same single bacillus. NASA and Convair raced to come up with a remedy and flight test it on the Agency's KC-135 aircraft. In December 1983, just two months before launch, the water bottles were finally flown and passed the test. A January team report noted that they needed to determine how quickly they could train young rats to drink from it.[10]

There were other lessons to be learned as well. In the event of a rodent death, the team wanted to know how many dead rats would be too much for the AEM odor filters, so Pfizer and Ames tested them with actual dead rats. The first batch of supposedly pathogen-free lab rats turned up with a pneumonia virus. AEM temperatures in the lab registered 88–92 degrees Fahrenheit (31.1 and 33.3 degrees C) in spite of a consistent room temperature of 72 (22.2 C). Observation revealed that the animals were sitting on the thermometer—not a problem in weightlessness but it alerted ground-control handlers to shield the thermal probe. For feeding, instead of food pellets, experimenters had to use food bars in the weightless environment, and had to autoclave then dry them first.[11] Backup teams would be needed for pre- and postflight activities, with written instructions for technicians who might have to care for the animals in case of an overseas landing. Weber's team created a prelaunch and postlanding timeline, with protocols for euthanizing very ill or injured animals, quick-freezing animal remains, postflight feeding and watering of animals, dissection at the landing site, and postflight disposal of rodent remains.[12]

The team also learned that it had no control over other payloads and might have to alter their own in response. A furnace experiment was added to the shuttle middeck, concerning for the amount of noise it generated. Weber's experiment depended on observations of the rats' in-flight movements and conduct. If they were stressed by the noise, the team wouldn't know if their behavior reflected that or discomfort from joint inflammation. Ames researchers tested the situation four weeks prior to launch and were relieved to see that "two of the three rats awoke and looked around, then went back to sleep."[13]

Lessons learned, *Challenger* successfully launched and landed the half-dozen rats, but the unexpected results ultimately did not answer Weber's original question. When supplier Charles River Laboratories could not come up with sufficient pathogen-free Lewis rats, the team had to substitute "gnotobiotic" rats—animals free of disease but not specifically Lewis-variety rodents. Unknown to team members, generic gnotobiotics took much longer to respond to the Freund's adjuvant. Neither the animals that flew nor the ground controls developed definitive symptoms for 14 days, several days more than it typically took Lewis rats to manifest symptoms, and the mission was only eight days long. The final report did not find fault but did point out that this fact had been described in an earlier journal publication (although not in the abstract).[14]

The twelfth SSIP experiment, scheduled for 1986 but delayed by *Challenger* until 1989, presented a massive public relations dilemma. In the study, designed by Andrew Fras of Binghamton (NY) High School, NASA was to drill holes into a nonweightbearing leg bone of two flight rats prior to launch, bisecting it, to see if healing was (as expected) slower in space than on Earth. (NASA would also bisect the leg bones of two ground-control rats.) The experiment did show that fracture repair would be delayed in weightlessness, useful data for space station and planetary mission planners.[15] However, news headlines about a rat's leg being "broken," conjuring up images of hammer-wielding astronauts, brought a vigorous negative response from the public and in the press.[16] The Agency reacted by establishing the Advisory Panel on Animal Care and Welfare, cochaired by manager of the Space Biology Program Thora Halstead and Associate Administrator (AA) for Space Science and Applications Charles Redmond. Members were managers from the four NASA Centers operating life sciences research programs, Headquarters General Counsel, the Security Officer, the Freedom of Information Act (FOIA) Officer, and representatives from the National Institutes of Health (NIH). The panel met twice weekly for six weeks prior to STS-29 "to ensure proper coordination and preparation" for the flight and wrote four white papers, a summary statement, and a question-and-answer sheet. The NASA Safety Officer coordinated activities "to prevent or respond to any adverse activities, such as picketing."[17]

The SSIP selection process was not weighted to ensure diversity among the students selected, as evidenced by Andrew Fras flying experiments on *two* missions, a study of brain cells being launched before his bone-healing project.[18] Only the handful (four girls and seven boys) who won an SSIP competition could take part in the actual design experience, too, and NASA records do not track an outcome in terms of their career choices.[19] Anecdotally, five years after his experiment "A Comparison of Honeycomb Structures Built by Apis millifera" flew on STS-41-C, Tennessee high school senior Dan M. Poskevich credited the SSIP experience with influencing his decision to study electrical engineering in college.[20] John Vellinger (who conceived of the "Chix in Space" experiment as an eighth-grader) flew incubated chicken eggs on STS-51L and STS-29 to evaluate embryonic development in microgravity. He used that experience to cofound Techshot, a Greenville, Indiana, development firm that specialized in biomedical technology.[21] Overall, SSIP was a slow and expensive way for any youngster to learn how to do a science experiment and in the worst-case scenario, a billion-dollar boondoggle for just one boy or girl.

After the STS-51-L accident took the life of the first Teacher in Space, Christa McAuliffe, K-12 student participation became more constrained physically. The SSIP became the Space Science Student Involvement Program, which centered on experiments "that could theoretically be conducted on the Space Station, in a wind tunnel or in a zero-gravity research facility." Another goal was to broaden participation by including "students interested in space, [but] not necessarily in scientific research," via art and writing competitions.[22] In the late 1990s, teams in grades three through five designed and built a "futuristic aircraft or spacecraft," and grades six through eight took part in a "Mission to Planet Earth." There was an "intergalactic art" competition for grades three through twelve. First-place winners went to Space Camp in Huntsville, Alabama, and with their teachers to the National Space Science Symposium in Washington, DC. High school students could also propose a Mars experiment as individuals or apply for internships at Centers to work on microgravity, wind tunnel technology, astronomy, and other studies.[23]

In the early 1980s, NASA had also implemented a separate small-payload program, the Get Away Special (GAS). Experiments had to fit into a standard-size sealed canister carried inside the shuttle or in the cargo bay and thus not require in-flight handling. Flying living specimens was still tricky. In June 1983, some New Jersey high school students and their teacher, along with two Temple University biology professors, launched a colony of 151 carpenter ants aboard *Challenger* during STS-7. The students had chosen carpenter ants because they were observably social beings, possessed an exoskeleton that made them hardier than some biological specimens, had feet that could grip surfaces even without gravity, and lived in confined places. The G-012 GAS experiment package consisted of ants, food, water, and soil, plus a cooling system and a camera. Unlike two control colonies on the ground, the space ants died. The students replicated flight conditions to investigate the cause of death, considering the length of time the colony sat on board the orbiter before liftoff (two months) and were left on board after landing (eight days). They concluded that "the ants died (within a few days of being placed in the canister) from dehydration, due to the purging of the canister with dry air (required by NASA to prevent condensation buildup in GAS canisters)." In other words, NASA had launched dead ants.[24]

GAS was geared more toward groups, and foreign youngsters, college students, private individuals, and corporations could all take part. In fact, they were encouraged to collaborate. Goddard Space Flight Center (GSFC) organizers invited participants to annual GAS symposia from 1984 through 1988 at Goddard (except for 1988, held at Cocoa Beach, Florida) and though the presenters were the adult leaders and organizers, over 300 students from high schools and universities attended.[25] Ultimately shuttles flew over a hundred K-12 GAS canisters until budget constraints pertaining to George W. Bush's Vision for Space Exploration led Administrator Sean O'Keefe to cancel GAS and the revamped SSIP in 2005.[26]

Many schoolchildren and college students worldwide took part in SSIP and GAS, and certainly they learned about the scientific method of inquiry and about the processes that real-life scientists go through to fund and carry out research. Exactly how many cannot be told, as NASA's own record-keeping was not rigorous regarding student activities. A draft of a 1989 report stated that 17 experiments had flown out of 57 SSIP competition winners. An additional 23 experiments were also "feasible for flight."[27] A 1991 Agency document reported that "24 student experiments ha[d] flown," but is not clear whether all 24 had been SSIPs. Through the National Science Teachers Association, it added, over one million students used NASA materials in the classroom and fifteen thousand students submitted proposals.[28] However, a document written more than two years *earlier* had claimed that over *six* million students and teachers had been introduced to the SSIP.[29] A 1997 brochure about the revamped, less science-centered SSIP claimed only ten thousand participants per year, a rate that would have required a century to involve just the one million youngsters reported six years earlier and 600 years to recruit six million kids.[30]

Arguably, K-12 student programs did meet the "major goal" of promoting "interest in science and engineering among US students by encouraging the natural excitement that many young people have about space-related subjects."[31] Still, it is valid to ask whether student experiments actually advanced a particular life-sciences line of inquiry

STS-107/Research-1 configuration

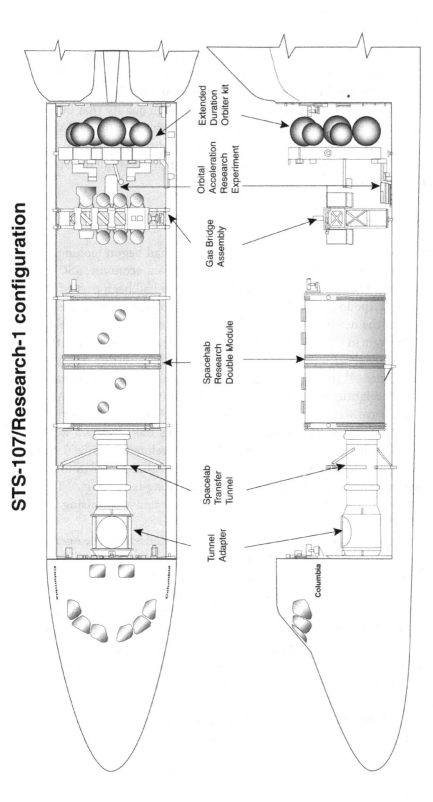

Extended Duration Orbiter kit

Orbital Acceleration Research Experiment

Gas Bridge Assembly

Spacehab Research Double Module

Spacelab Transfer Tunnel

Tunnel Adapter

Columbia

Columbia

Drawing shows location of GAS bridge assembly inside the shuttle payload bay. Photo credit: NASA-MSFC.

important to NASA, since they quite likely cost a professional researcher an opportunity to fly one. How many were one-shot experiments that no one followed up? How many students or groups published their results?[32] If published, was the work of high enough quality that others could cite them as references?[33] Was anyone tracking how many GAS or SSIP students went on to careers in science, medicine, or technology? How could one calculate whether the expense of flying student experiments was ultimately worth it?

Biomanufacturing in Space

Beginning in the mid-1970s, the space agency had begun looking for ways to integrate the shuttle into the mainstream economy. JSC Director Chris Kraft appointed Glynn Lunney, who had been part of the Apollo and Apollo-Soyuz Test Program (ASTP) management teams, to be head of the newly created Shuttle Payload Integration and Development Program in 1975. Kraft charged Lunney with reaching out to other government agencies and private industry to sell the shuttle's transportation services.[34] In 1977 Headquarters commissioned industry studies of the weightless environment as a commodity to which NASA offered exclusive access until the European Space Agency's (ESA) Ariane launched December 24, 1979. The United States was already a world leader in both pharmaceutical research and development (R&D) and production, and there was a commitment within that sector to invest money in new technologies on an ongoing and long-term basis.[35] Thus manufacturing biomedical products in space became a promising line of inquiry.[36]

An opportunity to pursue that line already existed within the Agency. Materials scientists at the Astronautics Laboratory at Marshall Space Flight Center (MSFC) in Huntsville, Alabama, had experimented with electrophoretic separation of biological materials aboard Apollo 14 and Apollo 16 and during the ASTP mission.[37] Nineteenth-century German chemist F. F. Reuss, working at Moscow University, had first demonstrated the process, in which an electrical charge separated desired molecules from a carrying medium (liquid or gel), and its application's potential, such as isolating specific proteins from blood serum, was intriguing.[38]

By 1972 Marshall was looking for new direction. In 1974 it had commissioned studies for materials processing by Auburn University,

Battelle Columbus Labs, General Electric, Lockheed, and the IIT Research Institute. Auburn's engineering, economics, and veterinary medicine schools were specifically asked to look at market demand and cost-price ratios.[39] By 1977 MSFC had commissioned a study of the possibility of commercial pharmaceutical production in orbit. Since drugmakers created something small in size, production could be automated, and there was potential for high profits and global market potential. In the case of insulin or human growth hormone, consumers would use the product for decades, even for life. A selling point for garnering corporate participation would be the link between the product and NASA, in the end user's mind, generating favorable public relations domestically and abroad. The Agency gave a contract to study the concept of commercial electrophoresis in orbit to McDonnell Douglas Astronautics Company (MDAC) of St. Louis, Missouri.[40]

McDonnell Douglas was then looking for ways to expand its product line and its market because of the volatility of the commercial and military aerospace sectors. Employment had dropped to about 57,000 in 1976 from a high of 140,000 in 1967.[41] Having worked successfully with NASA Headquarters as prime contractor for the Mercury and Gemini space capsules and with Marshall in developing the shuttle's aft propulsion system during the 1970s, MDAC vigorously pursued the manufacturing-in-space idea, drawing in part on internal resources left over from earlier aviation-related research.[42]

Searching for both an application for electrophoresis technology and a business model for pairing with nonaerospace firms, MDAC engaged the accounting firm of Price Waterhouse to perform market research on the pharmaceuticals industry and on ways to effectively approach, sell to, and team with likely partners. From an initial group of some twenty companies, MDAC ultimately partnered with the Ortho Pharmaceuticals division of Johnson & Johnson to build a space-worthy version of a device to carry out electrophoresis in the near-weightlessness of low-Earth orbit (LEO).[43] It would differ from existing electrophoretic equipment in that the flow of carrier medium would be continuous, generating separate streams of each product sought. Based on fluid dynamics equations, designers predicted that the throughput in space would be 500 times that of the best ground-based machines and with four or five times the purity, because the lack of gravity would eliminate friction and sedimentation they believed were responsible for low output and contamination on Earth.[44]

At the time, the companies kept the name of the product to be refined a very close secret. Only after the 1986 *Challenger* accident grounded MDAC and terrestrial advances in biotechnology made the trips into space unnecessary did the companies reveal the identity: erythropoietin. A hormone produced by the adrenal glands, it signals bone marrow to produce red blood cells. Without it a person develops anemia, a chronic condition affecting tens of millions of people and a side effect also of some cancer treatments. In 1980 dollars, Ortho estimated the market for a pure form of erythropoietin to be about a billion dollars a year.[45]

In January 1980, the director of MDAC's Materials Processing Study Group, James Rose, was able to negotiate a landmark Joint Endeavor Agreement (JEA) with NASA that initially called for the MDAC-Ortho apparatus, now called the Continuous Flow Electrophoresis System (CFES), to fly on a Spacelab mission in late 1982 or early 1983.[46] The Materials Processing in Space Division (Code E), Spacelab Mission Integration Division (Code S), and Space Transportation Operations (Code O) were in charge of implementing the agreement between NASA and MDAC. There was no mention of official life sciences sectors within NASA having a role. The emphasis was on proving the engineering and on marketing. By the summer of 1980 both sides had concluded that CFES could and should be modified to fly on the shuttle middeck but changed the agreement from six to three flights.[47]

MDAC and Ortho had already put two-plus years of work and millions of dollars of their own funds into the Electrophoresis Operations in Space (EOS) program by the time they signed the joint agreement with NASA. Terms called for NASA to provide the launch and support services free of charge, while MDAC contributed electrophoresis equipment and know-how, material to be processed, and analysis, and would train astronauts to operate the equipment in flight. MDAC would receive access to space, and NASA would receive a guarantee that one-third of the in-flight electrophoresis time would go to the Agency for its own experimental use.

In the early 1980s, University of Alabama researchers interested in "rational drug design" had also approached Marshall with the concept of custom-building a pharmaceutical molecule for greater precision and effectiveness. This required them to first crystallize a protein molecule. They would use the electrophoresis equipment in that process under a "proof of concept" contract that implied future addition of flights and services. With this contract, biosciences became the first endeavor to make commercial use of the Space Transportation System (STS) in flight.[48]

MDAC's initial CFES task was to develop a ground-test model. Primary among the challenges was simply to keep the biomaterials (enzymes and hormones) viable for long periods of time and under less-than-optimal environmental conditions. The samples were a very delicate balance of chemicals and had to be maintained at specific temperatures and pH levels. Shuttle flights were six or seven days in duration, and the machinery and materials to be processed were loaded onto an orbiter long before flight. Launches might be delayed as well.[49]

The second greatest challenge was money. According to Payload Specialist Charles Walker, then a McDonnell Douglas engineer and later a marketing executive for MDAC (a Boeing Company after 1997), cash was even more limited in the private sector than in government. This was in part because the profit that MDAC and Ortho were investing in such far-out R&D was subject to stockholder scrutiny, which was extra intense for MDAC because most of that profit had come from government projects. Taxpayers and Congress felt free to express opinions about how the corporation spent their dollars. Ultimately Walker estimated that MDAC invested $20 million between 1978 and 1986, in 1980 dollars, and Johnson & Johnson spent another $15–20 million from 1977 through 1985.[50]

The first CFES unit flew aboard STS-4 in June 1982, testing the device using a proprietary tissue culture medium and rat and egg albumin.[51] Instead of human hormones, the equipment processed a simulant suggested by MSFC: microscopic styrene beads of different colors, allowing photography to help verify the results. Even though the machinery was automated, the mission schedule called for pilot Henry "Hank" Hartsfield to run a dozen and a half samples between days L+0 (launch day) and L+3, to change out the collector multiple times and to engage in nearly 30 ground communications. These tasks alone required him to devote eight or nine hours of undivided attention to the CFES device and a significant number of other time blocks to set up the experiment, document it, and film the machinery in operation. His success in carrying out the experiment was a testament to Hartsfield but also to NASA, McDonnell Douglas, Johnson & Johnson, and contract personnel who established meaningful yet feasible experiment protocols, designed and constructed equipment that "anyone" could operate—and

Protein and virus crystallization research focused on determining the exact shape and structure of the molecules in order to engineer pharmaceuticals keyed to specific diseases and medical conditions. Photo credit: NASA-MSFC.

in a weightless environment—and trained Hartsfield to accomplish the mission goals. It also essentially proved part of the payload specialist concept: that a well-trained, well-equipped surrogate could perform another investigator's research in space and thus maximize efficiency of a mission.[52]

Results showed that over 400 times as much sample could be fractionated (split into component particles) over a specific time in LEO as could be done on the ground. Part of that improvement was that MDAC and MSFC designers kept upgrading the CFES. Additionally, weightlessness allowed for the use of a more concentrated sample.[53]

"Block 2" of the EOS program was STS-6. MDAC flew it and STS-7 in part to verify that increased output was a factor of greater sample concentration, and also to learn more about actual functioning of CFES gear in space. Designers improved the cooling process and applied a larger electric field for more flow and separation. They also changed a few variables in the liquid, including the electrochemically conductive buffering agent. The sample mix became hemoglobin and pneumococcal capsular polysaccharide on STS-6, then polystyrene latex particles of varying size and color, suspended in liquid, on STS-7.[54] A fourth equipment test flight, STS-8, came in the summer of 1983.

Results showed that some earlier models and predictions of particle behavior had been incorrect. For example, the basic premise of electrophoresis in space was that removing the gravity-induced problems of thermal convection (heat caused by the friction of the moving sample) and sedimentation would optimize the process. However, electrohydrodynamic problems that occurred on the ground turned up in space as well, distorting the flow of the sample stream.[55] Also, the increased sample concentration achieved in space changed the behavior of the liquid when the electric current was applied. A third observation was that certain mixtures of buffer and sample caused the entire batch to be differently charged at different points within the container.[56] These unexpected results motivated MSFC engineers to create new equipment designed specifically for gravity-free operations that would eliminate flow distortions and, more importantly, do so in a way that did not require extra attention from a human attendant.[57]

On a casual basis, Walker and Rose had begun to hint as early as 1979 that MDAC would be interested in adding its own payload specialist—the first-ever privately sponsored astronaut. It took three years for that to happen. The opportunity arose because an engineering group—the MSFC team, not a life sciences group—had a need for more data to enable theoretical work, specifically predictions of material behavior in microgravity. In the summer of 1982 MDAC was allowed to submit Walker's payload specialist application, provided he met the same medical fitness standards as astronauts, and in September Headquarters approved it.[58]

Payload specialists were intended to be trained in a particular set of experiments for a specific mission and were usually not NASA employees. As the first of a string of itinerant astronauts that would include a senator, a congressman, and a Saudi prince, Walker also became a guinea pig for the Agency's attempts to determine just how much and what sort of training to provide to ensure crew safety and mission success. If something benefited crew cohesion, Walker suggested in a 2004 interview, then do it.[59] Training the outsider broadly would reassure the commander and crew that the nonastronaut would not endanger their safety. Mission commander Hartsfield requested training for Walker on operating the Canadarm (Remote Manipulator System) and performing wastewater dumps. When management denied this, the rest of the crew invited the MDAC astronaut privately to look over their shoulders as *they* trained.[60]

During this time, the space agency also upgraded its capabilities in ways that allowed EOS to operate more efficiently. Foremost was putting the first Tracking and Data Relay Satellite (TDRS) into orbit, allowing increased ground-orbiter signal transmission. On the first few EOS flights, Walker, on the ground, could only communicate with the astronauts he had trained when the orbiter flew over ground stations, and then only if enough airtime remained after more critical communications. Astronauts would verbally read out numbers from the machine and listeners on the ground would quickly write them down. Getting NASA to tape record the received transmissions was a major victory, Walker recalled.[61]

Testing on ground-based controls always accompanied the work in LEO, as did mathematical modeling. The latter was needed to predict the effectiveness for other biological materials and to facilitate future mass production of pharmaceutical products. Roberts Associates of Vienna, Virginia received the contract to do this, and step one was simulating the hemoglobin experiment, that is, mimicking the exact results obtained in order to develop accurate algorithms needed to write predictive software. Even unexpected negative results, like those obtained on STS-6 and STS-7, were useful because they provided more input for these models.[62]

Other technical challenges included sterilizing CFES gear to prevent bacterial contamination. Ground crew managed this chemically at KSC about five days before launch and added another chemical to prevent the formation of bubbles. The cleansing was inadequate, however, to prevent a contamination of the product aboard STS-41D in August 1984, the first mission that Walker flew as payload specialist. As a result, MDAC and Ortho were able to obtain only physical and chemical data, not biological, and the pharmaceutical maker was unable to proceed to the next stage of research: animal testing.[63]

STS-41D pointed out how human-intensive CFES was, processing the sample for about 100 hours through hundreds of tubes just a millimeter thick. MDAC requested a second shuttle flight for Walker, agreeing to share his time and equipment with MFSC. Aboard STS-51D, in April 1985, MSFC used its one-third portion of Walker's time to operate a hand-held protein crystallization experiment, a precursor to more sophisticated automated experiments studying the structure of proteins for pharmaceutical manufacturing applications. The Commercial Development Division of the Office of Commercial Programs (OCP) and the Microgravity Science and Applications Division (MSAD) of the Office of Space Science and Applications (OSSA) cosponsored the study.[64]

The success that MDAC already had in proving the CFES concept aboard STS-4, -6, -7, and -8 in 1982 and 1983 and in recruiting private investment from nonaerospace firms like Johnson & Johnson had helped persuade NASA that President Ronald Reagan's call for the private sector to lead the way in space commercialization was valid. Analysts and observers speculated that by 2000, commercial use of space could be generating $65 billion in annual revenues for US companies. In mid-1983, NASA established a task force to look at "the opportunities for and impediments to expanded commercial activities in space," and in September 1984 Administrator James Beggs created the Office of Commercial Programs at the AA level.[65] It was meant to forward development of new commercial high-tech enterprises such as CFES and new profitable applications of existing technology. The White House issued a policy statement on commercial use of space and in October NASA released its own document, the "NASA Commercial Use of Space Policy."[66]

By then the Agency had under its belt only a dozen orbiter flights, three with NASA-tended CFES gear and one with MDAC's Walker aboard, and no marketable product in terms of pharmaceuticals or anything else. Still, NASA was ready to declare optimistically that "space technology is ripe for its transition from exploration to major exploitation, from experimentation to expanded profitable commercial uses." The statement's preamble predicted that if the new policy were implemented, it would "expedite the expansion of self-sustaining, profit-earning, tax-paying, jobs-providing commercial space activities."[67]

Given the technical success of EOS, the promise of on-orbit protein crystal production test flights, and overtures from other bioindustries to MDAC regarding joint ventures in space, MDAC and other corporations also made bold predictions. MDAC president John F. Yardley forecasted a 45 percent market share and 15 percent profit (or more) just three years after getting "big factory spacecraft" up and running, once the anticipated small Leasecraft (Fairchild Industries' astronaut-tended satellite) paved the way. He and others in private industry expected Japan and Europe to be significant additional markets but also likely partners. James Rose, still chief EOS program engineer at MDAC, thought that if his company's unit flew on "every shuttle flight available to it" the program could see 30 days a year of production, enough to "begin to generate a positive cash flow."[68]

According to a policy statement on commercial space applications, NASA would "expand its traditional links with the aerospace industry and academia to also embrace other industries such as new high-technology entrepreneurial ventures and the financial and non-aerospace industrial and academic communities."[69] In a 1984 *Aviation Week & Space Technology* (*AWST*) interview, Rose pointed to a pact with Washington University in St. Louis to swap an on-campus electrophoresis device for live canine pancreatic cells for flight and ground research and a discussion to do the same with the University of Texas Medical Branch (UTMB). Negotiations were under way with Summa Medical Corporation and Lovelace Medical Foundation, both of Albuquerque, and with Scripps Institute in San Diego.[70]

Confident that electrophoresis held strong promise, the MDAC-Ortho team still expected to have a product on the market in 1988. For Walker's second flight, STS-51D, the mission goals were to evaluate the flow of the material within the equipment to look for points at which it might become contaminated; evaluate the optimum processing rate and sample-to-buffer ratio; and see if software changes improved the machine's efficiency. Walker would be unable to make corrections in flight.[71] The new sterilization process proved effective, and the CFES equipment and Walker flew seven months later in November 1985 as part of STS-61B. This time, after purifying a liter of the biological sample material, Walker was to reconfigure the CFES to retest the process on several different concentrations of sample versus buffer.[72]

Because EOS flew so many times in quick succession—seven shuttle flights in just three and a half years—it was a good example of the kind of in-flight life sciences experimental program the OCP was seeking. It was able to build, evolve, and grow, rather than fly to serve a political or diplomatic function. Walker explained that quick turnaround served a definite scientific and engineering need. "The Space Shuttle [and] the crew . . . [are] such [an] expensive and valuable asset that you don't fly a second time to just do the same things you did the first time," he said. "You advance. . . . You prove what you wanted to prove, or you find out you've got a problem. . . . On the next flight, you already have plans to go further in terms of the scientific or the technical investigation. You do more. You stretch the envelope. . . . Also, if you had problems the first time around, you fix those, and [then] you demonstrate the fix." The CFES unit on STS-61B was actually a preproduction device, with manufacturing equipment tentatively manifested for an STS-61M in the summer of 1986.[73] The mood was buoyant as to the shuttle living up to its promises, including the pledge of taking private industry into space.[74]

Instead, the EOS program disintegrated with *Challenger* in early 1986, but only partly due to the accident. The space agency did not, in the aftermath, immediately cut every nonagency program—the Teacher in Space, Journalist in Space, or EOS. At least over the next few weeks—perhaps even months, while the Rogers Commission deliberated—records imply that NASA intended to keep those going, and commercial space endeavors still had a modest priority with the shuttle and a future with the space station program.[75] It is unlikely that anyone knew less than a month after the accident that it would be almost three years before STS returned to flight status. In June 1986 NASA cancelled the journalist program, with its 40 "regional nominees."[76] Teacher in Space remained in an odd limbo, with backup teacher Barbara Morgan finally flying in 2007 as a full mission specialist.[77] As for EOS, a weekly report from the Payload Specialist Liaison Office at the Space Operations Directorate at JSC on February 19, 1986, said merely, "The Electrophoresis Operations in Space (EOS) has been put on hold. This 61-M payload and the payload specialist activities have been discontinued by McDonnell [Douglas]."[78]

That came about because Ortho had terminated its contract with MDAC four months before the *Challenger* accident to team with Amgen Inc., a bioengineering firm that had developed a productive, clean method of making large quantities of erythropoietin on Earth, obviously at much less cost and less risk.[79] MDAC's board had voted to push on alone while it searched for a new partner for space-made drugs. The industry people had done their best, as had NASA, but ultimately all took a gamble on for-profit space bioprocessing in the 1980s and lost.[80]

The absence of "traditional" life sciences personnel in the electrophoresis and protein-crystallization projects and the continuing promotion of commercial space bioprocessing after 1986 is evidence that NASA's goal in these endeavors had been not so much the science as the potential for support for a space station. Congress would have noticed and approved of the creation of jobs across a wide range of US industry. Director of International Affairs Kenneth S. Pedersen wrote a memo in 1982 describing other ways in which such "potential partners" could help politically. "Foreign industrial support can help expand the overall industrial interest in a Space Station and willingness to fund space R&D that can contribute to Space Station utilization," he wrote. "Thus, a larger corps of domestic industry (besides aerospace) may visibly support the Space Station."[81]

Support abroad would also lower the cost to the US government, possibly make the project more bulletproof politically, strengthen US industry by directing foreign brain pools into NASA, and underscore the point that the time was right for a permanently crewed space station. Pedersen offered to scout "conferences that include potential foreign users of a Space Station: scientists, business groups, and applications-oriented groups." After visits to Europe, Japan, and Canada four years later, Administrator James Beggs wrote to his foreign counterparts inviting them to appoint observers to the new Space Station Science Advisory Committee and a forthcoming Industrial Committee, "established to ensure that the Space Station maximizes the commercial opportunities of space."[82] By 1988, his successor James Fletcher was also using microgravity manufacturing as a selling point to motivate ESA in particular to share the cost of station construction. A memorandum of understanding (MOU) between NASA and ESA that year predicted that the station "and its evolutionary additions could provide for a variety of capabilities, for example . . . a research and manufacturing capability in space, where the unique space environment enhances commercial opportunities; an infrastructure to encourage commercial investment in space."[83]

Space Industries, Inc.: The Industrial Space Facility (ISF)

The possibility of capitalizing on life sciences in space also spurred, albeit abortively, a niche market: the orbital factory. Former JSC Director of Engineering and Development Maxime Faget, who by 1983 was heading Space Industries, Inc. (SII), began negotiating with MDAC and Johnson & Johnson about building an unpressurized orbital platform for electrophoresis production. Before the end of 1983, MDAC had signed for Phase I action on design and fabrication of an EOS-committed free-flyer.[84] Human-tended, it could be fitted with pressurized containers to make it "habitable" for brief periods but would not be equipped with costly life-support systems. It would dock with the space station and astronauts would access its onboard equipment, changing out raw material for finished product. NASA and SII signed an MOU in 1985 for two 45-foot-long (13.7-meter) free-flyers, the first to be launched in 1989 (three years before expected station deployment), the other in 1992.[85]

In late 1986, SII teamed with Westinghouse to manage its new Wespace subsidiary, the latter's experience with automated operations in hazardous nuclear environments and its electronics base making it a natural partner to construct small, solar-powered, automated space modules. In 1987 Boeing formed Commercial Space Development Co. to "invest in, own, or develop individual or joint ventures in commercial space." It signed on to design and build the ISF docking/berthing mechanisms.[86] The two NASA Centers for the Commercial Development of Space (CCDS) involved in protein crystallography for medical research and pharmaceutical production joined with four others interested in materials processing and reported that an ISF-type orbital facility "would provide valuable flight experience." Ames began constructing a full-scale mockup to test autonomous intelligent systems and do other research.[87]

Faget asked NASA for any space station schematics NASA had so his engineers could design a docking adapter, which they did, and he patented the device a few years later. At the same time, a competitor, Fairchild Space Company, began to explore the option of Leasecraft, basic platforms built for a variety of customers, including MDAC. A third

Maxime Faget. Photo credit: NASA.

company, Spacehab, Inc., was examining (with MDAC) payload-bay modules for commercial and government microgravity activities, including cleaner animal experiment modules. General Electric Astro Space became a fourth contender, updating its Department of Defense (DOD) photorecon film reentry vehicle as a microgravity lab. MSFC engineers worked on a pallet for their own electrophoresis projects, initially only accessible by astronauts in Manned Maneuvering Units (MMU). If bioprocessing proved feasible on the expected scale, the pallets could be adapted later for habitability.[88]

This burst of interest in commercial space platforms came during the Reagan years (1981–1989), when vigorous political conservatism created a strong push to privatize and downsize government. NASA under James Fletcher became the target of outspoken congressional criticism in 1987–88 for failing to support space privatization, including Faget's orbiting facility (or its competitors). Lawmakers accused NASA of being "reluctant." A NASA Advisory Council (NAC) task force on international relations stated that the Agency had "not perceived its mission to include the promotion of US industry generally." It compared the United States unfavorably in that regard to ESA and Japan, claiming that "those who make US policy need to understand that its inability to win a competition for a $100 million launch contract is equivalent in economic terms to the import of 10,000 Toyotas." The House Appropriations Sub-committee linked its support for the station in fiscal year (FY) 1988 to NASA's enthusiasm for the ISF and even tacked on an unrequested $25 million to lease space on it.[89]

In a January 7, 1988, meeting, the White House Economic Policy Council asked NASA to become the "anchor tenant" of 70 percent of the ISF, paying SII's consortium $140 million annually for five years, per a congressional directive of the previous week. NASA would have to launch it in 1991 or 1992. In response, NASA stepped up its campaign against new mandates on how it operated. "A NASA official" claimed in

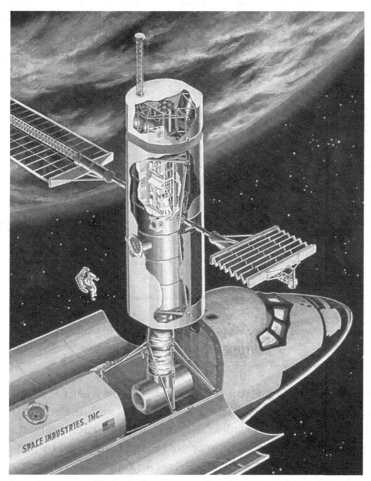

Artist's concept of Space Industries, Inc. free-flyer. Image courtesy Max Faget Collection, University of Houston–Clear Lake.

Table 2.—Space Infrastructure Platforms[a] That Could Be Serviced by Shuttle or an Orbital Maneuvering Vehicl

	Unpressurized coorbiting platforms (serviced by means of extravehicular activity)				Pressurized platforms (serviced internally while docked)	
	SPAS	MESA	LEASECRAFT	EURECA	Space Industries' Platform	European Modified Spacelab
Date available (now, or approximate, assuming start in 1985)	Now	Now	1986	1987	Late 1980's	1989
Cost[b] (billions of fiscal year 1984 dollars)	0.005	0.01	0.2	0.2	0.3	0.6
Characteristics						
Power to users (kW)	0.6	0.1	6	2	20	6
Pressurized volume (ft³)	None	None	None	None	2,500	3,000
Nominal crew size	None	None	None	None	1–3 only when docked	3
Miscellaneous	3,000 lb Payload	200 lb Payload	20,000 lb Payload	2,000 lb Payload	25,000 lb Payload	20,000 lb Payload
Capabilities[c]						
Time on orbit	10 days	8 months	Unlimited	6 months	3–6 months	Unlimited
Laboratories for:						
Life sciences	No	No	Modest	Modest	Modest	Moderate
Space science/applications	Modest	Modest	Modest	Modest	No	Moderate
Materials science	Modest	No	Modest	Modest	Moderate	Moderate
Technology development	No	No	Modest	Modest	Moderate	Modest
Observatories	No	No	Modest	Modest	Modest	Moderate
Data/communication node	No	No	No	No	No	No
Servicing of satellites	No	No	No	No	No	No
Manufacturing facility (materials processing)	No	No	Considerable	Modest	Extensive	Considerable
Large structure assembly	No	No	No	No	No	No
Transportation node (assembly, checkout, and launch)	No	No	No	No	No	No
Fuel and supply depot	No	No	No	No	No	No
Response to reasons advanced for space infrastructure[c]						
Maintain U.S. space leadership and technology capability	No	No	Modest	No	Modest	No
Respond to U.S.S.R. space activities	No	No	Modest	No	Modest	Modest
Enable long-term human presence in space	No	No	No	No	No	No
Attention-getting heroic public spectacle	No	No	No	No	No	No
Extended international cooperation	Yes	No	No	Yes	No	Unclear
Promote U.S. commercialization of space	Unclear	Modest	Considerable	No	Considerable	No
Maintain vigorous NASA engineering capability	No	No	No	No	No	No
Enhance national security, broadly defined	No	No	No	No	No	No
Space travel for non-technicians	No	No	No	No	No	No

[a]Listed platforms are illustrative examples; the list is not exhaustive.
[b]Costs include design, development, and production; launch and operational costs are not included. Some costs are estimated by the Office of Technology Assess ment; others were provided to OTA.
[c]Clearly judgmental.

This chart compared unpressurized versus pressurized platform options for Congress. From Office of Technology Assessment, *Civilian Space Stations and the U.S. Future in Space*, OTA-STI-241 (Washington, DC: US Congress, Nov. 1984), 8.

the trade press that ISF's real price tag would be double what Faget and SII partner, former shuttle astronaut Joseph P. Allen, said. The proposal could cause the Agency to "consider canceling current station contracts and going out with new proposals" and would "hand much more responsibility" for the station to the private sector. A proposed R&D review board would "take manifesting decisions away from NASA."[90] "Administration officials" were said to be angry enough to start talking about taking away NASA's authority for ISF and giving it to the new oversight group as soon as the lease was signed.[91]

During the final year of the Reagan presidency, new policy crafted "with heavy influence" from the DOC, DOT, and Treasury, plus clashing agendas, killed SII's bid to move from drawing board to LEO. Fletcher and company argued strenuously against a commercial free-flyer, initially on the grounds that the technological infrastructure, including robotics for tending it, were not yet in existence. A few in the press thought NASA was just overly cautious, not obstructionist, a result of negative experiences with recent satellite deals. Others speculated that NASA feared having a facsimile of a station would cause Congress to cancel the real one rather than use free-flyers to transition to a permanent crewed station that would advance US capabilities in science, technology, and commerce. The interagency Economic Planning Council, which included the Department of Commerce (DOC), allegedly saw great potential for resurrecting the payoff potential of the *human* space program. Amidst the tussling over money for the station, Associate Administrator Andrew Stofan stated publicly that if its budget were cut he would rather cancel the US station than see it turned into another Mir or Skylab.[92]

At the same time, some scientists within the Agency were arguing *for* the ISF, Spacelabs, or any other device for putting more bioprocessing and materials science experiments into orbit. The Microgravity Materials Science Assessment Task Force headed by scientist-astronaut Bonnie Dunbar recommended to Congress a Commercially Developed Space Facility that "coupled with an aggressive US program in technology advancement will allow US researchers to compete effectively with international groups."[93] Deputy Administrator Dale Myers gave to Congress an OSSA "NASA Microgravity Strategic Report" prepared for internal use, essentially a wish-list. As things stood, NASA had no strategic plans for the sort of research described and had been relying "entirely upon unsolicited proposals" for the past decade.[94] These words, from NASA's own scientists, contrasted with repeated statements from Fletcher, including congressional testimony, that "commensurate firm requirements have not been identified either within or without the federal government."[95] His emphasis would have been on the word "requirements."

Along with the argument that there were no plans or requirement for a free-flyer, which "a Space Industries official" retorted was "patently incorrect and NASA knows it," the Agency's Deputy General Counsel

Table 1.1. This Office of Technology Assessment report contrasts the capabilities of a NASA-only infrastructure vs. what free-flyers could accomplish

Comparison of Some Options for Low-Earth Orbit Independently Operating Infrastructure

	Shuttle Orbiter	Extended Duration Orbiter: Phase I	Extended Duration Orbiter: Phase II	Free-flying spacelab (developed as permanent infrastructure)	NASA infrastructure aspirations	
					Initial operational capability	Mature, fully developed
Date available (assuming start in 1985)	Now	1988	1990	1990	1992	1996–2000
Cost[b] (billions of fiscal year 1984 dollars)	None	0.2	0.5	2–3	8	20
Characteristics						
Power to users (kW)	7	7	20	6	80	200
Pressurized volume (m3)	60	60	100 (with Spacelab habitat)	100	200	300
Nominal crew size	6	5	5	3	8	20
Miscellaneous	Can accept Spacelab	No new technology	New technology required modest laboratory space;	Modest crew accommodations	Orbital maneuvering vehicle plus two free-flying unpressurized platforms	Reusable orbital transfer vehicle plus several more platforms
Capabilities[c]						
Time on Orbit	10 days	20 days	50 days	Unlimited (60–90 day resupply)	Unlimited (90 day resupply)	Unlimited (90 day resupply)

Laboratories for:						
Life sciences	Moderate	Moderate	Considerable	Extensive	Extensive	Extensive
Space science/applications	Modest	Modest	Modest	Modest	Extensive	Extensive
Materials science	Some	Some	Moderate	Moderate	Extensive	Extensive
Technology development	Modest	Modest	Some	Moderate	Extensive	Extensive
Observatories	No	Modest	Modest	Modest	Extensive	Extensive
Datal/communication node	No	No	No	No	Considerable	Extensive
Servicing of satellites	Modest	Modest	Modest	Modest	Considerable	Extensive
Manufacturing facility (materials processing)	No	No	Modest	Modest	Considerable	Extensive
Large structure assembly	No	No	No	Modest	Moderate	Extensive
Transportation node	No	No	No	No	Moderate	Extensive[d]
Fuel and supply depot	No	No	No	No	No	Considerable
Response to reasons advanced for space infrastructure						
Maintain US space leadership and technology capability	No	Modest	Modest	Modest	Considerable	Extensive
Respond to USSR space activities	No	Modest	Modest	Modest	Considerable	Extensive

(continued)

Table 1.1—*Continued*

	Shuttle Orbiter	Extended Duration Orbiter: Phase I	Extended Duration Orbiter: Phase II	Free-flying spacelab (developed as permanent infrastructure)	NASA infrastructure aspirations	
					Initial operational capability	Mature, fully developed
Enable long-term human presence in space	No	Modest	Modest	Considerable	Extensive	Extensive
Attention-getting heroic public spectacle	No	Modest	Modest	Modest	Modest	Modest
Extended international cooperation	Modest	Modest	Moderate	Moderate	Moderate	Moderate
Promote US commercialization of space	Modest	Modest	Modest	Considerable	Considerable	Considerable
Maintain vigorous NASA engineering capability	No	No	No	Modest	Extensive	Extensive
Enhance national security, broadly defined	No	No	No	Modest	Unclear	Unclear
Space travel for nontechnicians	Modest	Modest	Modest	Modest	Considerable	Considerable

Source: US Congress, Office of Technology Assessment, *Civilian Space Stations and the U.S. Future in Space*, OTA-STI-241 (Washington, DC: GPO, 1984), 8.

a Listed options are illustrative examples; the list is not exhaustive.

b Costs include design, development, and production; launch and operational costs are not included. Some costs are estimated by the Office of Technology Assessment; others were provided to OTA.

c Clearly judgmental.

d Including launch to the Moon, Mars, and some asteroids.

opined that NASA did "not now have the legal authority to enter such an agreement." The Administrator maneuvered to change the focus and put giant aerospace contractors on alert and on his side by declaring the whole free-flyer question to be really a matter of open versus directed contracting. *AWST* quoted "a congressional aide" who said sole sourcing was a "technicality." Congress *did* have authority to grant such a contract, and agencies asked it to do so all the time. A "senior space station official" next threatened to cancel the four prime station agreements outright.[96]

In the years since Reagan's 1984 State of the Union announcement about building a space station within a decade, 45 companies had expressed interest. They would not be able to bid on the free-flyer if it were an uncontested contract awarded to someone else. The Committee's action came at budget-submittal time, too, and Administrator Fletcher had already been planning to approach Congress with a request for some 30 percent more money to spend, so even contractors not interested in bidding on a free-flyer likewise saw opportunity passing them by if the budget got hung up in a squabble over the orbital platforms.[97]

Then, in response to a Reagan directive on space commercialization in early February, Fletcher did a seeming about-face.[98] NASA's objection to flying a private-industry Spacehab module, which it had signed an MOU to do in 1985, "was resolved recently" when shuttle upgrades increased launch/return weight limits. The whole objection there, "NASA officials" said, had been center of gravity (CG) and now they had "disposed" of the problem. The Agency would now "study" the idea of lofting a Spacehab (specifically) in a shuttle cargo bay. This seemed to contradict a two-month-old NASA study on potential uses of ISF that concluded it would "increase the risk of failure during payload operation and meet few scientific requirements of the agency."[99]

Lawmakers were extremely suspicious of such a rapid reversal. A troika of joint appropriations committee members, Senator William J. Proxmire (D-WI) and Representatives Edward Boland (D-MA) and Bill Green (R-NY), held up some $90 million in station funds appropriated in FY 1988. They had favored keeping an early noncrewed capability in the timeline, which they believed could have ensured that NASA got its station. ISF seemed a natural fit. NASA supporter Senator Bill Nelson (D-FL) quickly called hearings on the whole matter before his Space Science and Applications Subcommittee. In a display of partisan politics, other Democratic members of the House Space Science and Technology

Figure A-12.—One Contractor's Summary of Infrastructure ("Space Station") Payoffs

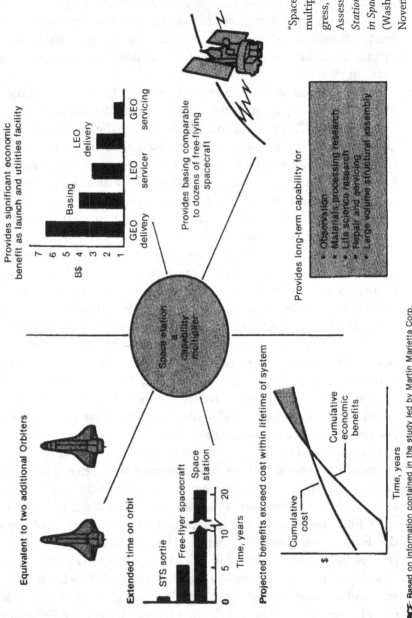

Equivalent to two additional Orbiters

Extended time on orbit

STS sortie

Free-flyer spacecraft

Space station

Time, years

0   5   10   20

Provides significant economic benefit as launch and utilities facility

B$
7
6
5
4
3
2
1

GEO delivery   Basing   LEO servicer   LEO delivery   GEO servicing

Space station a capability multiplier

Provides basing comparable to dozens of free-flying spacecraft

Provides long-term capability for

* Observation
* Materials processing research
* Life science research
* Repair and servicing
* Large volume structural assembly

Projected benefits exceed cost within lifetime of system

$

Cumulative cost

Cumulative economic benefits

Time, years

"Space station a capability multiplier." From US Congress, Office of Technology Assessment, *Civilian Space Stations and the U.S. Future in Space*, OTA-STI-241 (Washington, DC: GPO, November 1984), 154.

SOURCE: Based on information contained in the study led by Martin Marietta Corp.

Committee complained to NASA about granting a contract without their approval. Senators Donald Riegle (D-MI) and Ernest Hollings (D-SC), and chair of the House Committee on Science, Space, and Technology Robert A. Roe (D-NJ), none from states with huge NASA employment, wrote to Fletcher and asked him to put the Request for Proposal (RFP) on hold until Congress had legislated a procedure for the deal.[100] At a March 1988 Nelson subcommittee hearing, Deputy Administrator Dale Myers was hazy about how NASA would divide funding among the station, ISF, Spacehab, or some combination of those. NASA was "exploring" the idea of shifting money, meaning it might proceed with the work and pay for it in the "mid-1990s" with "new funds" appropriated now, possibly some of the $700 million increase NASA officials requested for the year. That money was essentially a request for government guarantee of a lease contract, which was anathema to free market advocates. Conservatives and liberals alike balked at what amounted to "buy-now, pay-later," as the administration had probably expected. Representative Kenneth Hood "Buddy" MacKay (D-FL), a member of both the budget committee and space subcommittee, called Myers's proposal "the epitome of nonplanning," pointing out that it set free-flyer charges on a collision course with the ever-climbing station costs. MacKay put NASA "on notice" that "we will have to drop one of the programs."[101]

The House finally voted to allow NASA to release a commercialization RFP, but Congress reserved the right to approve the project. The Senate essentially tabled the controversy by calling for a National Academy of Sciences (NAS) report on the issue, to be completed a year hence.[102] That April 1989 study—which NASA had commissioned the National Research Council (NRC), a part of the NAS, to do—advised the space agency to forego leasing a commercial free-flyer. The arguments were suspiciously familiar: it was not needed, would not bring about station or space manufacturing any sooner, microgravity was too immature a science, and an extended duration orbiter could do the same thing. Furthermore, leasing an ISF would cost three times what anyone predicted, the NRC declared. In response, Faget pointed out that the NRC had concluded just the reverse in 1988. The difference, he said, was that the most recent panel "was composed of experts in basic research disciplines, instead of having a commercial inclination." Another unnamed industry official pointed out that the second NRC group had studied a particular group of experiments to see if they needed the higher power and better microgravity a free-flyer could provide and concluded that

Table 1.2. Reasons for a space station

	Goals						
OBJECTIVES	Increase space activities' efficiency; reduce their net cost	Involve the general public directly	Derive economic benefits	Derive scientific, political, and social benefits	Increase international cooperation	Study and explore the physical universe	Bring life to the physical universe
1. Establish a global information system / service re natural hazards	N	N	P	Y	Y	N	N
2. Establish lower cost reusable transportation service to the Moon and establish human presence there	Y	P	P	Y	Y	Y	Y
3. Use space probes to obtain information re Mars and some asteroids prior to early human exploration	N	N	N	Y	Y	Y	N
4. Conduct medical research of direct interest to the general public	N	N	P	Y	P	N	N
5. Bring at least hundreds of the general public per year into space for short visits	N	Y	Y	Y	Y	N	Y

Goal						
6. Establish a global, direct, audio broadcasting, common-user system/service	N	P	Y	Y	N	N
7. Make essentially all data generated by civilian satellites and spacecraft directly available to the general public	N	Y	Y	Y	N	N
8. Exploit radio-optical free space electromagnetic propagation for long-distance energy distribution	N	Y	P	Y	N	N
9. Reduce the unit cost of space transportation and space activities[a]	Y	Y	Y	N	N	N
10. Increase space-related private sector sales	Y	Y	N	N	N	N

Source: US Congress, Office of Technology Assessment, *Civilian Space Stations and the U.S. Future in Space*, OTA-STI-241 (Washington, DC: GPO, 1984), 16.

Notes: Y: Yes; N: No; P: Perhaps; depends on how carried out.

[a] This would advance the prospects of successfully addressing all other goals.

they did not, but the experiments they considered had already been optimized for the shuttle. The Committee did not ask how good the experiments *could* be made to be. When questioned, another official admitted that the group "had adopted NASA's viewpoint and disregarded other information that presented the facility in a more favorable light."[103]

Congress asked NASA for a reply to the NRC report and held hearings into the matter the following month. Nothing of substance resulted. In the intervening year, a new president and Congress were elected, and a new administrator took charge of the nation's space agency.

According to then Associate Deputy Administrator Philip Culbertson, after Johnson & Johnson opted to make erythropoietin on the ground and Fairchild gave up, and as SII's customers "kind of vaporized," Faget had no way to prove to potential investors that he could get clients, in spite of government assurances of support for his space platform. Also, the more designers studied the problems associated with free-flyers, the more problems they discovered and the concept became that much more expensive. Joe Allen, who had joined Faget's firm in 1984 to work on the ISF, said it would still have "provide[d] a permanent Space Station for a cost about 50 times less than the station NASA was planning."[104] As late as 1987, NASA Ames was still working on the full-scale ISF mockup, to be completed by May of that year.[105] Allen advocated for commercial space facilities as a member of the Advisory Committee for the Future of the US Space Program in Houston in 1990, and the House, at least, communicated its interest in an SII free-flyer to the Committee.[106] Faget's company moved on to other projects and other government agencies, such as the DOD, in search of business.[107]

Some put blame on the space agency for its initial overly optimistic estimates of how fast it could turn around the shuttle and get it back into space. Expecting in 1982 to be able to soon launch an orbiter nearly every two weeks, NASA helped create a false aura of ease-of-access and of a launch frequency rate that would have pushed launch prices low enough to be very attractive to all sorts of industries.[108] After *Challenger*, the impact of this struck home, and for some in the space agency it was as though the proverbial scales had fallen from their eyes.

In a 1991 interview, former Administrator James Fletcher placed the blame on the Office of Management and Budget (OMB), which by the late 1960s had modeled its procedures after the Department of Defense's systems-analysis approach. "I had a very low opinion of that group," Fletcher said. "They pulled some really crazy things. . . . [OMB] forced NASA to do that cost-effective analysis to show that . . . the development cost was worth the return, and reduce the cost of the Shuttle." After bringing in analysis from industry and the DOD, he added, "you put all that together and . . . it showed, leaving out DTMO (Development, Test, and Mission Operations), that twenty-five [flights would] get our money back. It wasn't [NASA's] idea at all. I didn't like it because this was a new way of getting to and from space, especially *from* space, never tried before. . . . We had a lot of arguments over that. Over the assumptions."[109]

2

Working in
the Space Environment

Over the course of the shuttle program, the astronaut corps became more diverse in background, gender, age, and nationality with the inclusion of women and astronauts from other countries. The size of crews expanded from a commander and pilot to a complement that included mission and payload specialists. This created a broader pool of human test subjects, making space research more applicable to Earth medicine. It also presented new challenges as the Agency worked to equip and maintain flight crews and manage programs carrying out increasingly ambitious research.

Anthropometry and Astronauts

In 1978, ESA selected three astronauts to train with the United States, and West Germany's Ulf Merbold flew in 1983 as part of the STS-9 crew, working on the joint US-ESA Spacelab 1. Pleased with the success of the Canadarm remote manipulator project, in 1982 NASA invited Canada to take part, and the Canadian National Research Council selected six astronauts, including a neurologist (Roberta Bondar) and a physiologist (Kenneth Money) in December 1983. Physicist Marc Garneau was the first Canadian to fly as part of the STS-41G mission in 1984.[1] In both instances the relevant international agencies did their own medical screening, but it was NASA's policy to reevaluate payload specialist candidates and it maintained final say over an individual's fitness.[2]

While nationality presented no medical obstacles to the astronaut selection process, sex and corresponding body type and size differences

ESA's Ulf Merbold, a German payload specialist, working at the Gradient Heating Facility on the Materials Science Double Rack on STS-9, Spacelab-1. NASA-MSFC image.

The first six Canadian Space Agency astronauts, chosen 1983: (*back, L to R*): Ken Money, Marc Garneau, Steve MacLean, Bjarni Tryggvason; (*front, L to R*): Robert Thirsk, Roberta Bondar. © Canadian Space Agency, 1983.

were an issue from the start. With the larger shuttle and a crew not drawn from the military jet test-pilot population, the life support systems and gear all had to be designed to fit people much bigger or smaller than the average military pilot in any dimension. This included height, reach, sitting height, grip strength, circumference, mass, center of gravity, shoe size, facial width, and dozens of other measurements. All of these needed to be taken into account to design escape systems, seating, and extravehicular activity (EVA) suits, and place everything a crewmember might need within reach. The military's height requirement had been a minimum of 5'4" (64 inches, 162.5 cm), but the average pilot actually measured just under 5'10" (69.9 inches, 177.5 cm) and weighed about 172 pounds (78 kg), at the time some 6 inches (15 cm) and 40 pounds (18 kg) more than a typical American woman.[3] To keep early space capsule weight down, NASA actually had a *maximum* height restriction of 5'11" (71 inches, 180 cm).[4]

Since a space station was such an enormous undertaking, foreign participation was being solicited and one key partner would be Japan. Statistically, Japanese people were shorter than Americans, making for an even wider range of heights, to cover both US and Japanese body types. The Agency extended the size range, from the bottom fifth percentile of Japanese females to the top ninety-fifth percentile of US males (civilians included, race unspecified). Human factors engineers would have to deal with astronauts varying in height from 58.6 to 74.8 inches (148.9–190.1 cm)—roughly 4'10.5" to nearly 6'3"—and from 90.4 to 217.2 pounds (41–98.5 kg). Other body measurements would naturally vary correspondingly. Controls, stowage, and equipment also had to be reachable, accessible, and usable within any amount of gravity, from zero to three g's. Because gravity no longer gave an assist, it required a bit more strength to reach *downward* in space but was easier reaching up.[5]

NASA biometricians learned with Skylab that given a space large enough to move around, body measurements change noticeably in zero-g. Spine-related measurements, such as standing and sitting height and upward reach, increase an average of 3 percent (about 2 inches, or 5.08 cm) during the first few days. NASA did not want astronauts to be unable to get back into their pressure suits or to be sitting up too high in their seat to strap in for reentry. The "body strike envelope"—the volume of space that limbs and heads of seated restrained crew might hit during high accelerations, as with launch and reentry—was affected,

The first female NASA astronauts, shown with a Personal Rescue Enclosure, 1980. (*L to R*): Margaret R. (Rhea) Seddon, Kathryn D. Sullivan, Judith A. Resnick, Sally K. Ride, Anna L. Fisher, Shannon W. Lucid. Photo credit: NASA.

so crew seating was arranged to prevent flailing extremities from hitting control panels. Fluid shifting shrinks leg girth and expands upper extremities, manifesting as facial puffiness, pertinent to oxygen mask and eye protection design.[6]

Since astronauts would be exposed to more radiation than they would be on Earth, NASA also needed to know the body surface area. This was also important in designing the temperature control system, dealing with a much greater variation than on Earth. Body surface area varied by about 30 percent between the 5th and 95th percentile adult males.[7]

Being fully prepared for "ordinary people" in space also mattered given the numbers then expected. In a 1984 interview, Lt. Gen. James Abrahamson, Associate Administrator for Space Flight, predicted that "by the year 2000 we'll have over five thousand human beings that have gone into space. . . . Out toward that turn of the century, when we get a larger module, there may be in fact, twenty, thirty, or forty people up at a time."[8] A huge station was not Abrahamson's personal pipe dream. An undated committee report from 1979 or 1980 projected crews of 500-plus after 2000, with orbital missions of one to two years.[9] Boeing and JSC planners carried out a 1982 study of an orbiting "Space Operations Center" staffed by 109 "astronauts," including three cooks, five secretaries, and one "morale officer," with janitorial staff "TBD." Another 1982 projection prepared at KSC for the Operational Medicine Office at Headquarters referred to "current NASA Space Station studies" (by Boeing) that projected "up to one hundred crew."[10]

An ad hoc group in Houston and Florida comprised of astronaut William Thornton, John T. Jackson, and Gene Coleman had decided early on to create an anthropometry database and from 1977 to 1979 gathered statistics on astronauts and applicants. No one at NASA collected data from 1980 to 1984, but Lockheed Engineering and Sciences in Houston received a contract to do so from 1985 through 1991. Both studies classified participants as pilots, mission specialists, payload specialists, and "observers," and collected data on 14 different body dimensions. The cost of accommodating literally everyone "could be astronomical," the report read. Thus it was "necessary to have percentile info that is specific to a particular assignment or position to which an astronaut has been designated." In other words, only US citizens, mostly military, would pilot US shuttles so designers could safely assume a narrower range of heights.[11] The Air Force still required a minimum standing height of 5'4" (64 inches, 162.5 cm) for pilots, so that would be NASA's baseline as well.

Size also mattered for lunar and planetary habitats with their partial gravity. In the late 1980s and early 1990s, a team from the Lunar and Mars Exploration Project Office (LMEPO), part of the System Engineering and Integration Office at JSC, began work on the Lunar Base Accommodation Study and considered habitat design in relation to body size. The goal of their "construction shack" project was to design a small building to serve as an office, meeting area, and living quarters for workers on the Moon's surface.[12] It would be made from a "Freedom-size module" of 14.7 feet by 45 feet (4.5 meters by 13.7 meters), with one airlock. Crews would work at the shack on rotating 30- to 60-day missions.[13]

Because human beings "walked" differently in one-sixth gravity than

This cutaway artist's concept from 1989 depicts the dust removal annex for an inflatable lunar habitat, attached at right. It would have included an air shower and a place to change out of and store the surface suit. Photo credit: NASA-JSC.

they did in Earth gravity, floors would be covered with something like Astroturf, for traction. Ceilings would need to be taller, architects believed, but padded because lunar workers would likely adopt the loping, bouncing gait of the Apollo moonwalkers. Likewise, stairs might be 1.5 meters (almost 5 feet) high because people could jump so much higher on the Moon. Perhaps inhabitants needed just a fire pole, as on Skylab.[14]

The group evaluated size factors and other physical details of moving around in lunar gravity. Team member Martha Evert took a 50-parabola ride in the KC-135, a four-engine turbojet whose flight path induced weightlessness. She performed the motions a typical day would require, including jogging and running, stopping and starting, reaching, and tasks to assess, balance, control, and speed.[15] The group drew up a schematic for the construction shack, and then six team members, two women and four men, spent a day inside a mockup, going through activities such as eating, sleeping, exercising, dressing, photography, recreation, and care and transport of an injured crewmember. One person, at 5'2" (62 inches, 157.5 cm) the shortest, noted that size made a difference in terms of "roominess" of the enclosed bunk bed compartments. Small people could undress and dress in a bunk but their taller comrades had to use the hallway. Two people could sit on a bunk for a private conversation but three was a crowd. Fitting all six crew around the dinner table was tight, too, and the walkway needed to be redesigned because the "injured" party would not fit around corners on a stretcher.[16] Body size variances also meant disparities in the ability to lift objects in reduced gravity. This came up in another KC-135 study that tested reach and lifting capabilities in simulated one-sixth and one-third gravity. Center of gravity, which would change with height (standing or seated), made a significant difference in the ability to lift weights varying between 50 and 150 pounds (23 and 68 kilograms).[17]

Given the per-hour cost of a human in space, sluggishness was of concern. Size and gender differences produced different calorie intake requirements, so NASA contracted for a study of nutritional needs for Space Station Freedom by the US Army Soldier Systems Center in Natick, Massachusetts.[18] Charles Winget of the Ames Biomedical Research Division studied desynchronization of circadian rhythm in response to disruptive light-dark cycles, as part of a bed rest study in the mid-1980s. His group found that a woman's heart rate was higher at any time of day than a man's and men had greater variation in body temperature. These differences and factors such as being a "night owl" as opposed to an early riser made a difference in the body's ability to adapt to space.[19]

Other sex-based differences showed up in the 1990s. One was susceptibility to decompression sickness, or "the bends," most likely to occur in connection with EVAs. Nitrogen bubbles from the blood become painfully trapped in joints and in the most severe instances can cause death. A USAF School of Aerospace Medicine (USAFSAM) study reviewed results of decompression chamber tests of women between January 1978 and December 1988. Among 81 records they noted "a significant inverse linear relationship . . . between the number of days since the start of [the] last menstrual period and the incidence of [decompression sickness.] . . . The conclusion is that women are at higher risk of developing altitude-related decompression sickness during menses, with the risk decreasing linearly as the time since last menstrual period increases." The underlying mechanism for the correlation remained unknown.[20]

Researchers at NASA and in the military also found female subjects to be more susceptible to orthostatic intolerance (OI), passing out when attempting to stand after a landing. Deconditioning to g-force tolerance was 50 percent greater in the females than among males. They were also significantly less able (15–61 percent) to buffer against this with physical conditioning regimes or pressure suits, but they recovered more quickly than the men.[21] One study of males in the early 1990s indicated that OI in men was affected primarily by height and blood volume. The taller the astronaut, the more likely he would be to pass out. Researchers wondered, then, why short women had lower tolerance than tall men. One explanation was that a female's lower blood volume caused a faster reduction of the rate at which the heart refilled with blood between heartbeats and as a consequence lessened ability to maintain blood pressure levels. They also noted that blood pooled in a woman's pelvic area rather than her legs by as much as six times the rate it did with men. Females might be helped with pressure suits that squeezed the pelvic area more and the legs less.[22]

The Agency wanted to know if aerobic training could lessen recovery time.[23] To study that and bone loss, Ames Research Center investigators employed a sample population of athletes and others who used one limb or one set of limbs much more than the other. An Olympic gold medal shot-putter, an iron worker, and a rugby team took part. Researchers found a 20–25 percent difference in bone density between the much-used limb and the less-used one. The study also tested a bone stiffness analyzer as a way to detect osteoporosis earlier than x-rays, which required at least 20 percent bone loss to reveal damage. The device used a very low-dose X-ray beam and became the DEXA (Dual Energy X-ray Absorptiometry) commonly used in the United States decades later to detect age-related bone loss.[24]

One physiological difference no one had envisioned themselves testing was extreme age disparity. The flight of former Mercury astronaut Senator John Glenn in 1998 drew much attention to geriatric medicine, both as it pertained to the 77-year-old's flight and as an illustration of how space medicine research impacted large segments of the population. The elderly and the bedridden, and even younger sufferers of diseases that mimicked rapid aging, such as progeria, often had bone loss issues and in some cases hearing, vision, and balance problems.

Glenn had gained fame February 20, 1962, when he became the first American to orbit the Earth. In a memoir he recalled that while serving on the Senate Special Committee on Aging he had come across a book describing physiological changes in spaceflight and was struck by how similar they seemed to the effects of aging. That spurred Glenn to read more on the subject and talk with Arnauld Nicogossian, Sam Pool, and Carolyn Huntoon at NASA, then with physicians at the National Institute on Aging (NIA), and finally with NASA Administrator Daniel Goldin. In December 1995, Glenn offered himself as a medical guinea pig for as many experiments as NASA could design. It took time to convince Goldin that flying him would be valid science and a fruitful use of taxpayer money. Finally, after cautioning Glenn that he would have to pass the same medical tests as astronauts four decades younger than himself, Goldin agreed in January 1998 to a shuttle flight.[25]

Along with the expected charges of political favoritism, the announcement of Glenn's upcoming flight created a call to send an elderly *woman* into space. The population of female senior citizens was much larger than that of males. The National Organization for Women, the National Women's History Project, senators from California and Oklahoma, the American Association of University Women, and individuals from all over the United States rallied to the cause of one woman in particular—Oklahoman Jerrie Cobb. She had been the first of a group of experienced female pilots tested between 1959 and 1961 for fitness to fly in space. The leader of the study, Dr. Randolph Lovelace II, had been in charge of the medical screening of NASA's first astronauts, and this, to the 20-plus women subjects, had implied that NASA had requested the study. However, NASA had not directed anyone to find women astronaut candidates. Congress called hearings on the women's charges of sexism in the summer of 1962, and the star witness against female astronauts was Mercury hero John H. Glenn. Supported by his testimony, NASA was able to stick to its policy of hiring only military jet test pilots until the shuttle era, dashing the hopes of the baker's dozen who passed the Lovelace tests (the "Mercury Thirteen"). Glenn's spokesperson said the senator was being made a scapegoat for the attitudes of society decades earlier. The 68-year-old Cobb met with Daniel Goldin in 1999, but he refused her offer to duplicate Glenn's STS-95 flight. The Agency continued to publicly market Glenn's flight as a contribution to geriatric medicine and reportedly "encouraged Cobb to apply," but at the same time in press statements treated the idea of flying Cobb only as a "consolation" prize that was "not going to happen."[26]

During the 1998, flight Glenn assisted with some of the more than 80 experiments manifested. Goldin had taken Glenn's offer to be a medical guinea pig seriously, so he was a subject for several experiments, as well as some of the DSO (Detailed Supplementary Objective) tests.[27] DSO 603C, "Orthostatic Function During Entry, Landing and Egress," asked if a septuagenarian would be able to maintain consciousness given the g-force of liftoff and reentry and whether countermeasures would prevent OI.[28] Similarly, DSO 626 took stock of physiological responses to the act of standing, before and after the flight. A third, DSO 630, consisted of Glenn wearing a Holter cardiac monitor to see if his heart rate was less variable in microgravity and whether any electrocardiographic changes occurred in flight. All crewmembers took part in DSOs 497 and 498, blood tests to evaluate the functioning of the immune system. With three other crew members, Glenn was a test subject for DSO 627, which consisted of pre- and postflight DEXA scans to document lean body mass changes, coupled with resistance exercise to measure endurance and reaction speed of specific muscles in the leg, knee, ankle, and back. He took doses of two amino acids each day, alanine by capsule and histidine via injection. Another group experiment evaluated the inner ear's functioning in maintaining balance before and after the mission.[29] A protein turnover experiment used blood and urine samples from Glenn and another astronaut, and the pair also had to record all food consumed. Data would provide insight into whole-body and skeletal muscle protein metabolism, important for understanding muscle atrophy in space and among bedridden and wheelchair-bound people.[30] Glenn and others also had MRIs before and after flight, and filled out a questionnaire daily to study "muscle, intervertebral disc, and bone marrow changes" in connection with reports of back pain in weightlessness.[31] A sleep study required Glenn to swallow a "pill" developed at Ames that measured and transmitted his body core temperature to a device worn on a belt. He slept in a head net and an instrumented vest with 21 monitors that measured the volume and rate of respiration, eye movements, brain waves, heartbeat patterns, heart rate, pulse, and blood oxygen content. After sleep he took reaction and cognition tests that consisted of responding to numbers flashed on a computer screen.[32]

A search of the National Library of Medicine/NIH's PubMed database revealed only a handful of journal articles based on research conducted on John Glenn as part of STS-95. Two specifically identified a subject as having been Glenn in their abstracts or titles. Both indicated that there were marked differences between the astronaut in his late seventies and his younger crewmates, but concluded that they were not negative disparities. Glenn had kept in shape in the three and a half decades since his three orbits around the Earth with power walks, free weights, and stretching and still flew his own plane, a two-engine propeller aircraft. A Phoenix (Arizona) Heart Institute study of Glenn's cardiac performance indicated that his heart kept up with the average astronaut during and after spaceflight.[33] The senator acknowledged in his memoir that STS-95 did point out where age had crept up on him: flexibility. NASA physical therapists helped him with that.[34]

In the end, Glenn's flight proved that an extremely fit, geriatric ex-astronaut could survive a multiple-g centrifuge ride, get around in 80 pounds (36.29 kg) of gear, drive the tank-like pad escape vehicle, learn new tasks and procedures, master the tiny body motions needed to propel himself in weightlessness, and reliably operate sophisticated gear required for in-flight experiments.[35] Investigators did no truly new science instigated by his presence and with a test cohort of just one, experiments on Glenn did not generate statistically useful data.

What the mission did point out, and what was of immediate benefit to the average taxpayer, was Glenn's message of the importance of staying fit. NASA would not have flown Glenn if he had not met its astronaut fitness standards. The NIA enlisted Glenn for help with a national campaign to increase exercise among the elderly, and he took part in the advertising for an institute book, *Exercise: A Guide from the National Institute on Aging,* and wrote an essay crediting his lifelong exercise regime and healthy eating for making his participation in STS-95 possible.[36]

Extravehicular Activity (EVA) Suit Research at Ames Research Center and Johnson Space Center

In preparation for space station assembly EVAs and planetary surface operations, NASA began to look at a totally new space suit. It needed to be more comfortable and for a much longer period, maintainable at remote locations, tolerant of lunar and Martian environments, and cost effective. Because of its expertise in human factors, Ames Research Center got the assignment.

Their approach was to reexamine "hard suits," modular garments with a mostly hard exoskeleton and built-in life support. Early designers had tested the notion in anticipation of the 1970s Apollo 18–21 lunar missions, but those flights had been canceled. Ames engineer Hubert C. Vykukal had developed the "AX series," working in the 1960s with the

AiResearch Division of the Garrett Corporation (part of Honeywell after 1999). By the 1980s the series was up to model number five. Contractors were Micro Craft, Inc. of Tennessee; Air-Lock, Inc. (a Connecticut subsidiary of the David Clark Company); and ILC Dover of Delaware.[37]

AX-5 had rotating bearings in the joints to allow for a fuller range of motion, was milled from solid pieces of aluminum, and also had some steel parts. There was no cloth at all, a departure from past suits, and the garment was modular, with 15 major parts that could be swapped out for repairs or to accommodate wearers ranging from 4'10" (147.32 cm) to 7'0" (213.36 cm) tall. It had a double shell to protect the wearer from micrometeroids and would be coated with an extremely thin layer of gold—against hydrazine corrosion, for protection against single-atom oxygen, and to offer some thermal shielding. Underneath, astronauts would wear a liquid-cooled garment made by ILC Dover, which had made a similar undergarment for shuttle suits. AX-5 was heavy at 185 pounds (84 kg), but in LEO weight would not matter as much as it would on the lunar surface or on Mars.[38]

One goal was to eliminate the current need for "a major overhaul on Earth" after each mission or 30 to 50 hours of wear. A second was to allow wearers to don and doff them in seconds rather than minutes. Another goal was achieving a consistent level of internal pressurization that maintained the suit's volume, thus reducing the fatigue of maneuvering in a suit and eliminating the need for time "prebreathing"—purging the blood of nitrogen by breathing pure oxygen—before EVAs. Spacewalks involved exposure to significant changes in pressure, risking decompression sickness.[39] Fatigue and degradation of muscle and bone on long missions was a vital issue. Crews that had spent months in weightlessness traveling to Mars might not have enough physical strength or endurance to don a 185-pound (83.91-kg) suit.

In the spring of 1989, Ames personnel brought an AX-5 to the JSC Weightless Training Facility (WETF) and had four crewmembers act as test subjects. The suit had been undergoing KC-135 tests at Edwards AFB for ease of donning, removal, and translation (linear movement forward, backward, or to the side), and also had undergone a year of underwater testing at the small Ames Neutral Buoyancy Test Facility. No one had previously gathered metabolic data during simulations of actual use, however, and that required the use of astronauts in the large-scale underwater facility at JSC. The AX-5 also went to the Anthropometrics and Biomechanics Lab there for testing during strenuous activity.

Right: Underwater testing of the AX-5 suit in Ames' Neutral Buoyancy Tank. Photo credit: NASA-ARC.

Below: JSC RATS testing their Mark III hard suit in the Arizona desert. Photo credit: NASA-JSC.

Subjects used Cybex exercise machines wearing an AX-5, then the shuttle suit, while researchers collected metabolic information. In April, JSC installed a three-dimensional video motion-tracking system, as well as electromyography equipment and a force/torque plate. These allowed the simultaneous integrated collection of data on the electrical activity of skeletal muscles, reaction forces, and 3-D motion graphics. Designers could then calculate the acceleration, velocity, displacement, centers of gravity for each limb and for the total suited body, and joint angles and torques. In December, tests of both the AX and Mark series suits were still under way, but as the fortunes of Space Station Freedom declined so did those of the AX suit, although later designs absorbed some of its components.[40]

The Mark series was a homegrown JSC suit design overseen by project engineer Joseph Kosmo of the EVA Systems group within the Planet Surface Systems group. The latest model, Mark III, met the same performance requirements and was a zero-prebreathe suit, but only the upper torso was hard. The Mark III prototype weighed 154 pounds (70 kg) and was 65 percent metal, modular, and the wearer entered from the rear.[41] Durability and repairability were crucial since replacement parts would be difficult to obtain, even on station, so designers set a goal of 1,248 hours of use, nearly triple the 462 hours expected for shuttle suits.[42]

JSC's target suit weight was 80 pounds (36 kg), so designers tried to shave off ounces. They kept after physiologists for a firm number on an acceptable level of carbon dioxide (CO_2) inside the helmet. Being able to set levels higher than what NASA then allowed but no more than what undersea divers considered safe would allow them to use a hollow-fiber membrane, enabling a smaller CO_2 scrubber and a smaller fan. Likewise, being able to accurately predict the metabolic rate of an astronaut working on the Moon or on Mars would also impact the heat removal requirements. The group also pleaded for the development of "reasonable and practical REM (radiation) level requirements based on overall mission phases of activity." Radiation shielding was obviously necessary but its weight, if NASA mandated "extraordinary radiation protective measures," would aggravate fatigue, impair mobility, and slow the wearer down.[43] Neither could the suits be so cumbersome that an astronaut who tripped could not get up again.[44]

Gloves were actually the biggest problem because pressurizing them could make it impossible to bend the fingers. "High-operability suits," including "highly dexterous gloves," were a critical near-term goal.[45]

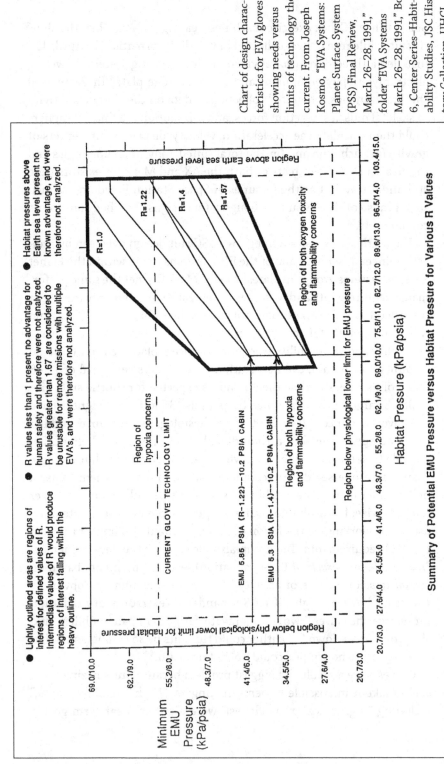

Chart of design characteristics for EVA gloves, showing needs versus limits of technology then current. From Joseph Kosmo, "EVA Systems: Planet Surface System (PSS) Final Review, March 26–28, 1991," Box 6, Center Series–Habitability Studies, JSC History Collection, UHCL.

● Lightly outlined areas are regions of interest for defined values of R. Intermediate values of R would produce regions of interest falling within the heavy outline.

● R values less than 1 present no advantage for human safety and therefore were not analyzed. R values greater than 1.67 are considered to be unusable for remote missions with multiple EVA's, and were therefore not analyzed.

● Habitat pressures above Earth sea level present no known advantage, and were therefore not analyzed.

Minimum EMU Pressure (kPa/psia)

69.0/10.0
62.1/9.0
55.2/8.0
48.3/7.0
41.4/6.0
34.5/5.0
27.6/4.0
20.7/3.0

Region below physiological lower limit for habitat pressure

Region above Earth sea level pressure

R=1.0
R=1.22
R=1.4
R=1.67

Region of hypoxia concerns

CURRENT GLOVE TECHNOLOGY LIMIT

EMU 5.85 PSIA (R–1.22)––10.2 PSIA CABIN

EMU 5.3 PSIA (R–1.4)––10.2 PSIA CABIN

Region of both hypoxia and flammability concerns

Region below physiological lower limit for EMU pressure

Region of both oxygen toxicity and flammability concerns

Habitat Pressure (kPa/psia)

20.7/3.0 27.6/4.0 34.5/5.0 41.4/6.0 48.3/7.0 55.2/8.0 62.1/9.0 69.0/10.0 75.8/11.0 82.7/12.0 89.6/13.0 96.5/14.0 103.4/15.0

Summary of Potential EMU Pressure versus Habitat Pressure for Various R Values

Lunar astronauts had reported that it was very difficult even to open one's hand and the constant motions of opening and closing led to aching hands and arms, even bleeding under the nails.[46] The University of Wisconsin and Astronautics Corporation of America had been working in the late 1980s on an EVA glove "tactile feedback system," which had built-in "electrocutaneous stimulators" in an internal ("comfort") glove, and were researching a newer generation of "vibrotactile simulators" small enough to be integrated into the glove.[47] One simple solution was to have multiple pairs of gloves for fine manipulation, handling large bulky items, or for specific tasks.[48] Fingers were a high-wear area and the Russians resolved the possibility of fingertip leaking with a pressure-sealing arm cuff at the wrist.[49] It was also a challenge simply making a garment so small and intricate. In the early 1990s, fabricators made casts of astronauts' hands in various positions and then individually measured the critical bending points. NASA hoped that new technology, laser scanning, could speed up that process.[50]

With suits to be used for lunar and planetary surface excursions, designers expected dust to be a major problem. Apollo astronauts had observed that dust on the Moon's moistureless surface remained suspended once stirred up, and, due to the static electricity that dryness caused, tiny, sharp-edged, unweathered regolith particles adhered to everything. Astronauts could not help tracking it into the landing modules, and once inside and able to remove their suits they discovered that the Moon had an unpleasant odor no one would want to live with for months in a sealed habitat.[51] The Planet Surface Systems team called lunar dust "the single most limiting factor to the continuation of extensive long-term EVA operations." The team worked to develop concepts for keeping the malodorous grit out of the living quarters, as well as the connectors, bearings, and mechanical linkages of the suits. One option was an "air shower." Adapted from similar units in clean rooms, the device could generate air flow at the rate of 9,100 linear feet (2,774 linear meters) per minute. The suited astronaut would stand on a special filter mat inside an airlock for at least 15 seconds while the laminar flow of ionized air coming from multiple directions would blow everything that could be removed downward onto the filter.[52]

Mars, unlike the Moon, has an atmosphere (albeit of carbon dioxide) in which dust would blow. In fact, the Red Planet has fierce dust storms. Martian particles would not be sharp but rounded, and were expected to be as much as a hundred times smaller than lunar grit. It was crucial to astronaut health that it be removed, as the smaller the particle, the more easily it could become lodged in airways and lungs.[53]

Ultimately, for a combination of political and economic reasons, NASA would opt to use two *different* suits on the ISS: a Russian Orlan when egressing via the airlock on the Russian module, and the shuttle EVA suit for egress through the US segment's airlock. Suit designer Hamilton Standard leased an Orlan in 1994 to study its design for advantages, especially in prebreathing time. Work went on for suits the Agency hoped would someday be used on a return to the Moon or to Mars.[54]

Closed Environmental Life Support Systems (CELSS)

Ames, JSC, KSC, and the John C. Stennis Space Center in Mississippi were all engaged in research on the use of plants for recycling, purification, and food production. Contractors and universities, mainly Utah State University, carried out some of the studies as part of programs that went by various names, such as "Biological Life Support Systems" or "Regenerative Life Support Systems." Overall, the technology and concept were referred to as CELSS ("sells"), for Closed Environmental Life Support Systems. ("Closed" systems did not require replenishment from Earth with chemicals, mechanical parts, and filters.)

Ames was the lead Center for machine-human interface, station interior design, and plant research and in 1983 had a contract with Boeing to examine means to cultivate plants on Freedom or in planetary habitats. Astronauts could supplement their diet with fiber and vitamins from fresh fruit and vegetables, and the sight and smell of a garden, plus the activity of gardening, could create a more Earthlike environment and provide psychological benefit.

Pursuing the cost-savings angle, Boeing looked at Agency forecasts postulating six typical missions: a low-inclination space station, a high-inclination station, a military post in LEO, a lunar base, an asteroid base, and a Mars mission. Then its engineers calculated how much recycling would be possible for each depending on the amount of sunlight, population, and mission duration. Given a six-year low-inclination mission, Boeing figured that regenerating 50 percent of the astronauts' diet would be the break-even point—that is, less costly than resupplying them with food from Earth. For an eight-year mission, astronauts could produce 97 percent of their food, raising fish along with plants and grinding container-grown wheat to make bread and pasta. The savings in 1983 dollars would be $68 million over 15 years. The Boeing team reckoned the break-even points at 5.5 years for a high-inclination station; 10 years for a military post housing 4–24 people; and 5.5 years for a lunar base with 12–48 occupants. The Mars mission was problematic in that a 50 percent system would not pay off for five years, by which time the crew would have returned to Earth. Likewise, the asteroid mining mission would have to use the 97 percent recycling system and would break even at 2.5 years, about the time it would take a resupply mission to get there.[55]

Ames and KSC opted to push CELSS for its nutritional and psychological benefits and the "cleanness" and ease of plant-based environmental recycling (no moving parts). That meant knowing how much food could be produced, of what variety, what nutrients it would provide, how many and what types of impurities it would remove and how, and what could be done with inedible parts of the plant. It wasn't even clear at first that plants would grow fully in microgravity. The 1985 Spacelab-2 flight of a Plant Growth Unit (PGU) with pine seedlings revealed that lignin production, which gives structure and strength to cell walls and also acts as part of the vascular system of fluid transport, was impeded in space.[56]

Boeing in 1989 proposed CELSS as a "total life support system" and started slowly to prove the concept. Initially, it would have been "merely a small cabinet-like facility in the space station's laboratory devoted to plant research." A system of seeds embedded in "accordion-type trays in a vertical stack, with plant foliage and roots growing out the sides" was already in the patent process. Bare roots would be fed by a misting system (aeroponics) that would be self-recycling. If the concept were successfully proven, next might come a half-module-size, CELSS-devoted area full of PGUs, "each about the size of a laundry basket." In step three, "around the year 2020 or so," Boeing's CELSS unit would be a separate module connected to the space station, roughly 45 feet long (13.7 meters) and 14 feet (4.3 meters) in diameter.[57]

Plants would clean the air and water and feed the astronauts, but by then Boeing had decided they would no longer serve as a psychological diversion. Humans carried too many viruses, fungi, and bacteria to which plants were susceptible, and raising food was "not necessarily the most productive thing the astronauts can do." Thus the contractor's concept included two robots: "Tracy," to handle the seed/plant trays, and "Harvey," who would harvest the useable biomass (plant matter). Both box-shaped robots, mounted on tracks running the length of the module, would monitor the spray system and inspect plants for disease. Light would come from a system of solar collectors, fiber optics, and light emitting diodes (LEDs), but "only those frequencies useful to plants would be collected by 'Fresnal' [sic] lenses." The Japanese were testing Fresnel lenses inside buildings and underwater, blocking out infrared and ultraviolet light, and cutting down on radiation and heat. Boeing gave a figure of $740 million for a "ready-for-launch CELSS module for the space station," saying that it would feed four and pay for itself in five to eight years. (At the time, the station was to be in orbit in the mid-1990s.)[58]

KSC had undertaken to develop and test prototype PGUs in 1986 to gain experience with technology that might be employed by astronaut-gardeners and to test environmental variables.[59] Set up inside the Biomass Production Chamber, originally a tank for testing Mercury and Gemini capsule airtightness, each PGU measured 2,825 cubic feet (80 cubic meters) and could hold 64 trays. Roots were maintained in a "thin film" of water and nutrients selected to give maximum yield to that specific crop.[60] As the focus was on larger scale space "farming," researchers hoped to see whether the system produced sufficient biomass and to solve problems created by the systems themselves, such as removing waste heat generated by the incandescent lighting then used.[61] As part of its Project Breadboard experiment, KSC scientists also had to find a way to use or dispose of dead plant material.

The KSC researchers tested potatoes, sweet potatoes, dwarf wheat, soybeans, sugar peas, snap beans, safflower, rapeseed (Canola), cowpeas, and peanuts. These met a variety of nutritional needs, such as carbohydrates, fiber, protein, and fat, and gave the greatest yield for the least light and nutrient solution and with the shortest growing period. Using a liquid growth medium, controlled lighting and temperature, and dwarf varieties, a team of Utah State agriculture scientists and KSC biologists were able to multiply wheat yield to four times that of traditional farms and achieve an edible biomass of 45 percent.[62]

Their research progressed through multiple generations into the mid-1990s. The Breadboard team was able to study every facet of plant farming, including the amount of equipment repairs and person-hours required to raise the crops. As their mechanical setup matured, they tried different lighting sources and more sophisticated water and nutrient delivery devices. They carried the work through the final harvest, including threshing the grain. The task was "probably the most labor-intensive activit[y] associated with any of the four species," wrote the authors of one study comparing wheat, potatoes, soybeans, and lettuce. "The threshing and seed sifting created a great deal of air-borne chaff and dust. . . . Mechanization of planting and harvesting and dust containment systems should be a high priority for future studies with wheat."[63]

Acreage was a trade-off in terms of the energy-consuming lighting required to successfully and reliably raise crops, even in a biologically ideal environment. Using only natural lighting on the Moon would require an estimated 40–50 square meters (431 to 538 square feet) of growing space per person. In a high-density hydroponic chamber, Utah State's chambers grew sufficient food in an area just 12–13 meters square per crewmember (129–140 square feet), thanks primarily to controllable artificial lighting and atmosphere. Project Breadboard scientists learned that the best yields for wheat and potatoes were in a CO_2-rich environment, cool temperatures, and a lot of light. The cost for those crops had to be weighed against the output of rice and soybeans, which prefer more darkness but might not yield as much food.[64] Crops with vertical leaves, such as wheat, were efficient at high levels of light, while crops with horizontal leaves, such as potatoes, were efficient where light levels were low. Nutrient uptake varied over the plant's lifecycle, so different amounts of plant food had to be added at different periods. Longer "days" made wheat grow faster, but resulted in fewer seeds per head. Higher CO_2 caused more photosynthesis but decreased the concentrations of minerals needed to prevent sterility. Higher temperatures also were a factor in producing enough seeds per head, the edible part of the wheat.[65] Algae separately came under scrutiny at KSC for its nutritional and bioreactive properties. It, too, could be manipulated to improve yields.[66]

By 1989 Ames, with its prototype Crop Growth Research Chamber, was also evaluating specific food crops like wheat, lettuce, soybeans, and potatoes for the amount of biomass attainable in a weightless, hydroponic (or aeroponic), artificially lit, temperature-controlled growth chamber. They measured the growth time from planting to harvest, the germination percentage, size of the vegetable, and the percentage of the plant that was edible. The uniqueness of the Chamber was that it tested the closed-loop aspects of space crop production. Each environmental factor could be manipulated for maximum growth rate and usefulness (edibility, oxygen production, pollutant removal) by changing the amount of plant food in the water, variety of hybrids used, quality of light and atmosphere, temperature, density of planting, position of the seeds in the germination medium, and possibly the amount of g-force to which the growth chamber would be exposed.[67] Researchers even experimented with a simulated Martian atmosphere, and the results were promising, although not stellar. Supplementing it with oxygen, biologists found that seeds had a 70 percent germination rate, although they grew at only 30 percent the rate of Earth plants.[68]

Within a year the CELSS project researchers in Ames had a definite growth chamber design in mind and gave it a name that NASA thought would sound appealing: the "salad machine." Seeds would go into orbit prepackaged in "cassettes," be allowed to sprout, and then astronauts would move them to a soilless "nutrient-delivery" system inside a small, sealed chamber. Engineering constraints for Freedom meant room for just one growing rack measuring 36 by 41.5 by 80 inches (91.4 by 105.4 by 203.2 cm), or about 69 cubic feet (2 cubic meters). It would have to use less than a kilowatt of power and produce minimal waste heat, meaning LEDs not incandescent lights. The University of Wisconsin and Automated Agriculture Associates in Dodgeville worked to overcome the key stumbling block: reproducing light in the blue portion of the spectrum.[69] The goal for NASA's machine, per Project Manager and Principal Investigator (PI) Mark Kliss, was to produce three salads per week per person for a crew of four, plus clean the air and water. With no weeds or insect pests in space, the cost in astronaut time to plant and harvest the salads would average just 15–20 minutes every few days.[70]

JSC also became involved in CELSS, and its Regenerative Life Support Systems underwent tests in early 1991. The Engineering Directorate's Crew and Thermal Systems Division planted Waldman's Green Lettuce seeds in 480 receptacles and then harvested the crop. This was part one of a four-part study, using a substrate to give roots an anchor in less-than-Earth gravity and prevent them from becoming waterlogged.[71]

If lunar or Martian regolith contained the right elements, that would lessen the need to transport plant nutrients from Earth. JSC studies of Apollo lunar samples showed that the Moon had the "precursor primary minerals of terrestrial soils" but some key nutrients, like nitrogen and phosphorus, were missing. Lack of weathering and prevalence of meteorite activity had created regolith with a preponderance of glasses and minerals such as olivine (chrysolite), a magnesium iron silicate; pyroxene, a silicate mineral typically found in volcanic lavas; and anorthosite, a kind of feldspar crystallized from magma. These were chemically soluble in the acidic environment that growing plants created, and magnesium, iron, calcium, and a small amount of potassium could be dissolved out in the presence of such acidic water. However, the amounts of potassium were insufficient for plant health, two heavy metals (nickel and chromium) might be released into the biosystem, and repeated plantings could reduce the pH to a level at which a toxic amount of aluminum would be present as well. Legumes such as peas and beans could fix critical nitrogen in soil but not enough to avoid having to import nitrogen from Earth, though researchers found that enriched CO_2 levels increased nitrogen fixing. Extra nitrogen, in turn, symbiotically increased uptake of carbon dioxide generated during waste recycling. With time, the waste recycling system would generate some phosphorus, an essential human nutrient and plant fertilizer. CELSS could not exist successfully as merely a salad machine, but would have to be part of a fully integrated system that produced food *and* cleaned the air, water, and solid waste by separating out impurities either present or added by the CELSS process.[72]

Research at Stennis, ongoing since the 1970s, made a strong argument for CELSS purification of air and water. Although most studies there centered on aquatic plants used on a large scale, Stennis scientists performed controlled tests, fertilizing tomatoes and string beans with a weak (0.5 percent) urine solution. Human urine is sterile, 2.5 percent urea (a fertilizer ingredient) and 2.5 percent other elements including hormones, amino acids, and enzymes. After pumping the solution through containers holding the roots, 15 minutes on and 45 minutes off every 24 hours, and replenishing the solution every week over four months, plants flourished. Cherry tomato vines grew 11 feet (3.35 meters) high and bore an average of 210 fruits per plant. Stennis researchers also tested a system using household wastewater on plants embedded in a container of gravel and oyster shell and found that tomatoes and string beans not only thrived in the wastewater, but that they also removed nearly all of the particulates and toxins within a week.[73]

Some Stennis plants, because of their symbiotic relationship with the naturally occurring microbes in the rhizome (soil area adjacent to roots), could accelerate the breakdown of industrial contaminants like trichloroethylene, used in adhesives. Roots drew in other contaminants such

'Yecora Rojo'
83 days old

NASA CELSS research
focused on using food crops
for nutrition and environmental cleaning both. Photo
credit: NASA-MSFC.

as benzene, cyanide, and phenol, where they were broken down within the leaves. Some could also remove heavy metals such as cadmium, lead, and mercury from wastewater and concentrate it in the roots. Water hyacinths removed 92 percent of the cadmium and mercury, 95.5 percent of the lead. Scientists grew plants in soil contaminated with cesium, strontium, and cobalt and included food crops that might be grown by astronauts. Tomatoes in particular could take in huge amounts of these radioactive elements.[74]

After a name change to Advanced Life Support System, which incorporated physicochemical environmental scrubbing, JSC research continued into the late 1990s, carried by the momentum of President George H. W. Bush's Space Exploration Initiative (SEI). Plant-only recycling remained more or less permanently under study, but something of a stepchild within planetary or space station programs. The Life Sciences Data Archive listed the Growth Apparatus for the Regenerative Development of Edible Nourishment (GARDEN), the Variable Pressure Growth Chamber, and the Life Support Systems Integration Facility in the 1995–1997 Lunar-Mars Life Support Test Project (formerly the Early Human Testing Initiative).[75] The next decade brought UW's Advanced Astroculture test unit with Expedition 2 to the ISS, flown again with the Biomass Production System and the Photosynthesis Experiment and System Testing Operation (PESTO) experiments on Expedition 4, and the Plant Generic Bioprocessing Apparatus on Expedition 5, to study the structural components of plants. NASA manifested Optimization for Root Zone Substrates for the 2007 ISS Expedition 15.[76]

In spite of the 40 percent budget increase that SEI provided and the promising research by industry, academia, and NASA scientists, CELSS research elsewhere stalled in the 1990s. The NASA Specialized Center of Research and Training (NSCORT) for CELSS at Purdue University opened in 1990 and closed in 1995. NASA's chief visionary for CELSS, planetary terraforming, and environmental life sciences in general, Mel Averner, retired from the Agency in the late 1990s. As CELSS program manager, he believed the best applications for space farming were inhabited lunar and Mars bases, yet noted that there was "no NASA approved mission" for CELSS.[77] The "no mission" argument that had grounded women astronauts in the early 1960s and killed Max Faget's free-flyer in the 1980s would doom hard suits, CELSS and CERV in the 1990s. In 2005, with the second President Bush's Vision for Space Exploration

(VSE), it would also eliminate a crucial piece of space station research equipment, the Centrifuge.

Organizational Restructuring and Station, 1980–1992

As NASA moved from Apollo to STS and the station, organizational changes in life sciences areas were inevitable. New vehicles, a new astronaut cadre, and new types of missions represented completely different environments and ways of working, the thinking went, and consequently required a fresh management approach. That meant internal structure changes at multiple levels and places within the Agency.[78]

In September 1981, Administrator James Beggs formally made Associate Administrators (AA) the "major line officers of NASA," responsible for "program content and execution." Instead of 11 Centers and 19 other entities reporting directly to him and a Deputy Administrators, Center directors would report to a specific AA, who would also be responsible for that Center's "program and institutional resources." The Office of Space Science Applications would be comprised of the former Office of Space Science and the Office of Space and Terrestrial Applications, due to program reductions and similarities in technology.[79] The Life Sciences Division, under biologist Gerald Soffen, was part of OSSA.

By June 1982, head of the OSSA was Associate Administrator Burton I. Edelson, an engineer who favored robotic missions over crewed ones. AA for the shuttle program was then-Maj. Gen. James A. Abrahamson.[80] They were preparing to run an ambitious shuttle operation that would be flying to a permanently manned space station by 1990. This plan did not include a major role for life science research as a paying passenger or an internal NASA priority, but Beggs's appointment in late 1983 of Arnauld E. Nicogossian, then Chief of Operational Medicine and a former JSC flight surgeon, to be Director of Life Sciences did seem to show an emphasis on astronaut (and passenger) health, safety, and productivity. When astrophysicist Lennard A. Fisk, later Chief Scientist for the Agency, replaced Edelson as OSSA head in 1987, that appointment proved to be a move in favor of the life sciences, as Fisk supported such work and was able to increase funding during his tenure.[81] Fisk, though, explained the support in a 2010 interview as being primarily an outcome of an early 1980s agreement between Beggs and the chair of the National Academies of Science's Space Studies Board, Thomas Donahue, who was then "leading the charge of the science community against

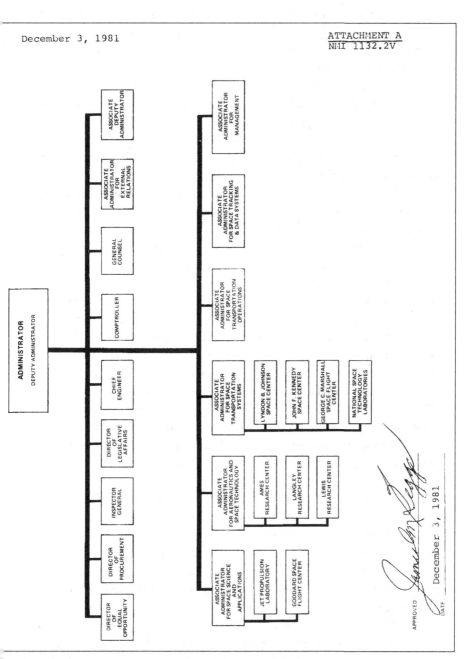

eadquarters organizational chart, Dec. 1981. From "NASA Organizational Charts," *NASA History*, NASA History Division, https://history.nasa.gov/orgcharts/63HQorg81-12-3.pdf.

the Space Station." The agreement was brokered by NASA Chief Scientist Frank McDonald, Fisk said, and essentially promised that space science would "in effect" receive 20 percent of the annual NASA budget if Donahue and the Board backed off on the protests. A 1982 Beggs interview supports this, with a remark that NASA's overall R&D budget was increasing 13–14 percent in FY 1983 "but the space science budget will be up almost nineteen percent. So it has received the major increase and the major emphasis." In the early 1980s OSSA did not have to pay for the launch vehicles, tracking, and some other facets of a project so a steady 20 percent guarantee was a very meaningful number.[82]

The Life Sciences Division was responsible, in concept, for strategic planning, prioritizing missions, and overseeing Centers' proposal, peer review, and selection processes.[83] The Centers, however, still had strong, independent life sciences groups, a potential source of conflict. The 1982 Space Station Life Science Research Facilities study project, headed by William Bishop, was run such that people in Houston charged with astronaut care were designing the crew health, safety, and productivity-related components of the space station. Life scientists at Ames were designing work areas and space suits, overlapping with some JSC functions and Ames also expected to research medical needs of astronauts, another JSC task. Three groups worked on habitats: the MSFC conceptual personnel looking at the way modules fit together structurally worked independently of designers at JSC (whose chief concern was safety and ergonomics) and Ames personnel studied interior design for psychological enhancement. Although Bishop was leader, the project was heavily dependent on the initiative of Centers cross-checking each other's needs.[84]

By November 1986, Ames's leadership seemed more hesitant about its role in station development, saying it was "still being defined" but was principally "technology development," specifically autonomous systems in line with its work in computers, space suits, human factors, and enhancing a facility's livability and productivity. Ames was also "a science center," doing research in bone demineralization, muscle atrophy, and cardiovascular deconditioning.[85]

JSC did not have the luxury of taking a back seat but seemed unclear about its role or roles with the space station. A February 1987 organizational chart showed 18 offices and directorates reporting to Center Director Aaron Cohen, with two devoted to space station work. Director of Space and Life Sciences Joseph P. Kerwin had no obvious station

Ames Research Center organizational structure, 1982. From Richard G. Mulligan, *The NASA Organization,* November 1984, in folder "The NASA Organization, November 1984," box 14, Organization Series, JSC Files, JSC History Collection, UHCL.

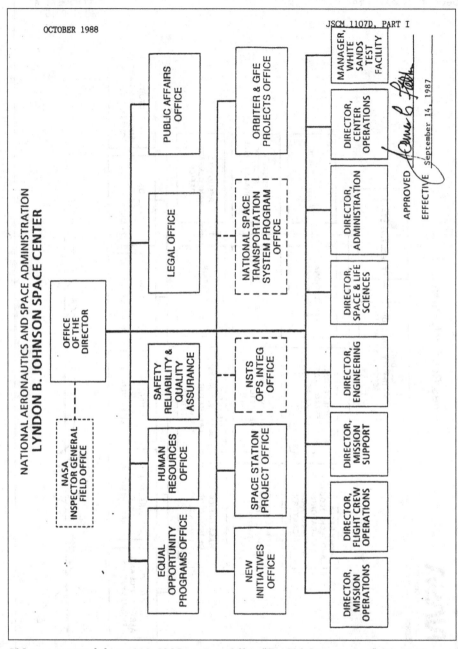

JSC organizational chart 1988. JSC Directives Office, "The JSC Organization," Oct. 1988, folder "1988, 1990," box 14, Organization Series, JSC Files, JSC History Collection, UHCL.

responsibilities, while the shuttle, engineering, and administrative organizations each were in a box designating a direct role in the station. Acting Manager of the Space Station Program Office, John W. Aaron, another of the 18 people under Cohen, had boxes for systems engineering and integration, international and external affairs, customer integration, data management and operations, technical and management information systems (MIS), and program management. His peer Clarke Covington, Manager of the Space Station Projects Office (SSPO), had avionics, structural elements, management integration, user and international relations, systems, test and logistics, manufacturing, and engineering. There was no overt mention of safety or ergonomics. In spite of a Medical Sciences Division, Life Sciences Project Division, and Man-Systems Division within the Space and Life Sciences Directorate, there was no obvious link between those responsibilities and space station design and functioning at JSC.[86]

As for Marshall Space Flight Center, biology was a payload that needed to be managed like any other. Positions for Spacelab management, science payload projects, and microgravity projects fell within the Payload Projects Office, the head of which reported to then-Director of Marshall J. R. Thompson Jr. Another unit that reported to him, the Space Station Project Office, headed by L. E. Powell, employed personnel who dealt with habitability and laboratory modules.[87] Associate Director for Space Systems, E. R. Tanner, who reported to the head of the science and engineering group J. B. Odom, held responsibility for experiments and instrument development and Spacelab payload integration, along with work on the Hubble Space Telescope, the Advanced X-ray Astrophysics Facility (AXAF), and other general, nonmedical, and nonlife sciences station and spacecraft tasks.[88]

By the end of 1987, as the lines of responsibility for the space station became less clear, the OSSA reorganized again. The announcement to employees implied that this was because station planning had progressed, and the need for individuals with more specific responsibilities in various scientific disciplines had become obvious. Associate Administrator Fisk sent out an organizational chart with new positions in space physics, astrophysics, microgravity sciences and applications, solar system exploration, and Spacelab. While Fisk could cite individual astronomy and physics projects by name—Hubble Space Telescope (HST), Solar Max, the International Solar-Terrestrial Physics balloon, and rocketry programs—he specified nothing for the life sciences because there

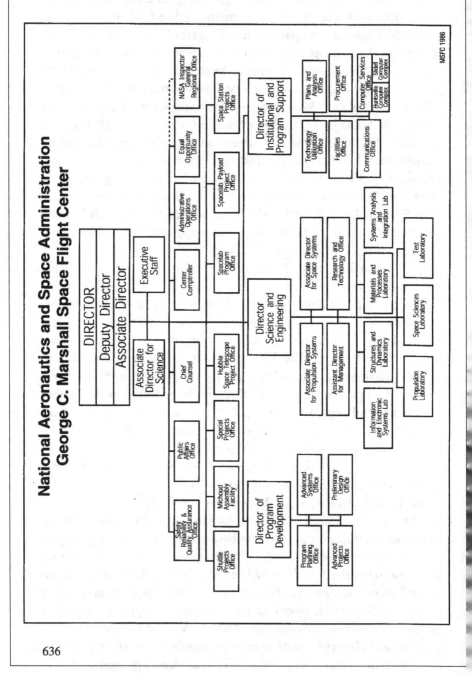

Marshall Space Flight Center organizational chart, 1986. From Andrew J. Dunar and Stephen P. Waring, *Power to Explore: A History of Marshall Space Flight Center 1960–1990*, SP-4313 (Washington, DC: NASA, 1999), 636.

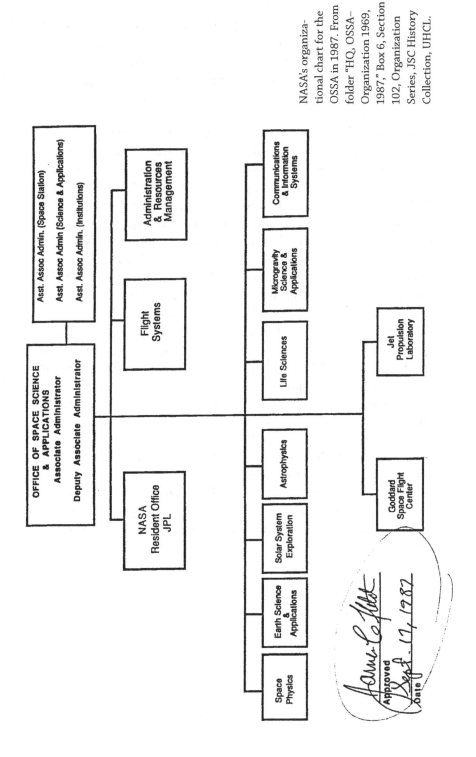

NASA's organizational chart for the OSSA in 1987. From folder "HQ, OSSA–Organization 1969, 1987," Box 6, Section 102, Organization Series, JSC History Collection, UHCL.

was no particular mission dedicated to it under way or in the immediate pipeline. (EOS was "materials processing.") However, the chart did contain a box marked Life Sciences as one of seven generic disciplines, such as Space Physics, Earth Science and Applications, or Astrophysics.[89]

In the early 1990s, when Congress came within one vote of canceling the station, NASA employees at all levels had to redesign, rethink, and replan for a scaled-down station. The cutbacks eliminated astronomy, satellite repair, and other functions and so left basically an orbiting laboratory for biomedical research on the effects of the space environment and for (bio)materials processing studies. Because life sciences was at the heart of the argument over whether to retain the station or cancel it, there was some thought about refocusing the division, moving the research component into the Office of Exploration, and thus committing it to the Moon and Mars ventures George H. W. Bush proposed in 1989. However, others, including the Bush administration itself, said the redesign need not emphasize science at all, but just demonstrate the United States' continued *technological* superiority.[90]

That had to sting, at least in the short term. During the Clinton-Goldin years (1992–2000) would come significant restructuring, including organizational consolidation for life sciences and overall "reinvention" of NASA. Projects and programs were able to make headway, increase funding modestly in some areas, and strengthen relationships with the scientific community. A new Office of Life and Microgravity Sciences and Applications (OLMSA) would be dedicated in 1993 to space life sciences and report directly to the NASA Administrator. Joan Vernikos noted that OLMSA "had no institutional authority over a Field Center and no coordinating authority over the life sciences programs elsewhere in NASA."[91] However, its managers, programs, and projects, did have the ear and the attention of the new administrator, and that would allow life sciences to flourish, if only for a time.

3

Safety, Science, and Operational Medicine

Shuttle and Station in the 1980s and 1990s

Others have written extensively about the 1986 *Challenger* accident and its causes and outcomes, so it will not be the focus here. However, that event, along with increasing emphasis in the late 1980s on the upcoming space station, underscored the need for better preventive measures and for safety and life support systems, including escape mechanisms, to be designed into vehicles and platforms. It also gave rise to thinking about alternative access to LEO to fill extended gaps during a system-wide shuttle grounding—in this case, two years and eight months of lost time.

Shuttle Safety Improvements Post-*Challenger*

NASA realized that shuttle astronauts could not escape from every emergency scenario, but there were measures it could take to improve the odds of their survival. These included new protective garb and reexamining the question of ejection seats. North American Rockwell had equipped *Columbia* with pilot and copilot ejection seats for the first two missions. Launch abort plans included attempting to steer the orbiter, intact, to a low enough altitude for a survivable ejection should pilots be unable to coax it to a landing strip. There had never been any such plan for passengers later, however, so NASA ordered the two seats removed from *Columbia* after the final two-person R&D mission, STS-4, and did not have them installed in later orbiters.[1] The subsequent launch abort plan called for attempted ditching if an orbiter could not return to a landing strip. The seats had a 16-g crash rating, but Chief of the Astronaut Office John Young, a veteran of six Gemini, Apollo, and shuttle missions, thought ditching would not be survivable and told Houston management so in 1984. He advocated a tractor-rocket extraction system, which was basically an ejection device to propel astronauts out the side of the vehicle.[2]

The JSC Engineering and Development Directorate reexamined the ditching issue in 1985 and found cause for optimism. Engineers from the Spacecraft Design Division reviewed 1:20 scale ditching tests done at Langley Research Center (LaRC) in 1974 and JSC analysis from 1981. These indicated that the external shape of the orbiter promised "good ditching geometry." Similarly, the long nose and absence of engine nacelles on the wings reduced the "potential of diving" and the smooth lower surface theoretically would keep an orbiter stable in the water. However, structurally the bottom of the shuttle was likely too fragile to withstand the impact, and in 50 percent of the tests the crew cabin compartment broke loose inside the vehicle. JSC designers concluded that crew survival was "highly dependent upon structural damage. With minor damage, the flotation attitude will initially be rather high. With major damage—cargo bay and wings flooded—a watertight crew cabin can provide flotation but the water level may cover the egress hatch."[3]

After the *Challenger* accident, the Engineering Directorate first revisited water ditching, then investigated bailout (rather than ejection). McDonnell Douglas conducted a ditching study in 1987 and, after modeling a number of scenarios, concluded that the mid-fuselage would need triple the number of frames and the payload bay aft bulkhead also would require such extensive modification that redesign would be "prohibitive."[4] In February and March 1988, test drops of dummies, then live jumps by military parachutists from a C-141 at Edwards AFB "up to the Orbiter 200 knot [230.16 mile/370.4 km per hour] range," demonstrated the feasibility of crew bailout, albeit only at certain altitudes.[5] Astronauts would deploy an extendable escape pole, exit via a harness system that connected a parachute pack on their backs to the pole, slide out, and drop back to Earth. The pack also contained a bailout oxygen system, water-activated one-person life raft, sea rescue dye, and 2.11 quarts (2 liters) of drinking water. A flotation device would keep an unconscious person's head out of the water and an extra drogue parachute helped ensure the astronaut would land feet-first. An automatic

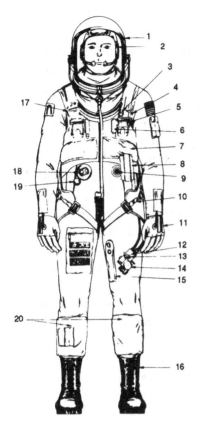

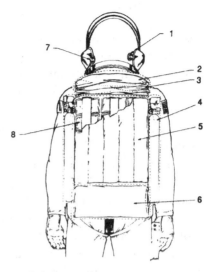

1 Antisuffocation device
2 Drogue chute
3 Pilot chute
4 Emergency oxygen system
5 Parachute pack • automatic opener
• main canopy
• locator beacon
6 Life raft (LRU-18) • sea dye marker
7 Radio communication line (to headset interface unit)
8 Emergency water supply

1 Helmet
2 Communications carrier (comm cap)
3 Parachute arming lanyard (red apple)
4 Rip cord handle (D-ring)
5 Upper parachute attach fittings (frost fittings)
6 Sea water activation release system (seawars)
7 Life preserver unit (LPU)
8 Carabiner
9 Suit ventilator valve
10 Lower parachute attach fittings (ejector snaps)
11 Suit gloves
12 Suit oxygen manifold
13 Suit oxygen on/off valve
14 G-suit controller valve
15 Shroud cutter
16 Boots
17 Pole lanyard attach ring (harness ring)
18 Suit controller valve
19 Emergency O_2 system activation lanyard (green apple)
20 Survival gear
• survival radio
• signal mirror
• chem lights
• pen gun flare kit
• smoke flare
• motion sickness pills

Sketches of new egress gear, front and back. From James W. McBarron II, et al. "Individual Systems for Crewmember Life Support and Extravehicular Activity," in *Space Biology & Medicine*, Vol. II., edited by Nicogossian et al. (Reston, VA: AIAA, 1996), p. 284.

parachute release was activated by contact with seawater so the sodden chute would not pull the jumper under. The C-141 tests, performed by Navy parachutists in both the tucked and head-first prone positions, and with poles of two different lengths, demonstrated that such egress need not cause uncontrollable swinging or spinning, or overstress the load capacity of the exit pole. Crew could get through the escape hatch quickly and smoothly, and safely clear the orbiter wing.[6]

The pole would need to be stowed in pieces and the astronauts would set it up shortly before reentry. Vibration tests showed that its connecting pins would become deformed after 30 minutes, making them difficult for the test team to remove for stowage, and perhaps impossible to redeploy should astronauts ever need to do so twice on the same mission. Quality Assurance was still working on the problem in late August 1988, planning another 10 "dummy drops," followed by 20 more dummy/pole integration jumps. By September 1988 the parachute system had been redesigned again, retested, and was ready for final review and certification.[7]

Retesting revealed another complication. The crew first had to equalize cabin pressure so the door would open. They accomplished this by means of an emergency vent, before jettisoning the hatch. The plate that covered the vent might blow into the orbiter and damage control cables, however, so JSC engineers came up with a "debris-capturing device" to snag the plate before it could destroy critical orbiter components. They managed to design, fabricate, and ship a device within three weeks for bonding over the vent panel. With the new device, shaped charges on the panel would "blast the fragments out into the payload bay at an estimated four hundred feet [121.92 meters] per second," to be caught "in a Kevlar net with 'rip away stitching' to absorb energy." The Kevlar net blew completely off in tests so they tried again with a higher-strength cold-bond epoxy. A month later engineers added an aluminum honeycomb plug under the new net "to catch the smaller shrapnel . . . previously unidentified and first experienced in the TTA [Thermochemical Test Area] test." Higher-priority testing delayed a lab evaluation, however, so it was not ready for the first return-to-flight mission.[8]

JSC personnel were having less luck with an emergency egress slide, reporting "the problems with the release of the outboard cover [have] matured to the status of a hazard." The escape parachutes also caused some concern when several of them deployed "at a higher than desired altitude," theoretically exposing the jumper to a long ride through an

Left: The new escape pole can be seen to Mission Specialist John Grunsfeld's right. Photo credit: NASA.

Below: CSA Mission Specialist Julie Payette simulates a water egress using a mockup of the new pole escape system. Photo credit: NASA.

extremely cold, oxygen-thin region of the atmosphere. Cable modifications "apparently fixed the problem," the Flight Crew Operations group added, but not well enough to certify the equipment. In fact, canopies failed to deploy at all in two tests. An oxygen system that gave an astronaut 10 minutes of breathing time during descent was unreliable, too. During a test run in October 1988 at the White Sands Test Facility (WSTF) the oxygen regulator leaked. In January 1989, the group reported that none of the regulators they had tried was yet working.[9]

Extremely high-altitude bailout also presented a problem in that lack of air pressure could be fatal. At 63,000 feet (19,202 meters), about 11 miles (17.7 km) high, human blood boils due to the low pressure, and so the post-*Challenger* orange Shuttle Launch and Entry Suit (LES), part of the Crew Altitude Protection System (CAPS), was a counterpressure/anti-exposure garment. It had bladders that would apply enough pressure to various parts of the body for the wearer to survive at altitudes of 100,000 feet (30,480 meters) for about 40 minutes.[10] In the winter of 1988, the USAF School of Aerospace Medicine at Brooks AFB in Texas conducted nearly two dozen explosive decompression tests on the suit. In February, the Naval Air Development Center in Pennsylvania conducted cold water tests for survivability on an engineering model of the suit. The Brooks Human Use Committee approved it in March 1988 for testing on people.[11]

The Space and Life Sciences Directorate under Carolyn Huntoon began planning in March 1988 to add a connection for electrocardiogram leads and created a working group to get the suits ready in two weeks for the first Return to Flight (RTF) mission, STS-26. It proved extremely difficult to rapidly modify the suit to include a "penetration" through which wires could be fitted without compromising the integrity of the pressure system. A newly formed Biomedical Instrumentation Port Working Group met with the suit manufacturer, the David Clark Company of Worcester, Massachusetts. They debated the pros and cons of various spots such as the front, side, or leg; portals such as a "generic hole"; or wires snaking in through a ventilation tube. In December, the group had to go before a JSC board to request authorization to install a half-million-dollar hole in their new Launch and Entry Suits. Meanwhile, reach and visibility tests made with the assistance of the STS-26 crew exposed difficulties for pilots in operating some switches and buttons while wearing the entire CAPS ensemble. Rockwell would have to move their seats 3 inches (7.62 cm) aft to compensate for the thick para-chute and life-raft pack strapped to the pilots' backs.[12]

The JSC Space and Life Sciences Directorate also managed some of the orbiter safety upgrades. Rockwell was attempting to meet a require-ment for a 16-g load on the crew, and then the Orbiter and GFE Proj-ects Office in Houston told the Flight Crew Support Division to check out lighter weight seats, possibly in time for STS-26 in September.[13] In October, the Man-Systems Division (MSD) organized the Orbiter Seat Strength Working Group, composed of representatives from the Federal Aviation Administration (FAA), Navy, Army, Air Force, Langley, Rock-well, the commercial aircraft industry, and seat manufacturers, to de-termine the actual crashworthiness requirements. By December, MSD engineers were test fitting the seats in launch position with astronaut volunteers of various heights and found the new seats needed modifica-tion to protect their knees. Designers moved mission specialist seats 5.6 inches (14.2 cm) aft. Commenting on the crew compartment upgrades, the Working Group called it "quite robust, significantly better than most aircraft," but still noted that "the seat strength is significantly less." They "felt very strongly that the seats should be as strong as the supporting structure." All of these improvements, JSC personnel estimated, now could not be implemented before STS-29 in March 1989.[14]

While studying ideas for a possible shuttle follow-on in the mid- to late 1980s, ejection seats came up again and John Young (after May 1987 Special Assistant for Engineering, Operations, and Safety at JSC) lobbied Administrator Richard Truly to include them or an escape pod.[15] The National STS Payload Safety Review Panel also had formu-lated policy regarding emergency evacuation of Spacelab modules inside an orbiter. The panel allowed a maximum of five minutes for stowing and securing all payload elements that might "present an uncontrolled penetration hazard to the Spacelab module." Astronauts responsible for Spacelab would then exit the orbiter with the others if needed.[16]

There had always been a contingency for the shuttle to land at emer-gency backup sites: Morón and Zaragoza in Spain, Dakar in Senegal (changed to Banjul, Gambia, 1988–2002), and Casablanca in Morocco (at the Ben Guerir Air Base near Marrakesh in 1988). Runway lengths post-*Challenger* were not long enough, so NASA took the Rogers Com-mission's recommendation and installed arresting nets to avoid over-runs. All-American Engineering Company (later part of French-owned

Zodiac Group, S.A.) installed the nets in February 1987, but they were never used.[17]

Records from JSC archives also reveal discussion about destruction of the shuttle stack on the pad in case of fire or after liftoff if voice and telemetry communication failed. The assembly had a destruct mechanism, originally shaped charges attached to the solid rocket boosters (SRBs), that could only be set off by the Range Safety Officer (RSO) at the Cape Canaveral Air Force Station at Patrick AFB. According to Young, they were added when ejection seats were included in shuttle design, on the theory that a pilotless vehicle was uncontrollable and might crash into an inhabited area.[18]

However, studies in the mid-1970s had shown the SRBs' shaped charges to be insufficient to break up the huge external tank (ET) and noted that the tank's "large quantities of LOX [liquid oxygen] and LH2 [liquid hydrogen] present a considerable potential explosion hazard upon ground impact should the ET fall to the surface relatively intact." Dispersing the two gasses by exploding the tank while in the air would "reduce the impact area hazard by reducing the explosive yield." The 1976 report recommended addition of linear shaped charges to the external tank's operational instrumentation tray to assure destruction at any point in the first 100 seconds of flight. This was called the Triplex Command Destruct System. NASA asked study authors at the Naval Surface Weapons Center (NSWC) to reexamine this. The Navy's 1981 report noted that in most instances this would also destroy the orbiter.

Auto destruct was also considered for instances when something caused the stack to explode or come apart. Among these, the NSWC estimated the probability of an "SRB case/nozzle failure" to be greatest. "Eastern Test Range Personnel estimate that a few seconds may elapse before it is definitely determined that the cluster has lost a major component," their report read. "Additional seconds may be consumed in consolidating and evaluating all information pertinent to the loss. As a consequence, it may be 10 to 15 seconds after the initial loss that the command destruct signal is given."[19] A casing failure and ground-instructed SRB destruction happened with *Challenger* in 1986, after the initial catastrophic breakup.

Another study in 1987 looked at scenarios in which the entire stack might "hammerhead in" if one SRB were lost but concluded that it would not.[20] Wayward motion of a damaged SRB would quickly cause it to part from the remainder of the stack, which "simulations show . . . continues to climb until instability causes angular rates to exceed structural allowables or until RSS [Range Safety System] action can be implemented. Since the separation of the failed SRB from the stack constitutes 'vehicle breakup,' detection of this event permits the RSO to initiate range destruct while the remaining stack is still climbing. Thus complete breakup of the stack will occur in either case before ground impact."[21]

To save an intact shuttle and crew moving uprange away from the Cape would require the malfunctioning SRB to move cleanly away and the pilot to rapidly jettison both external tank and remaining SRB before the Range Safety Officer hit the destruct button, then execute a contingency landing—or ditch. Range Safety personnel had a few months earlier mused that they "must treat the shuttle as an unmanned expendable launch vehicle. It will always be difficult, if not impossible, to treat the shuttle as an unmanned vehicle. Therefore, the intact high-velocity impact scenario must be considered credible."[22] In other words, if the orbiter had not shaken off both external tank and SRB, the military was telling NASA that it could not guarantee RSOs would kill astronauts rather than watch them "hammerhead in."

The Rogers Commission report on the *Challenger* tragedy, issued in June 1987, recommended NASA and the Air Force "critically examine" removing the external tank destruct system "since the current range safety system does not allow for selective destruction of components."[23] Range Safety was "reluctant to agree," and Marshall Space Flight Center's Shuttle Projects Office began studying a "compromise solution" of "safing" the external tank explosive system after SRB separation if no population centers were in danger.[24]

When astronaut Bryan O'Connor attended the first Crew Safety and Rescue Workshop at the 1989 Paris Air Show and asked the USSR manned program's Chief Designer Yuri Semyonov about range safety for their shuttle, Buran, the Soviet had replied, "Why would you want to destroy a manned space vehicle?" O'Connor shared this with the astronaut corps, plus a chart and graph he had made showing that roughly a million Russians lived under the path of the first Buran flight. Special Assistant Young forwarded it all to Headquarters. The Soviets were not unconcerned about safety of their citizens, he argued; they had simply realized that more damage would be done by "fragments and shrapnel," just as "cluster bomblets will produce a much larger lethal area than any single bomb." A system of gradually terminating thrust on the SRBs and shutting off the main engines, Young wrote, would improve the chance of an abort with intact return to KSC.[25]

He rebuked his superiors with the same points in 1990. *Challenger* "demonstrated that the Range Safety System is not required on the External Tank. When cryogenic propellant leaves the tank . . . dispersal will be instant. . . . When the liquid hydrogen and oxygen cryos flashed to gas, everything appeared to explode. . . . The tank cannot survive in ascent. Even if [it] did . . . the damage caused might well be much less than would result from Range Safety System detonation."[26] Young suggested sardonically that "we get legal guidance before we put a Range Safety System on the Advanced Solid Rocket Booster that when activated will kill the flight crew out right—*after* which it *might* be protecting life and property on the ground (according to a statistical analysis)." During 75 percent of the ascent phase, Young wrote, Launch Control could stop a wounded STS by simply shutting off main engines, terminating the SRBs' thrust, telling the crew to return to KSC or eject, and then let things fall into the Atlantic.[27] Young continued to press for changes he believed would increase crew survivability, such as a no-night launch/landing policy. Among the drawbacks to doing either in the dark were the RSO's lessened ability to physically see what was happening, decreased depth perception at night for the pilots, fewer runway options, and increased difficulty of water rescue in the dark.[28]

At the end of its investigation, the Rogers Commission agreed with the assessment of member and shuttle commander Robert L. Crippen: no known escape system could have saved the *Challenger* crew. Still, it insisted that "the vehicle that will carry astronauts into orbit through this decade and the next incorporate systems that provide some chance for crew survival in emergencies."[29] An undocumented online source stated that the destruct system was "no longer used" beginning with STS-79, "completely removed" by STS-88, and "has not been present on any tank since then," making it "no longer possible to destroy the vehicle during second stage ascent." However, a KSC press release for STS-103 in December 1999 stated that crews had replaced a "damaged range safety cable" that "runs from the right-hand SRB forward attach point, through the external tank and connects to the left-hand booster" and NASA media information as late as 2008 listed "Arm solid rocket booster and external tank range safety safe and arm devices" among tasks done at T minus 5 in the countdown to launch.[30]

This painstaking *Columbia* debris reconstruction in a KSC hanger involved thousands of recovered fragments. Photo credit: NASA-KSC.

After the *Challenger* accident, NASA maintained a perfect safety record through 2011 carrying astronauts into orbit. It would lose its next shuttle coming back to Earth.

Substitutes for the Shuttle? LifeSat and Expendable Launch Vehicles (ELVs)

Just three NASA satellites had been flown for the benefit of life scientists in the 1960s, all part of Ames' Biosatellite program. There had been one flown in the 1970s (the Orbiting Frog Otolith mission) and none in the early 1980s because the shuttle was intended make up the deficit.[31] Until *Challenger*, it did.

In 1987 William Gilbreath of the Materials and Physical Science Branch at NASA-Ames was on a one-year Headquarters assignment studying life science experiment possibilities for shuttle, expendable launch vehicles, and the hoped-for unmanned platforms. At the same time, Thora Halstead, Chief of the Space Biology Program and president of the American Society for Gravitational and Space Biology (ASGSB), was urging members to work for more recognition for their field by other scientists, politicians, and the public. She pointed out that astrophysicists and planetary scientists achieved recognition because they were involved in missions with "continuous access to space" and "a bevy of planetary probes."[32] They had successfully launched experiments aboard roughly 100 satellites during the same years that the life scientists had been allowed their four, plus another 2,000 sounding rocket missions.[33] Gilbreath discussed the use of satellites with Halstead, and soon Ames scientists were considering another basic biosatellite program, launching capsules on an expendable launch vehicle (ELV). Compared to the shuttle, putting inert biosamples into orbit should be relatively simple, fast, and inexpensive.[34]

Gilbreath announced NASA's consideration of LifeSat in the same issue of the ASGSB newsletter, saying that "the scientific community is being polled to assess support." It was intended to augment—not supplant—Spacelab, "to provide more frequent and controlled access to space, allow for extended times aloft, and permit special orbits to be reached (which might be of interest to radiation biologists, for example)." LifeSat would carry a 20 cubic foot (0.57 cubic meters), 450-pound (204.12 kg) payload: "Inflight manipulation of the experiment could be automated if desired and postflight analysis could begin within one hour of landing."[35]

Ames scientists—originally interested in studying the immune system—conducted the LifeSat Phase A feasibility study in 1988.[36] It would be launched on a Delta-II class ELV, orbit for 60 days, deorbit, and parachute to a soft landing on US soil. The reusable capsule would resemble a Discoverer satellite of the late 1950s, weighing about 2,800 pounds and measuring 45 inches in diameter and 36 inches in length. It could carry 525 pounds of payload—rats, mice, "lower life forms," and their life support equipment. They proposed flying three such capsules per year for 10 years.[37]

In August 1988, a month before RTF, NASA transferred LifeSat project management to JSC, which issued a Request for Proposal (RFP) for the Phase B design study in January 1989. By October, however, a requirement for artificial gravity had been added and rather than centrifuge just some of the rodents, it was stipulated that the entire module spin. This allowed for streamlined design and faster payload module access, but none of the biospecimens inside could be weightless controls. Still, ESA and the space agencies of Canada, France, Japan, and Germany expressed interest. NASA gave one study contract to General Electric Aerospace's Reentry Systems Division, supported by Lockheed Missiles and Space (designers of the original Discoverer) as subcontractor. Another contract, reportedly "much later," went to Science Applications International Corporation (SAIC), supported by Fairchild Space and Electronics. The Agency funded each for $1 million over nine months.[38]

Would-be participants formed a Science Working Group (SWG) by September 1989, with Emily Morey-Holton of Ames as chair and project scientist and Halstead as program scientist. Gilbreath was program manager, and Michael Richardson of the Flight Projects Office of the New Initiatives Office at JSC was project manager. It was around this point that the emphasis shifted from gravitational biology to radiation as LifeSat's purpose, reportedly due to "science community" input.[39] This shifted the goal from studying astronaut immune-system decline to obtaining data on how long one could stay in space before the cancer risk became unacceptable.[40]

LifeSat's usefulness for radiation research probably had roots in the intriguing results from Ames' 1967 Biosatellite 2, although that had used a controlled onboard radiation source. Radiation had promoted bacteria growth by 48 percent. Insect larvae chromosomes underwent lethal mutations and breakage, cutting survival rates of some species in half and increasing deformities by 50 percent in others. Amphibian eggs fertilized in space had produced malformed offspring. Some plant cells expected to show negative effects of radiation exposure had not done so in space, while others unexpectedly did. Spores had seemed unaffected.[41] Such puzzles begged further inquiry into the combined effects of radiation and reduced gravity.

A group of radiation scientists working mainly at JSC, Langley, and Rockwell studied four possible orbits for LifeSat suggested by the Lawrence Berkeley National Laboratory's (LBNL) Walter Schimmerling at a LifeSat workshop. Two were circular polar orbits at 200 km and 900 km, both of which would allow for fractional gravity, tracking, and precision landings. Two had "highly eccentric" elliptical orbits with 400 km perigee and a 36,000 km apogee, one polar and the other equatorial.[42]

General Electric submitted, as expected, plans for a Discoverer-based capsule, spinning on its longitudinal axis as the "traditional" means of creating gravity, and making soft landings via steerable parafoil parachute to hit the ground with less impact. SAIC's design was of a different shape but rotating end-over-end "using a deployable 100-foot [30.48-meter] AstroMast truss." This would result in lower RPM and less space sickness for the animals inside. It also recommended the round Apollo-style parachute reentry system for its proven reliability, "accepting a penalty both in footprint size and impact velocity." Both designs met the requirement for a maximum 15-g load and designated White Sands Missile Range in New Mexico as the landing point.[43]

In January 1991, Ames researchers chose a small Philadelphia firm, Micro-G, as Small Business Innovation Research (SBIR) vendor to make a "Variable-G Facility," meaning a centrifuge.[44] This implies that no longer would the entire satellite spin to provide artificial gravity. JSC assembled a design team from the Project Office at Ames, the Headquarters Life Sciences Directorate, the Engineering Directorate, and Safety, Reliability, and Quality Assurance (SR & QA) in Houston. Going against past practices, it incorporated SR & QA early to avoid additional costs and delays if redesign were necessary. (This caused clashes since SR & QA policy was based on crewed vehicles, with triple redundancy and other restrictions.)[45]

The SWG made other changes after a March 1991 meeting. The module should weigh less but hold 78 percent more mass because it now was to carry three shuttle middeck lockers and a centrifuge. Power requirements and the placement of power sources changed. Engineers had to recalculate LifeSat's center of gravity based on the weight of heavy new equipment, its placement within the capsule, and "the continuous transfer of angular momentum from the centrifuge to the spacecraft via internal aerodynamic drag and mechanical friction." LifeSats would have to operate in multiple orbits to allow varying degrees of exposure and encounters with more than one kind of radioactive particle. The SWG chose the highly elliptical polar orbit, varying from 350 to 20,600 km, in order to pass through the doughnut-shaped Van Allen Belts just once and avoid proton contamination of the galactic cosmic radiation (GCR) experiments as much as possible.[46]

In the four years since inception, LifeSat had become an engineering nightmare. "Due to the low dosage of GCR [galactic cosmic radiation] and the amount of proton contamination," one frustrated NASA engineer wrote later, "it is recommended that GCR missions only be flown much above the Van Allen Belt. To achieve a mission which leaves Earth, passing through the Van Allen Belt only once, receiving GCR for 60 days, then returning through the belts, requires an apogee more than twice the distance to the Moon."[47] That forced the inclusion of an independent, onboard navigation system for deorbit because the satellite would be out of range of the Global Positioning System (GPS). Orbital changes also constrained the amount of mass a LifeSat could hold, so mice, not rats, would have to be used to get the "n" (sample size) scientists wanted. Increased design complexity overall meant that LifeSat, judged "well within the state-of-the-art" in 1988 using off-the-shelf technology, became so heavy the Delta II would no longer be able to lift it.[48]

For political reasons, money had also become a bigger issue. Phase C/D of the competition was to have begun October 1, 1991, a year after George H. W. Bush and Congress agreed on the Budget Enforcement Act of 1990, instituting pay-as-you-go rules.[49] Senator Fritz Hollings (D-SC), Chair of the Commerce, Science, and Transportation Committee, called NASA's FY 1992 budget submission "extremely ambitious" in asking for a 13.6 percent increase over FY 1991. His committee was "compelled" to cut half a billion dollars from the request to save the National Aero-Space Plane (NASP) and because trusting NASA after recent failures with Hubble and the Galileo space probe was now an issue. NASA had just unveiled the latest Freedom redesign in March, and some in Congress were keen to see that accepted before funding something novel.[50]

Senator Albert A. Gore Jr. of the Science and Technology Committee spoke at length about NASA's Authorization Act. Congress chose to reduce funds for SEI, the Comet Rendezvous Asteroid Flyby (CRAF), the Search for Extraterrestrial Intelligence (SETI) program, and new launch systems and was denying *any* funding for LifeSat, Gore said, because it could not rely on NASA in budget matters. "At first blush," he said, "a $15 million program within a $15.7 *billion* budget request seems modest enough. However, nowhere in the budget does NASA state that, when all costs related to the LifeSat mission are calculated, it will cost the federal government some $870 million. This is a fact that only came out in questioning of NASA officials during subcommittee hearings." Gore accused the Agency of inserting a tiny "wedge" into its budget that it could later use to force in much larger amounts of money, possibly at the expense of higher-priority programs and projects (like Freedom).[51]

The mood was so mistrustful that the House inserted a stipulation that NASA annually submit budget figures on "the 5-year development costs of every project, mission, and program, as well as life-cycle costs associated with each." The Senate retained that wording.[52] A brief December 17 NASA Headquarters press release after the final vote noted that OSSA funding had risen 10 percent over FY 1991. About LifeSat it said only that its money had been deleted.[53]

Even if LifeSat had been approved, launches would not have been fast or frequent. OSSA's 1991 Strategic Plan had targeted LifeSat's first launch for 1996, ten years after *Challenger*. Launch two was manifested for a year later, number three for nine months after that (March 1998), and number four would be in December 1998. Penciled in were possible dates in June and December 1999 for the fifth and sixth LifeSat. There would be no rapid turnaround or ten flights a year.[54]

Records do not make entirely clear whether the outcome might have been different had a simple LEO design been maintained. They also do not shed light on the question of *why* the plans had changed and who bore responsibility for the increased mission complexity. Even knowing that the future held another 29-month loss of shuttle flights after the 2003 *Columbia* disaster and a nine-year gap after the retirement of the last orbiter in 2011, one can conclude from the LifeSat experience that using ELVs and satellites as quick and inexpensive replacements had been promising in concept but unlikely to succeed in execution.

Planning for Operational Medicine on the Station

Concurrent with the early years of STS operation, several Centers worked on safety, health, and medical care for crew of the anticipated space station as it evolved. One basic question was where to put the sick bay. Early concepts included combining the healthcare section with a life sciences research area in their own separate module or putting the medical facility among the core components of the station and having a separate biology module. A second issue was whether any one such module should host both human and nonhuman research.

The separate module idea was an early favorite. Life Sciences Division Deputy Director William Bishop wanted to lobby for it to be one of the first modules added to the basic habitat element.[55] The notion had several strikes against it: the Office of Space Science and Applications would have to foot the entire cost; it would reduce the Office's science and applications R&D capability by half since it would occupy one of the two science and applications docking ports; it would provide a fixed amount of space, whether or not that much was needed; and it would be "very expensive." A combined medical/health sciences research module would run $300 to $600 million in 1982 dollars. Option two, putting the module within the station core, would be less expensive, possibly safer, and more convenient but would freeze the working space at a small size, likely to be unacceptable before long, and it might expose other station residents to biohazards from sick or contaminated humans and/or animals. A third alternative, putting the healthcare facility in the core but making the research facility separate, would eliminate most contamination concerns but a separate module for just research might not be justifiable considering the costs of simply getting the station itself launched and assembled. Two other options combined nonhuman research facilities with those for humans but healthcare officials termed the mix "undesirable" and of scant benefit.[56]

The creator of the handwritten "Space Station Medical Operations Facility (MOF)/Human Life Sciences Research Laboratory (HLSRL) Concept and Development Plan," penned in 1982, recommended putting the MOF in the core elements with a separate human life sciences research lab in an attached module. OSSA and the Office of Space Transportation Systems (OSTS) could share the cost and NASA could start out cheaply and equip the module with increasing sophistication as research progressed, and even swap out the module for something better later. Should unexpected situations arise, medical personnel could respond immediately with on-site research. Parenthetically, the writer commented that "politics, money & time may force #5," referring to one of the "undesirable" options: integrating nonhuman research with the human module. The plan recognized the need to develop the medical and research capability on station "as a function of time, number of men in orbit, and mission duration," adding an instruction to "pace the system to the developing need." The writer expected funding to be a challenge and noted, "Every reason to believe no big change for the better—for the worse is more possible."[57]

JSC designers envisioned the station as a Space Operations Center with an engineering focus, a base for satellite repair and servicing of large spacecraft in orbit. Life sciences planners at JSC saw their facilities growing over three time spans. In Phase 1, two or three astronauts would live in a station consisting of one command module, one attached module, and a payload/upper stage assembly. Medical care would be a "SOMS-A Capability" (Shuttle Orbiter Medical System A) with "minimal human research capability." Residents would not carry out life sciences research because it was of "lower priority than MPS [main propulsion system], Earth/Solar observations, commercial or DOD." Phase 2 would have four to six astronauts, two command modules, one or two attached modules, the payload/upper stage assembly, space-based orbital test vehicles, and a logistics module. Phase 3, at some unspecified future time, would see eight people in orbit, adding two habitation modules, a docking tunnel, and one other attached module.[58]

The MOF would have "doctor's office capability with increased human research capability" spread across modules. There would be "hopefully one" life science facility for research. The MOF would have "emergency/surgical capability" in habitation module one, a clinic in command module two, and SOMS-A capability in command module one. Additionally, there would be "significant human research capability" in the MOF, and the HLSRL module would be a "full-up human/(non-human?) multidiscipline segregated system." At the time, they expected assembly to be just seven years away, 1989.[59]

JSC's space station Program Definition Working Group argued the case in 1982 for both a medical care facility and for human research on station, asking for a dedicated lab module to be berthed to the station core, rather than integrated into the habitation or command module, on the basis of being able to isolate it from biohazards. This research capability needed to be ready as soon as a crew size of four had been attained "because of the importance of understanding the effects of long duration space flight." Perhaps overly optimistic, the group asked that one crewmember be a doctor or surgeon with additional dental training, and one a payload specialist with 100 hours of medical training and some dental skills, and to be cross-trained in life sciences research in addition to their regular duties. Computers on station would store crew medical records and wherever NASA decided to put the Health Maintenance Facility with all its surgical and diagnostic equipment, the exercise gear would also be stored there.[60]

Medical operations personnel tried to predict the number and severity of illnesses and accidents likely to happen aboard a space station. These they readily admitted were "crude estimates." One early (1982) scenario assumed several 10- to 20-crew space stations in orbit and called for a "base station" with "sophisticated medical facilities" for an injured crewmember requiring evacuation but not able to handle reentry. Given expectation then of more and faster access to space via the shuttle, some also thought they might be able to send medical specialists *up* to the station in a crisis and have them return with the patient.[61]

Planners made their estimates based on analog situations that Antarctic researchers and the US Navy's nuclear submariners experienced.[62] They predicted that acute abdominal contingencies requiring surgery

SUMMARY OF HEALTH DATA COLLECTED DURING 10 YEARS OF
POLARIS SUBMARINE PATROLS[a]---20,960 MAN-YEARS

Disease/Condition	No. Cases[b]	Rate per MY	No. Cat.	Transfer At Sea	Deaths	Comments
Gen'l Surgery Referral	269	0.0128	32	6		70 appendicitis; 45 pionidal abscess; 23 burns.
Bone & Joint	264	0.0126	52	1		66 lumbosacral strain; 34 fractures; 2 amputations.
Gen'l Medical	240	0.0115	30	0		134 flu; 31 mononuc; 13 viremia.
Gastro-Intestinal	229	0.0109	19	6		155 gastroenteritis; 17 gastritis; 14 hepatitis.
Respiratory	185	0.00883	9	7		80 pneumonia; 43 URI; 36 acute bronchitis; 11 pneumothorax.
Ear, Nose, and Throat	165	0.00787	14	1		96 pharyngitis; 23 tonsilitis.
Urinary Tract	115	0.00549	19	3		39 ureteral calculi; 26 epid; 23 pyeloneph.
Psychiatric	58	0.00277	15	3	1	25 anxiety reaction; 13 neurotic depr.
Neurologic	53	0.00253	18	4	3	18 headache; 9 concussion; 8 migraine.
Dental	50	0.00239	9	1		28 periapical abscess; 13 pericoronitis+.
Eye	48	0.00229	16	3		18 corneal abrasions or foreign body; 16 conjunctivitis; 5 burns.
Cardiovascular	9	0.00043	5	2	1	3 Hypertension; 2 chest pain.
Total	1685	0.0804		37	5	

a. Compiled from data in: Tansey, W.A., J.M. Wilson, and K.E. Schaefer. 1979. Analysis of Health Data from 10 years of Polaris Submarine Patrols. Undersea Biomedical Research, Submarine Supplement, S217-S246.

b. Excludes transfer at sea and death; includes only cases resulting in 1 or more days lost from work.

(appendicitis, small bowel obstruction, hernia, diverticulitis, gallstones, and kidney stones) would be likely, as would heart attacks, mental disorders, tooth decay, or drug reactions. They attempted to classify the severity of illnesses and injuries, writing that "although the possibility exists for medical problems for which only highly invasive medical and surgical therapy would significantly increase patient survivability, the tendency to construct a 'Mayo Clinic in the Sky' must be avoided." Keeping a surgeon on board and outfitting such a facility on station would simply "not be cost-effective," so astronauts themselves would provide basic care. Until there was a larger and more sophisticated station in orbit, "the techniques and instrumentation necessary for major abdominal and orthopedic surgery in space should remain as a research priority rather than as an operational requirement."[63]

This contrasted with what the Program Definition Working Group was proposing, an idea that could be called a "Mayo Clinic in the Sky." The Working Group wrote that—while "it may never be practical or cost effective to treat such patients in space"—research could be carried out there on arthritis sufferers, paraplegics, people with slipped discs, heart patients, and burn victims. An appendix written by two Goddard Space Flight Center engineers added, "Eventually research can be extended to include direct human treatment in orbit, by the use of water tank immersion during launch, to allow g-abatement for patient transport." In other words, the ill or injured would first be encapsulated in a tube of water to lessen accelerative and decelerative forces and flown to an orbiting hospital. A presumably very well-funded NASA would treat them because space medicine had the power to "reduce human suffering, disfigurement and aging handicaps." Until that unspecified point in the future, the Working Group wrote, normal operational increases in crew size would at least lead to the need for advances in surgical and trauma treatment technology, and so it recommended a "separate attached module containing a small space hospital." Since hundreds of people would certainly be working in space by then, they reasoned, a few would likely be seriously injured and they could be relocated to this orbiting hospital. An astronaut physician/surgeon and an astronaut nurse would be standing by. When not taking care of patients they would be experimenting with new microgravity surgical techniques on resident animals. Such research would "minimize the risk of death [in space] and provide assurance to spaceworkers that if injury/illness should occur they will have a chance."[64]

One Working Group idea that quickly was nixed was a two-person emesis station. There would never be enough room to devote solely to vomiting, and the odds of two crew having to be shunted to a nonproductive area simultaneously and often were small. Radiation medicine specialist D. Stuart Nachtwey, who reviewed that section, noted that "the space sickness problem is a transient phenomenon and is being handled well enough by bags and wipes in the Orbiter."[65]

Some planners realized early on that traditional treatment procedures were gravity-dependent. Examples included intravenous delivery of fluids, gastric lavage (stomach "pumping"), fluid transfer (blood and urine), eye irrigation, cardiopulmonary resuscitation, control of vomited material, suctioning of blood during surgery, and containment of fluid and gastric contents during surgery. New equipment and techniques would have to be developed and tested before they could be used on missions that might last two years. NASA scheduled the prioritization of these needs, research, and the development of new tools and procedures for 1983 and 1984, aboard both the KC-135 and the shuttle, using both human and animal models, expecting station occupation to commence in 1990.[66]

Others worked to achieve and maintain operational health by making the station environment sanitary. There were obvious atmospheric issues such as temperature and humidity, but airflow rates aboard Skylab and CO_2 levels during the Salyut missions had created issues of both comfort and health. The presence of animals and plants onboard would mean the presence of extra microbes and particulates in the air and on surfaces. Microbial contaminants were a wild card because there was no way of predicting with certainty who would bring what bacteria or viruses into space, how many regular crewmembers might live on station, and how many visitors would be allowed and how often. Since 1977 NASA had maintained a list of banned pathogens (viruses and bacteria) for flight animals.[67] The experience on STS-41B, with the arthritic rats and their potatoes, proved that a sterile environment was impossible to achieve (see Chapter 1). Also, biological specimens would die or be killed in space, and dead animals would need to be isolated from both humans and other animals.[68] Humans might die as well, the Working Group conceded in a November 1982 analysis, noting, "Body disposition procedures must be considered."[69]

Particulates became a great concern in 1985 during STS-51B, which carried Spacelab 3. Servicing the food bars and waste trays in both

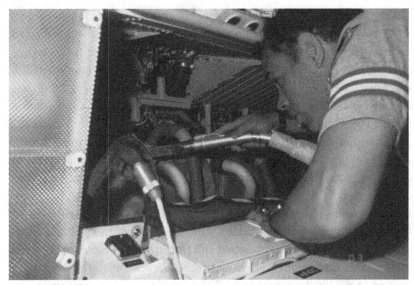

Pilot astronaut Frederick D. Gregory vacuums *Challenger*'s avionics air filters during the STS-51B Spacelab 3 mission in 1985. Photo credit: NASA-JSC.

primate and rodent cages resulted in the escape of debris particles, blamed on both "inadequate seal design" and "higher than expected vigor of monkeys, who kicked the material through the airflow of their zero-g into the Spacelab."[70] For the station, life scientists asked for "a restraint system for rhesus monkeys that will enable them to be continuously restrained in health and comfort for at least 90 days," plus "a method and hardware for the periodic, on-orbit cleaning and sanitizing of cage units."[71]

Restrictions on gaseous contaminants changed every time a redesign made the station smaller in volume but packed more gear into the tighter space. Existing industrial limits were not always consistent, either, with some exposures expressed as a duration of anywhere from a week to a month, others as parts per unit of air. Environmental scientists regularly added to the list of identified carcinogens too.

Station designers did not have enough information to fully understand issues of noise. In the past, cosmonauts, submariners, and space simulator inhabitants had all reported sleep interference, distraction, and annoyance caused by the sound of lab equipment, power generators, and ventilation. However, there were neither standards nor tests for noise tolerance, nor measures for predicting acoustics that could be used for the purposes of station design.[72]

By November 1985, plans called for the medical area to be equipped to handle burns, IVs, catheterization, minor surgery, local anesthesia, "bends" recompression, respiratory problems, medical imaging, and basic dental care for a crew of six. Each crew would have at least two licensed paramedics, and NASA would train all astronauts in first aid. An Earth physician would be on call constantly. JSC planners envisioned such a system being upgradable to handle a crew of 18. At the time, the Medical Operations Branch at JSC under James S. Logan was already providing each astronaut at least eight hours of CPR training, and at least two crewmembers aboard each shuttle mission had additional medical training enabling them to dispense pharmaceuticals, perform tracheotomies, open and maintain an airway, dress wounds, give electroencephalograms (EEGs), defibrillate a heart, and ventilate a patient who was not breathing.[73] There would be no Mayo Clinic, but there would be sufficient resources to treat common ailments and handle almost any foreseeable emergency.

Crew Emergency Rescue Vehicle (CERV)

The 1982 group had also declared that "no medical justifications for an emergency rescue capability exist." This was based on a solid faith in the redundancy of life-support and safety systems, crew training, and selection standards, and an assumed potential greater risk in an emergency flight back to Earth.[74] However, it was in direct contrast to Headquarters' plans, which called for a separate Crew Emergency Rescue Vehicle or CERV (pronounced "serve").

CERV was not a new idea. In the 1970s, General Electric had proposed the MOSES (Manned Orbital Space Escape System), a ballistic four-person vehicle shaped like a gumdrop. It was essentially a larger, manned version of their Discoverer space capsules, snagged in midair while reentering the atmosphere with cargoes of camera film or science specimens.[75] Schematics show that the four escape-suit clad MOSES occupants were to lie semirecumbent on Apollo-style seats, each at a right angle to both the person in front and behind, with the fourth to the side but facing in the opposite direction. Drawings also show a recovery parachute, retro engine, separable heat shield, and ballast and inflatable

A McDonnell
Douglas concept
for crew return,
circa 1987.
Photo credit:
NASA.

bumpers in front. It is not clear whether MOSES had the ability to land
on water, land, or both.[76] CERV ideas generated in the 1970s–1990s
would have influence later on the Orion Shuttle follow-on, in terms of
vehicle capabilities, purpose, and looks.

The 1997 near miss between the Mir space station and a remotely operated Progress supply vessel (see Chapter 8) would prove correct what
NASA planners of the early 1980s believed: most emergencies in orbit
would occur during the small percentage of time station crew engaged
in logistical operations. Thus a significant number would only require
the crew to get away from the station and wait for a shuttle to retrieve
them. Given that the shuttle program then called for dozens of launches
per year and a fleet of half a dozen vehicles, and that the Soviet Buran,
in development since 1976, was expected to be in use by the time a US
station was operational in the early 1990s, this should mean only a few

days' wait. Hence one escape concept of the early to mid-1980s was a small "orbit to orbit Shuttle," a cargo resupply vessel that looked something like the top stage of an Atlas rocket and flew alongside the station until needed. The remote manipulator arm would grab it, pull it close enough for the crew to board, then await rescue.[77]

After the *Challenger* accident, station designers realized they actually had no way to rescue stranded astronauts in under 28 days, the minimal time to prepare an orbiter for launch. Even that case assumed the entire fleet had not been grounded for safety reasons. Both the space station safety panel and the Critical Evaluation Task Force recommended a dedicated rescue vehicle, but Associate Administrator for the Space Station Office Andrew Stofan, calling it "highly desirable," said it was "not within [the] space station $8 billion program."[78]

Still, by October 1986, JSC personnel were reporting on what they called the Contingency Return Vehicle (CRV) in support of Space Station Freedom. Kornel Nagy led the 13-person study team, aided by another two dozen Center personnel from 13 organizations, including the Advanced Programs and Astronaut Offices, the Crew and Thermal Systems, and the Man-Systems and Medical Science Divisions. Although the team was laden with engineers, it included psychiatrist Patricia Santy, representing operational medicine.[79] Their task was to design, develop, and operate a CRV in time to service the crew of a station scheduled to be in LEO by 1994 and permanently crewed by 1996.[80]

Plans were fairly ambitious, with requirements for a backup CRV, shirtsleeve access to the vehicle, automatic reentry and landing (but with manual backup), on-orbit loiter capability, and the ability to land at preselected sites (with water landing as the baseline but recovery on land an option). To save costs, the team decided to make the vehicle "just adequate to the task," with "landing and recovery system . . . to be of minimum complexity." Under "Minimum CERV Design Ideas," the group also listed "Mono-stable," meaning it could enter the atmosphere at any angle and stabilize itself, and "pure ballistic," meaning that it needed no lifting capability (because it only went down) and the landing area could be readily predicted.[81]

Santy reported that accurate estimation of the incidence of illness and injury on station was impossible, in spite of analog studies, but most of the operational hazards had been identified "in a general sense." She recommended keeping in mind passengers might be already injured or ill, recovery forces and facilities might not be handy to the landing area, and the crew would still have to rely on what they had aboard their vehicle. Astronaut Steve Nagel listed "low [g] load" as a design requirement for the ill or injured, meaning that gravitational forces had to be as low as possible to prevent further trauma. It also required "adequate stay time on water or land" until rescue arrived—meaning a CERV would have to be watertight and float—and have a locator beacon for search and rescue teams.[82]

There would be medical ethics issues and decision-making problems with a rescue vehicle, Santy pointed out. Its presence would create an obligation to use it to save the life of even one individual who was critically ill or injured. Return would require the assistance of a healthy crewmember, putting the second party at risk, and, having adapted to microgravity, they might not be fit to help the ill or injured party once back on Earth.[83] Also, astronauts with certain illnesses would not be good risks for CRV travel. The g-load should be three, equivalent to the shuttle, or less, the study team advised, but the Advanced Program Office had a diagram indicating that only the technologically sophisticated (and expensive) lifting body designs could accomplish this. MOSES came in at around *nine* gs. Even modified Apollo capsules would pull four.[84]

Being able to find and retrieve the crew after return was obviously critical, particularly as they could not take off again or choose their landing site. Some vehicle designs were only marginally steerable as well, and the GPS was not yet fully operational. Neither was the Tracking and Data Relay Satellite System (TDRSS) coverage yet continuous.[85] This made an Apollo-style water landing look less unattractive. NASA's tracking systems were able to cover far fewer land target sites than water, and absolute accuracy was less important in the sense that a capsule aiming for the ocean was unlikely to hurt anyone because the population density there was essentially zero. A splashdown could also have a surprisingly quick recovery time. Of Apollo missions 13–17, the average target error was only 1.6 nautical miles (1.84 statute miles or 2.96 kilometers). The mean time required to get the crew out of the capsule and into a helicopter had been just under 37 minutes. Each of the Skylab recoveries brought the return module, crew still inside, onto the deck of the ship in less than 45 minutes. However, it had taken extensive US Navy resources to do so: an aircraft carrier, five helicopters, and three fixed-wing aircraft.[86]

The vessel closest to optimum under all these parameters, and which the team recommended, was a scaled-up Apollo capsule, reconfigured to do less but hold more—six astronauts instead of three—with an option for one strapped to a stretcher. A small increase in diameter from 13 to 14.5 feet (3.96 to 4.42 meters) would do it. Stripping it of all but the essential avionics and having little to no redundancy would save over 1,000 pounds (454 kg) and more than 41,000 cubic inches (671,869 cubic cm). However, had the guidance, navigation, and control system failed, the pilot might have had to literally eyeball the reentry through the capsule window, and miss-distances of up to 500 nautical miles (575 statute miles, 926 km) could have resulted. The Mission Operations Directorate viewed such procedures as "not highly demanding if reasonable vehicle margins exist," but assumed the pilot was very experienced.[87]

The estimated cost of design, development, engineering, and production in 1987 dollars was $315.7 million. Only $93.5 million of that would be NASA overhead, with the remainder paid out to industry. The space agency set Phase B definition for FY 1987.[88] NASA expected to then issue an RFP in the summer of 1988, stating the performance parameters but not indicating JSC's own preference for the Apollo-style vehicle. NASA would select a design and during the second phase, develop detailed plans with the goal of ultimately producing a test vehicle and two CERVs in time for the initial construction of Space Station Freedom.[89]

NASA Langley engineers in Virginia had separately been working on a lifting body design, the Cadillac of emergency return vehicles. In June 1988, the CERV group at JSC put together an initial CERV Operations Concept to present to Headquarters. Pending approval, they intended to get with Langley engineers to discuss the application of the CERV configuration and mission to the lifting-body project.[90] It was likely a foregone conclusion that a direct transfer would be impossibly unaffordable.

In April 1989, Acting Administrator Truly signed a memo to finally release the RFP for the CERV Definition Study, both Phases A and B, within weeks. There would be "several new system constraints." There would also be a Phase C/D competition, but even if a contractor lost in Phases A and B, it would still be allowed to spend its own money to stay in the race. Truly's successor as Associate Administrator for Space Flight would decide the winner.[91]

By August KSC was involved, studying four CERV issues with the University of Central Florida under a Universities Space Research Association grant. The issues were improving injured-crew removal features, integrating specialized medical gear, and "how to make CERV a good boat," as well as "systems requirements to support a healthy crew on the water."[92] Plans for sea landings indicate a movement away from the elegant lifting body and toward something that could get back the quickest, cheapest way possible.

Another indication of a Model-T approach came from JSC's Safety, Reliability, and Quality Assurance group. Trade studies had led the group to expect "a need for retrieving several astronauts during the life of the program," so they had looked at various ideas and reported that they might be ready to evaluate one basic method by STS-37, scheduled for 1990.[93] JSC also filed a patent application for an ACRV (A for "Assured") in December, a 1960s-style ballistic capsule with reentry parachutes.[94]

Meanwhile, Langley was proceeding with its lifting body, the HL-20 Personnel Launch System (PLS). It was to hold 10 people and be launched atop a Titan III booster, rendezvous with the space station, and then glide back to a runway landing. In October 1989, Rockwell won a contract to study the HL-20. Students at North Carolina State and North Carolina A&T universities, as part of a cooperative agreement with Langley, built a full-scale model for human factors research during the spring and summer of 1990. This consisted of visibility and emergency egress tests in both vertical and horizontal positions by Langley volunteers wearing unpressurized suits at first, then partial-pressure suits borrowed from JSC. Lockheed was charged with creating the prototype and an operating system in October 1991.[95]

For political as well as economic reasons, the United States chose in March 1992 to study the Russian Soyuz capsule as an alternative.[96] However, some CERV-like projects continued, particularly prior to the signing of the formal agreement by Daniel Goldin and Yuri Koptev, General Director of the Russian Space Agency (RSA), in July.[97] The WETF at JSC hosted a series of tests on the Buoyant Overdesigned ACRV Testbed (BOAT). The intent was to give meaningful data for three different designs, including the modified Apollo capsule and the Station Crew Return Alternative Module (SCRAM), a pre-MOSES design based on the Viking Mars landers of the mid-1970s. BOAT evaluation consisted of dry-ground testing of egress procedures, manned and unmanned Weightless Environment Training Facility (WETF) tests, and open-water rescue simulations at a wave-generating facility Texas A&M University used for examining offshore oil-drilling platforms.[98] In

February, even as Congress was urging the space agency to team with Russia, the Europeans, then planning for their own Columbus space station and Hermes crewed vehicle, approached NASA about cooperation on an ACRV. Informally, ESA asked to trade ACRV expense for Freedom costs, and the agencies got as far as engaging two contractors to study an eight-person capsule concept, then began to work on interface details with US counterparts. A tentative schedule called for an 18-month Phase B to start in October 1993, and if the capsule looked promising, a proposal to NASA in 1995 and a six-year Phase C/D to commence January 1996. ESA would do an orbital test flight in August 2001.[99] The end of the Cold War had made European leaders rethink Hermes, however, and in late 1991 they suspended it for a year of official "reflection." ESA modified and then suspended the program.[100]

According to UK space historians Rex Hall and David Shayler, the Soyuz idea came about in October 1991 when Freedom was "vastly over budget, seriously delayed, and close to Presidential cancellation" by George H. W. Bush. Energiya Design Bureau director Yuri Semyonov suggested Russia produce Soyuz TMs as a stopgap until the ACRV was ready. He saw the economic benefits of hard currency to his own program but also was "inspired" to help reduce Freedom's costs. The White House and Kremlin agreed and in June 1992 authorized NASA and NPO Energiya to undertake a $1 million study contract.[101]

The biggest technical stumbling block was the station's anticipated orbital inclination, very difficult to access from Kazakhstan, where the Russians launched. The Australian outback was the next choice and would have meant establishing a whole new search and rescue infrastructure for landings. A second critical point the evaluation team from NASA, Rockwell, and Lockheed quickly realized was that relying on a single Soyuz would limit space station occupancy to no more than three astronauts.[102] The amount of scientific research they could carry out would be severely curtailed, since the three would have to do all of the station maintenance alone. The Advisory Committee on Redesign of Space Station wanted to incorporate docking for two Soyuz vehicles into the configuration and increase crew stay time from six months to two years. Labor-intensive life sciences studies required longer stays and more crew to achieve productivity.[103]

Politically, the idea took some heat: using a vehicle based on a 1950s design would create complacency and retard domestic research on ACRVs. Committing to using a Soyuz TM would give Russia control over the progress and direction of station development. Eliminating research on an advanced US CERV could mean the loss of jobs for the aerospace industry in the United States, and NASA would be spending taxpayer money on overseas companies.[104]

There were safety concerns, too. Anthropometrics showed that 40 percent of NASA astronauts would be excluded from using a Soyuz (and thus from going to the station) because they were too big or too small for the Russian custom-molded seats. Energiya thought it could waive the weight limits but not the height restrictions, and returning astronauts would have to fly without pressure suits to fit into the capsule. According to Hall and Shayler, the Astronaut Office "was impressed by the measures demonstrated by NPO Energiya to ensure pressure integrity." The tragedy of Soyuz 11 in 1971, when three unsuited cosmonauts died on return after their capsule unexpectedly lost pressure, must have been on everyone's mind. Likewise, the Soyuz TM's oxygen-rich atmosphere concerned both parties, given the Apollo pad fire of 1967 that killed three US astronauts. Additionally, sick or injured space travelers would have to be strapped into Soyuz seats rather than laid on a pallet. That positioning precluded the return of spinal or leg injury patients. There was no room for medical equipment; taking station gear along would mean leaving the third crewmember behind. Since couch inserts were individually sized, there was talk of storing numbers of them on station, then quickly installing those that best fit the individuals evacuating. The Soyuz also required a one-hour predeparture leak test before embarking. These last two steps meant crucial time lost in a medical emergency.[105]

Still, the Agency proceeded with the Soyuz TM. NASA asked Boeing to see if it could be modified to serve as a vehicle with medical/life support for one, accompanied by one other crewmember, and the ability to get from the station to Earth within three hours. Landings had to be precise enough that a search and rescue team could get to the site and extract the occupants within one additional hour. NASA also wanted the adaptation done inexpensively, giving Boeing just $3 million to come up with a final design in 12 months.[106]

The matter was partially resolved in September 1993 when the United States and Russia issued a joint statement announcing cooperation on station and specifying a change in the orbital inclination to be "accessible by both U.S. and Russian resources."[107] NASA and the Russians continued to examine their options, and in 1995 a project called Lifeboat

Alpha, a joint Energiya-Rockwell-Khrunichev study, looked at the Russian Zarya ("daybreak") reentry vehicles of the 1980s as possible CRVs. In theory, they could be adapted to accommodate eight passengers. Zarya raised concerns, however, because it used a single small parachute during reentry instead of the large canopy the Soyuz employed, and it lacked a backup to the retrorockets that cushioned the landing. The acoustic levels of those rockets were also greater than NASA liked.[108]

In 1996, after anthropometry problems had caused crew changes during the Shuttle-Mir program, the Russians agreed to redesign the capsule, to be called the TMA, as an interim space lifeboat. Zarya plans would be scrapped in anticipation of NASA's X-38 lifting-body design.[109] As a possible CRV, the X-38 was under study by JSC with assistance from Langley and Dryden, and a feasibility study of European participation was under way. Over the next two years, NASA negotiated with ESA and individual European nations about expanding that participation to an operational CRV program.[110]

The X-38 was a seven-crew vehicle based on 1960s USAF research at Edwards AFB and on US Army parafoil drop tests at the Yuma Proving Grounds in Arizona. The program began in 1995 with one-sixth scale flights employing a Cessna and small parafoil. Scaled Composites, LLC of Mojave, California then built three 80 percent scale models and flight tested those in 1997, 1998, and 1999. A B-52 carried the test vehicles aloft and dropped them for a parafoil-aided landing on the dry lakebed at Edwards.[111]

Because the program relied on proven technology, was derived from past studies, and had the assistance of European partners, some estimates put the X-38's cost at one-quarter that of a $2 billion, ground-up CERV program. However, by FY 2000 it looked like it was going to cost $700 million, and NASA would take some of that money from the ISS. That part worried Congress, which would dig in its heels against President George W. Bush and NASA Headquarters when they called for elimination of the X-38, a station module, and nearly all ISS on-orbit research in budget requests to save sufficient money to simply keep the basic station program going. After spending $460 million of its $510 million flight test money, NASA eliminated the X-38 from its FY 2003 budget to save the station, which by then was billions of dollars more costly than anticipated. Ultimately estimates were that the X-38 would have cost $1.3 billion for the flight testing and three orbiters, but in the late 1990s, the Agency's various European partners (Germany, France,

The X-38 lifting body dropping away from its pylon beneath the B-52's wing and after landing on a dry lakebed near Armstrong Flight Research Center in California. Photo credit: NASA-Armstrong.

Spain, the Netherlands, Italy, Switzerland, Sweden, Austria, and Belgium) had been prepared to pay about half that. Japan had been lobbying JSC Center Director George Abbey and Dan Goldin to be allowed to participate. When NASA canceled the program in April 2002, it can be argued that the Agency saved only $210 million.[112]

The Russian Space Agency launched the first TMA to the ISS in October 2002.[113] Soyuz would continue to fill the role of shuttle alternative and space lifeboat for the remainder of the shuttle's operational lifetime.

4

Science and Scientists

Peer Review, the Extended Duration Orbiter Medical Project, Neurolab, and a Station Centrifuge

With STS-3 and its electrophoresis experiment, shuttle orbiters began to function as hands-on space laboratories. NASA geared up for what it expected to be hundreds of crewed missions and a full-time lab in LEO within the decade, supported by free-flyers and biosatellites. Organizing individual missions around a central platform (Spacelab) or theme (e.g., neurology) presented unique challenges and the potential for unmatchable results. If NASA could carry it off, a human-rated space station centrifuge held the tantalizing promise of a permanent presence in space becoming a reality.

Science and Standards

In the early 1980s, the NASA Life Sciences Advisory Committee (LSAC) held meetings at which members debated flight experiment proposals in closed sessions. To avoid the appearance of conflict of interest, those among the committee who had submitted a proposal absented themselves while the matter was discussed. Minutes of those meetings shed some light on how the selection procedure operated further up or down the chain of command.

A pharmacologist in attendance at a March 1981 meeting inquired why the Agency was flying rats instead of mice. The answer: NASA was not then developing cages or diets for mice. When a committee member asked if there were a policy on the use of countermeasures, a NASA scientist replied that it was "informal." The member objected to study-ing a countermeasure while at the same time studying the disease, but the minutes show NASA only ruling out one experiment for that reason. If no scientist submitted anything on a topic the committee deemed of interest, a member recommended "some data . . . be obtained even if no acceptable proposal in the area was available and on hand to be developed by NASA scientists." This opened the door for last-minute work by in-house experimenters and an advantage for them over ex-tramural PIs. Also, NASA could place experiments not approved on a "supplementary" or "rejected" list, then transition them back and forth from one category to the other. The committee could request that the scientist, even if working on a different species, be added to someone else's team as a coinvestigator or consultant, or that the experiment be incorporated into another.[1]

The LSAC made its recommendations based on its understanding of the science and NASA's expressed priorities at the time, but this intro-duced profound flaws into a system working years in advance of actual flight. For example, more than one LSAC member asserted that no phys-iochemical changes (e.g., insulin or calcium) would show up over the course of a shuttle flight, as it was too short a time. Some did not feel there was any interest in calcium loss in missions of less than 30 days and that even the existing information on decreased absorption and hormonal changes "might not be valuable." (None of that proved to be true.) A muscle specialist asked whether, "in the light of data obtained from studies of the Salyut missions, concerns or the problem of bone loss had diminished."[2] Soviet data had been difficult to obtain, NASA scientists did not consider it reliable, and certainly neither agency had developed completely successful countermeasures to all the medical challenges of spaceflight by 1981.[3]

Reviewing SLS-4 candidate payloads, LSAC meeting attendees in the spring of 1981 questioned the selection methodology. Two remarked that experiments in fields represented by a scientist on the committee were more likely to be approved. One reviewer argued for a second An-nouncement of Opportunity (AO) to maximize use of tissue samples, but another preferred instead to focus on doing the high-priority exper-iments well, "even if it meant discarding potentially useful samples." A NASA biologist encouraged researching by discipline to identify remain-ing questions, but an extramural neuroendocrinologist replied that such an approach could mean "thoughtless data-gathering missions."

Another outsider hoped the Agency "had a fail-safe mechanism" for halting even "highly recommended" experiments that made it past one committee if another found valid reason to nix it.[4]

Outside of the review committee setting, extramural jealousies simmering below the surface sometimes emerged. JSC physiologist John Charles recalled concerns of non-Agency scientists that NASA researchers delayed the flight of their experiments (due to the similarity of proposals). In a worst-case scenario, Agency science was substituted with no science at all. Charles had completed nearly all of a particular series of experiments, only to have the final one canceled because it had failed peer review. Ironically, it was too late to remove the equipment from the orbiter. Thus, he said, "It flew as ballast." Charles questioned how a NASA PI who had been trained by the same people as the extramural researchers, and specifically charged (as a job duty) with the task of doing in-flight science, could suddenly find his experiment "not good enough."[5]

Foreign experiments did not necessarily go through the same peer review, though they did receive the scrutiny of NASA technical personnel who verified that they would operate as planned, were not hazardous, and aligned with engineering objectives of test-flying a particular piece of equipment or platform.[6] NASA might launch foreign experiments for political reasons, domestic or international, or to allow a potential space station partner to gain needed practical experience. Student experiments went through a review process that compared their proposals to those of other high schoolers, not to any stated NASA scientific need.[7]

Perusing shuttle science manifests, the conclusion could be drawn that experiments were random. A significant percentage of shuttle time and space was spent on employing the vehicle for national policy reasons or as a commercial hauler.[8] It was as though NASA was operating like a technology and engineering factory rather than as a science laboratory.[9] Particularly in the early days, it was difficult to find any mention of an Agency need for peer review. A September 1981 press release, for example, announced that the OSS had chosen principal investigators for Spacelab 4, but made no mention of any process by which this might have been done.[10] A Space Science and Applications Notice dated September 1983, a general call for shuttle experiment submissions, said the "Space Biology Peer Panel, whose members are from the scientific community," would review proposals "for scientific merit," but gave no specifics as to the constitution of the review panel. Nor did scientific merit necessarily equate with something on the critical path of NASA's mission.[11]

The first meeting of the Life Sciences Peer Review and Policies Committee convened at Headquarters in early November 1988.[12] By the early 1990s, new peer review panels began evaluating and recommending experiments to fly on Space Station Freedom. Reviewers had to evaluate each proposal within their area of expertise, individually and as a group, and then justify the recommendation on the basis of how well the experiment would fulfill scientific need specified in the Research Announcement (RA) or Announcement of Opportunity.[13]

Outsiders reportedly still believed this process limited on-orbit time to a clique consisting mainly of Agency employees spending an outsize portion of the life sciences budget. Intramural NASA researchers, for their part, reportedly felt misunderstood and that nobody else could do as good a job with science as it related to astronauts.[14] After studying the situation in 1993, Earl W. Ferguson, Special Assistant for Research and Operations in OLMSA, told the NASA-NIH Advisory Panel on Biomedical and Behavioral Research that intramural peer review was a "major criticism of the existing system." Detailed Supplementary Objectives (DSOs)—experiments on astronauts in flight—were rare opportunities of operational access to crewmember medical data and were not open to extramural (non-NASA) investigators. To complicate matters, the same intramural scientists who had the monopoly on the unique environment and subjects were also the Agency's technical monitors of outside PIs investigating the same topics in a non-DSO situation.[15]

Dan Goldin's reorganization in March 1993 had put Harry Holloway at the top of the new Office of Life and Microgravity Sciences and Applications (OLMSA), and in response to some of the criticism Holloway decreed in May 1993 that everything "supported by Code U [OLMSA]" be peer reviewed up front and "at regular intervals" for "relevance and worthiness." Program managers could request "deviations," however, and nothing in Holloway's brief memo addressed the independence of the reviewers or outsider restrictions. Senator Barbara Mikulski (D-MD) of the Appropriations Committee wrote into the FY 1994 funding bill that NASA must "formally adopt the means of program administration of its life science program that is used at the NSF and the NIH," or lose its money. This included identical standards for intramural and extramural researchers, reviews every three years, and for the NASA scientists, reviews "by qualified individuals not associated with that particular program."[16] NASA responded quickly to Mikulski's big stick, putting up everything from 1993 for recompete. The Agency hired the American Institute of Biological Sciences (AIBS) to scrutinize more than 700 proposals.[17]

OLMSA's Life and Biomedical Sciences and Applications Division issued a revised Science Management Plan in February 1994 describing the new process and how it applied to both solicited and unsolicited proposals, intramural and extramural. Science programs were to have managers who would "assure that knowledge is transferred among the various intramural and extramural components" to build "appropriate advocacy for the science program within both the scientific community and NASA," and coordinate all related international and interagency projects. The division director (then Joan Vernikos) could appoint for special missions a program scientist who might "interpret" to all involved and design the strategy for experiment selection, but the new plan strictly proscribed any "real or apparent conflict of interest" on the part of the reviewers.[18]

Thus, every scientist was subject to a new peer review methodology and intramural PIs would lose oversight responsibility in areas of their own research. OLMSA would also disengage operational and basic science within the DSOs, allowing separate scrutiny of each. Goldin stated that he had been favoring scientists over engineers in high managerial appointments to indicate good faith on the Agency's part and to entice new research talent. The consensus was that improving the peer review process would increase the number of PIs submitting proposals, and that enlarging that population correspondingly would "increase NASA's external advocacy group."[19] By April NASA was making "major progress," with the new policy issued and procedures akin to those of NIH being "underway." Charles A. "Mickey" LeMaistre, president of M. D. Anderson Cancer Center in Houston and chair of the NASA-NIH Advisory Panel on Biomedical and Behavioral Research, wrote, "NASA is to be specifically commended for its forthright attitude and aggressive approach to reinvestigate its peer review process."[20]

Also affecting the quality of research was the fact that NASA's small cohort of test subjects (astronauts) usually meant very small sample populations ("n's"), perhaps even just one individual—an "n" of 1—so consequently experiment results were statistically questionable. That led the Agency to commission a study comparing its research protocols with the efforts of others working with select, very small populations, evaluating drugs or procedures to treat very rare illnesses.[21] The Institute of Medicine, part of the National Academy of Sciences, did the study and concluded that the answer was not to avoid the situation but to move forward with well-defined research questions. Experimental protocol should be tailored with alternative statistical design and analysis methods evaluated at every step. In light of the greater uncertainty a small population produced, several different analyses were needed to "evaluate the consistency and robustness" of the clinical trial. Also, great caution had to be exercised in interpreting results because potential for error, misinterpretation, and misapplication lay in extrapolation and generalization from small data sets. The Institute also issued a call for federal support for more research into mathematical modeling at both the theoretical and applied levels.[22]

The status of the ISS as both a national laboratory and an international partnership also led to a need to solidify human- and animal-subject review procedures. In 1997 Charles Sawin and Lawrence Chambers set up the first meetings in Houston with key Russian scientists whom the Institute for Medical and Biological Problems (IMBP) in Moscow had selected for what would become the first international board governing such experiments. In 1998 they brought in Canadians, Europeans, and Japanese. Russian scientists welcomed the new responsibility since it gave them more credibility and stature within their professional community. The Japanese were a challenge, because up to that point they had no "Common Rule" or standard protocol within their federal agencies concerning research on human subjects. By 2000, all the partners agreed to a standing committee that would meet to consider all requests for experiments on humans. The partners would evaluate proposals against the guidelines NASA had been using and refining throughout the shuttle program: the "Red Book," or *JSC Committee for the Protection of Human Subjects Guidelines for Investigators Proposing Human Research for Space Flight and Related Investigations*. Votes would be by consensus, so if anyone objected to an experiment it was to be rejected. With the establishment of this committee, each partner could then inject NASA standards into their own peer review process, elevating their nation's science in the eyes of global peers.[23]

After the *Columbia* tragedy curtailed ISS crew size to two six-month residents and a third short-stay visitor, Sawin worked to add a new step as "sort of a subset of the existing space station agreement," to avoid human experiment duplication. In such cases, NASA might realign congruent experiments to mandate experimenter teams and collective protocols among partners. "We don't have the crew time, we don't have the number of subjects, to afford [duplication]," Sawin said in a 2006 interview. "I think if we can do that we will learn more, we'll get better data, and we'll waste less, or not waste as much resources as we have."[24]

Extended Duration Orbiter Medical Project (EDOMP)

NASA backed off plans for orbiters that would stay in space for two to three months and began to consider a four-week option in the late 1980s. The Shuttle Program Office asked Life Sciences and the Astronaut Office at JSC about the feasibility of this. Although four-week shuttle missions never happened, the resulting research did gather data pertinent to station, Moon, and planetary missions.[25]

Operational medical personnel at JSC and the astronauts themselves knew that humans could function in space that long but questioned whether they could clamber out of the craft after return to Earth and function immediately. Would orthostatic intolerance be twice as bad if a mission were twice as long? How much fine muscle control might pilots have lost, just when they needed it for landing? How much more difficult would it be to exit a spacecraft while wearing a 75-pound (31.75-kg) backpack? To answer questions like these, NASA spent $40 million on EDOMP, using DSOs and led by Project Manager J. Travis Brown and Project Scientist Charles Sawin at JSC.[26]

Some earlier data had been contradictory, and assumptions and observations had proven to be questionable. Sample sizes had been too small for reliability, or information was simply unobtainable for various reasons, including incompatible or conflicting experiments. EDOMP researchers thus wanted to repeat earlier experiments or alter them and try new equipment and protocols.[27] Experiments selected covered many challenges of long-duration flight: cardiovascular changes and efforts at maintaining conditioning; the body's regulatory system for maintaining heart rate, blood pressure, circadian rhythms, and other physiological functions; the means of measuring performance; environmental safety; neurology; and human factors/ergonomics. Researchers carried out 36 experiments on 40 shuttle missions varying from 4 to 16 days, between January 1990 and January 1996 (STS-32 through STS-72), and including as few as five participants and as many as forty-two, with men and women, rookies and veterans, military and civilians, and pilots and mission specialists taking part.[28]

Orthostatic intolerance was the chief concern, as 5 percent of earlier crews had experienced dizziness, felt lightheaded, grayed out, or even fainted upon standing before egress, and another 5 to 6 percent experiences these issues during tests afterward. The post-*Challenger* Launch and Entry Suit had actually increased those numbers until a liquid cooling undergarment came online, and flight surgeons wanted to be prepared for the effects of future changes to the astronaut garb. Thus the project tested for OI susceptibility by age, gender, previous experience, and various physiological differences.[29]

Wearing Holter cardiac monitors and self-activating blood pressure devices during routine in-flight activities (every 20 minutes) and while asleep (every 30 minutes) provided more realistic feedback on cardiovascular condition than a test requiring astronauts to stop doing everything else. Data taken during, before, and after flight showed that heart rates and blood pressure changed at key points, such as liftoff, reentry, and landing. When astronauts reached Earth and stood, heart rate increased an average of 70 percent, in one case reaching 160 beats per minute, near the projected limits of tolerance. Blood pressure, however, dropped about 20 points in 22 percent of the astronauts tested, and in some cases was insufficient to retain consciousness. Echocardiograms of the left ventricle during flight showed that with each beat or stroke, the heart was pumping a smaller volume of blood than it would under normal gravity, partially explaining the rapidity of the heartbeat.[30]

Blood tests of the catecholamines responsible for controlling heart rate and blood pressure (e.g., epinephrine, norepinephrine, and dopamine) indicated, surprisingly, that individuals were consistent in their blood chemistry responses on the ground and in flight. Thus people could be divided into two groups: those who lost consciousness and those who did not. In medical terms, fainters had "low standing norepinephrine levels . . . which ultimately resulted in inadequate cerebral perfusion and presyncope. The failure to increase norepinephrine translated into lower peripheral vascular resistance, lower arterial pressures, and lower heart rate responses to decreasing systolic pressure." This suggested there might be a pharmacological countermeasure in the form of a "successful pressor agent" (a vasoconstrictor to increase blood pressure) and that it should be possible to predict who would be likely pass out and who would not.[31]

After 12 DSOs on dozens of subjects, the three cardiovascular PIs, John Charles, Janice Meck Fritsch-Yelle, and Peggy Whitson, concluded that physiological changes in zero-g "left the astronauts ill-prepared for the cardiovascular stresses associated with return to Earth." This was true even on four- or five-day missions, and crews took about a week to return to normal. The results revealed cardiovascular systems "under significant stress during nominal entry, landing, and seat egress . . . performing at or near . . . maximum capacity in a significant fraction (20%) of the study population. [The] swings in arterial pressure and heart rate indicate that they were unable to buffer arterial pressure changes as well as before flight. It is questionable whether sufficient reserve capacity remained to permit unaided emergency egress."[32]

Four countermeasure DSOs proved adrenal hormones, fluid-loading, pressure suits, and vasopressors not to be sufficient by themselves. The crew did not tolerate fludrocortisone (a synthetic adrenal hormone) well.[33] Experimenters had no say over wave-offs, when crew had already consumed the prescribed fluid but were told to make a few more orbits. Neither could researchers always get to the astronauts for postflight testing before they had consumed anything else. Sometimes missions were extended by a day after the astronauts had used the Lower Body Negative Pressure suit (LBNP) and then packed it away for landing.[34]

Encouragingly, echocardiograms showed that heart muscle did not deform in spaceflight. Holter monitors established that disrhythmias (changes in heart rhythm) did not increase. Ongoing study also revealed diurnal (daytime) cardiac rhythms, important in planning long missions.[35]

Loss of muscle mass and bone demineralization had proved to be related to the body's attempts to maintain the cardiovascular system's normal state. Skylab had shown that astronauts lost weight in space, much of it in lean body mass, promoting excretion of electrolytes, especially potassium, which affects cardiovascular function. After studying 13 males on six flights, the DSO 612 (Energy Utilization) team found that energy and fluid intake were lower in flight, resulting in a mean weight loss of 3–8 pounds (1.36–3.63 kg). DSO 610 (In-flight Assessment of Renal Stone Risk) verified that astronauts were excreting more calcium, which, combined with insufficient fluid intake, put them at increased risk for kidney stones. Although the sample size for the gastric emptying and intestinal transit time experiments was too small to be conclusive, previous experiments in bed rest studies had shown both to be faster with supine subjects, reducing nutrient absorption.[36]

DSO 484 (Assessment of Circadian Shifting in Astronauts by Bright Light) tested the utility of exposure to light of a certain intensity and duration prior to launch and of melatonin's use in flight to radically shift circadian rhythms by 9 to 12 hours over a short time period. Eight people took part during three missions, five men and three women. The protocol worked for seven of the eight, which was promising for future shuttle missions.[37]

Researchers wanted to quantify the physical requirements for reentry, landing, and emergency egress, then arrive at an optimal combination of preflight and in-flight fitness exercises to minimize muscle degradation. They hoped to create a timeline of muscle change, identifying critical stages at which it became deconditioned, then atrophied.[38] Eight DSOs measured the effectiveness of exercise countermeasures. Needle biopsies of the vastus lateralis thigh muscle on the day of return from space (R+0) and MRIs (on R+2 and R+7) of the soleus and gastrocnemius leg muscles determined that leg muscle shrinkage observed was due to actual atrophy, not just fluid shifting to the head. The Cross-Sectional Area of slow-twitch (endurance) muscles decreased 15 percent and in fast-twitch (sprinting) muscles by 22 percent. MRIs indicated a decrease of only 4 percent but seven days after return, subjects had regained nothing. Researchers could come to only general conclusions about intensity, duration, and timing of in-flight exercise but the consensus was that time in weightlessness invariably decreased physical performance.[39]

Overall, the more often crewmembers exercised intensely (estimated at 70 percent of maximum heart rate) during the mission, the less of a decrease in aerobic capacity they would experience, but not by a lot. The average decrease among exercisers was 18 percent, compared with 21 percent for the others.[40] Also, treadmill subjects showed less of a decrease than did those who used the upright cycle ergometer, and all regained their preflight aerobic levels within seven to ten days. Orthostatic intolerance was not affected by aerobic exercise, correlating with the cardiovascular study findings that it was individual blood chemistry that made one susceptible to OI. The question remained as to whether vigorous exercise late in the mission could make up for lack of exercise the rest of the time.[41]

Half a dozen DSOs anticipated the 1998 Neurolab mission, with tests on the effect of microgravity on the neurovestibular system responsible for balance, depth perception, and space sickness. The concerns were that pilots could not make critical visual perceptions and fine-motor hand motions vital for an emergency landing and crew could not stand upright or walk steadily enough for rapid egress. DSO 604 OI-3 (Visual-Vestibular Integration as a Function of Adaptation) examined "gaze," from visual "target acquisition" to "image stabilization" to "pursuit acquisition" of a moving target, along with head oscillations as astronauts kept their eyes on a test object. The five tasks of DSO 614 (the Effect of Prolonged Space Flight on Head and Gaze Stability during Locomotion) revealed that weightlessness had a "profound effect on sensory-motor function." It disturbed gaze during both target acquisition and locomotion when the subject walked on a treadmill toward a goal, but a positive link existed between prior time spent in space and performance, meaning this got better with experience. DSO 604 OI-1 (Visual-Vestibular Integration as a Function of Adaptation: Motion Perception Reporting) confirmed that zero-g provoked illusions of movement and position of the body in relation to other objects, particularly when the subjects' eyes were closed and they were unrestrained. During spacewalks, astronauts had reported disorienting perceptual and positional illusions. Small head movements made during the EDOMP tests created a perception of the surroundings moving in the same direction; bigger movements gave the illusion of motion in the opposite direction. This was more frequent with medium duration missions. Seventy percent reported this effect during flight, 80 percent during reentry, and 90 percent upon landing.[42]

DSO 605 (Postural Equilibrium Control During Landing/Egress) and DSO 614 verified previous US and Soviet observations of animals and humans, then quantified and calibrated the data. Each subject's vestibular system was unable to support upright posture immediately after landing or regain balance if perturbed. Subjects walked as though on a boat pitching up and down in a wild sea. With each step they thrust the foot with greater force and with increased angular amplitude at the knee and ankle. They were unstable when turning corners, increased their vertical acceleration, and tried to preserve stability by standing with their legs further apart, using their arms, and taking shorter steps. In jumping downward, as one might from a disabled orbiter, all were unsteady, one to the point of consistently falling over backward, which

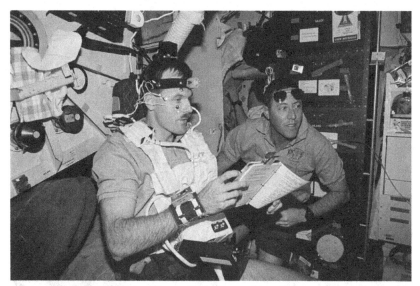

Astronaut Steven Smith wired for an EDOMP visual vestibular integration experiment, DSO 604 on STS-68 in 1994. Mission commander Michael Baker observes the data on a monitor out of sight. Photo credit: NASA.

researchers interpreted as meaning that the otolith-spinal reflex, which helps prepare leg muscles for the shock of landing, was "dramatically reduced in spaceflight," as was the perception of limb placement. These results emphasized how the visual system takes over in the absence of vestibular input and thus how the body integrates gaze and vision with movement. EDOMP researchers saw a need for sensory-motor countermeasures (including artificial gravity via in-flight centrifugation) and for preflight training so astronauts could recognize such effects and deal with them in various contingencies.[43]

The EDOMP experiments also evaluated overall functionality. A human-factors DSO examined the design and operation of the glovebox (GBX), general purpose workstation (GPWS), paper versus electronic procedures, touch screens versus track balls, stowage, noise, and lighting. The STS-50 (US Microgravity Laboratory-1) crew commented that the glovebox needed flexible armholes for better range of motion, accommodations for taller experimenters, foot restraints with knee support, and calibration markings to readjust it quickly. NASA redesigned and reflew GBX on STS-73 (USML-2). Astronauts critiqued the workstation on STS-58 and found it enhanced productivity, but recommended it be enlarged to hold two animals and two operators. They rated it much

EDOMP Waste Collection Device. Photo credit: NASA.

less restrictive than the GBX and said it caused no neck or shoulder pain, in spite of working hunched over for long periods.[44]

Astronauts noted that fine motor actions took less time than gross movements, malfunctions took more time to fix than expected, some tasks required more force or torque than others, and restraints and hand/footholds needed to be more or less robust depending on the activity and location. For stowage, they wished for more training with actual packing foam and in-flight configuration. They requested more Velcro and more handholds, that stowage lockers be easy to open with one hand, equipment stowed so that it neither jammed nor floated out, and that there be quick ways to restrain small items. NASA management wanted to move to a "paperless Shuttle," and astronauts found electronic interfaces, such as questionnaires and checklists, personally helpful because they did not have to keep finding their place again on a page.[45]

An environmental health team looked at airborne particulate hazards, chemical accumulation in the air, and microorganism and pathogen presence in water and air and on surfaces. Only formaldehyde was present at levels higher than the federal exposure limit. Of concern also were combustion products from overheating of electronics, particularly Teflon, Kapton, and epoxy resin components that released hydrogen chloride, hydrogen fluoride, hydrogen cyanide, and carbon monoxide.

Table 4.1. EDOMP environmental tests showed that shuttle air quality could vary greatly in type, amount, and contamination hazard

STS Mission	Contamination Concern	Analytical Results
28	Teleprinter cable	SSAS sample showed nothing unusual
31	High benzene in preflight sample	Benzene, found at 0.5 mg/m^3 in preflight sample, was scrubbed down to 0.01 mg/m^3 late in mission
35	Odor of burning electronics near the data display units (2 failures)	SSB sample showed 0.01 mg/m^3 benzene, which was later reproduced from ground pyrolysis of identical electronic components
37	Odors in galley area	SSB showed no unusual contaminants
40	Noxious odors from refrigerator / freezer	SSB sample showed no clear evidence of contamination. Ground studies of burned motor showed released ammonia and formaldehyde
49	Odor from airlock after EVA	Acetaldehyde (0.6 mg/m^3) was unusually high in SSB sample
50	Burning odor near American Echo Research Imager	SSB sample showed unusually high concentration of dichloromethane
53	Crew experienced nasal congestion possibly due to air contaminant	No unusual contaminants found in SSB sample
54	Odors in area of waste control system	Two incompletely identified organic compounds were found
55	Noxious odors from contingency waste container	Three dimethyl sulfides found at concentrations that would produce a noxious odor

Source: Charles F. Sawin, Gerald R. Taylor, and Wanda L. Smith, eds., *Extended Duration Orbiter Medical Project: Final Report 1989–1995*, SP-1999-534 (Houston: NASA-JSC, 1999), 4–8.

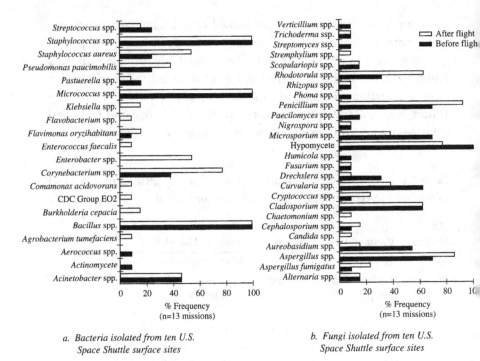

a. *Bacteria isolated from ten U.S.*
   Space Shuttle surface sites

b. *Fungi isolated from ten U.S.*
   Space Shuttle surface sites

Bacteria and fungi found on EDOMP-mission shuttle surface sites. From Charles F. Sawin, Gerald R. Taylor, and Wanda L. Smith, eds., *Extended Duration Orbiter Medical Project: Final Report 1989–1995*, SP-1999-534 (Houston: NASA-JSC, 1999), 4–11.

EDOMP documented nine "toxicological incidents" between STS-35 and STS-55, four from burning electronics.[46]

The EDOMP database originally hoped for did not happen completely. Researchers had to piece together findings from different missions. Some individuals and mission commanders proved more or less cooperative than others. Orbiters landed at different sites, with different ground support.[47] Subjects sometimes understood instructions differently, gear did not always work, and occasionally astronauts might, on their own initiative, "fix" an experimental protocol to give the researcher "the *right* experiment" and not the one approved by the review committee. This gave the frustrated investigator a new (and useless) data set of one, or an experimental cohort with one less participant, or it invalidated an experiment outright. From his experience, NASA physiologist John Charles concluded that the most productive subjects were the military pilots and "backseaters," trained to carry out all missions

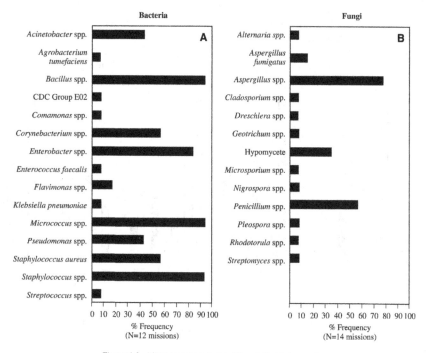

Figure 4-3. Microorganisms isolated from inflight air samples.

Missions measured the amount and type of illness-causing airborne contaminants. From Charles F. Sawin, Gerald R. Taylor, and Wanda L. Smith, eds., *Extended Duration Orbiter Medical Project: Final Report 1989–1995*, SP-1999-534 (Houston: NASA-JSC, 1999), 4–10.

exactly and completely and not to insert their own opinions. The worst were other scientists.[48] Mission commanders reportedly did not insist on crew compliance, either.

EDOMP, heavily dependent on cooperation in orbit, exemplified the debate over the extent to which astronauts should be *required* to participate in medical tests, not just *invited*. NASA's position was that in-flight experiments would be conducted according to international medical ethics governing ground-based human-subjects research. Participation was strictly voluntary; subjects could opt out at any time regardless of how that affected the investigator's ability to complete the experiment, and astronauts had to give informed consent prior to commencing. That meant PIs had to explain all possible hazards and discomforts, and the astronaut had to understand them. Astronauts had the right to take part only in experiments of benefit to themselves. There also were rules

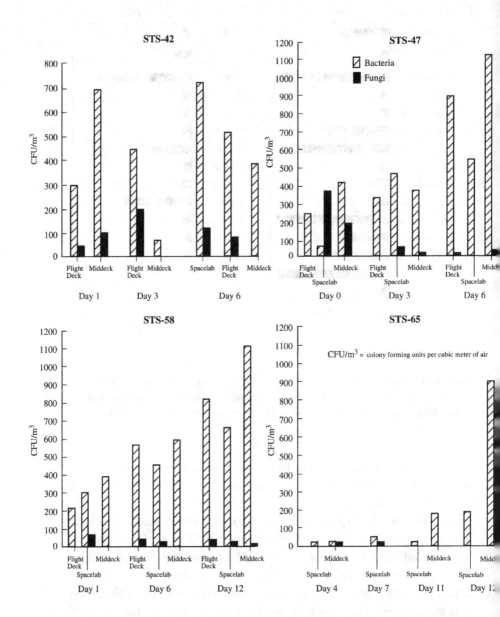

Charts illustrate the accumulation of data on sickness-causing contaminants by mission. Bacteria were overwhelmingly more present than fungi. From Charles F. Sawin, Gerald R. Taylor, and Wanda L. Smith, eds., *Extended Duration Orbiter Medical Project: Final Report 1989–1995*, SP-1999-534 (Houston: NASA-JSC, 1999), 4–9.

Table 4.2. Comparison of Sound Levels for STS-40, STS-50, and STS-57

Flight	Source Location	Conditions	Actual Decibels
FLIGHT DECK: DESIGN LIMIT 63 DECIBELS			
STS-40	Flight Deck (Center)	Nominal systems (ECLSS)	61.8
STS-50	Flight Deck (Center)	Nominal systems (ECLSS)	64.0
STS-57	Flight Deck (Center)	ECLSS + SAREX	72
		ECLSS + A/G	62
MIDDECK: DESIGN LIMIT 68 DECIBELS			
STS-40	Middeck (Center)	Nominal systems (ECLSS)	63
STS-40		ECLSS + AEM	64.7
STS-40	Middeck (1 foot from AEM)	ECLSS + AEM + OR/F	67.6
STS-50	Middeck (Center)	Nominal systems (ECLSS)	59.9
STS-50	Middeck (Center)	ECLSS + EVIS + Bike	67.9
STS-50	Middeck (Center)	ECLSS + Vacuum Cleaner	79.9
STS-57	Middeck (Center)	Nominal systems (ECLSS)	63
STS-57		Nominal systems (ECLSS)	62
SPACELAB/SPACEHAB: DESIGN LIMIT 68 DECIBELS			
STS-40	Spacelab (Center)	SR/F—one compressor on	69.7
STS-40	Spacelab (4 feet from SR/F)	SR/F—both compressors on	72.6
STS-50	Spacelab (Center)	Nominal systems (ECLSS)	61.6
STS-50	Spacelab (Center)	Nominal systems (ECLSS)	61.2
STS-50	Spacelab (operator)	ECLSS + DPM	64.7
STS-50	Spacelab (operator)	ECLSS + GBX on	61.0
STS-50	Spacelab (operator)	ECLSS + STDCE	63.8
STS-57	SpaceHab (Center)	ECLSS, fans off	63
STS-57	SpaceHab (Center)	ECLSS, fans on	66

Source: Charles F. Sawin, Gerald R. Taylor, and Wanda L. Smith, eds., *Extended Duration Orbiter Medical Project: Final Report 1989–1995*, SP-1999-534 (Houston: NASA-JSC, 1999), 6–7.
Notes: Sound levels sometimes exceeded design guidelines with experiments and equipment in operation.
ECLSS: Environmental Control and Life Support System
EVIS: Ergometer Vibration Isolation System (exercise bike)
A/G: air-to-ground
GBX: Glovebox
SR/F: Spacelab refrigerator/freezer
SAREX: Space Amateur Radio EXperiment
EDOMP: Extended Duration Orbiter Medical Project
STDCE: Surface Tension Driven Convection Experiment

Table 4.3. EDOMP studies covered a range of environmental health and ergonomics issues that might arise on a four-week mission and affect the crew's ability to work and return to Earth safely

Topic	Description	STS Flight
1.	Comfort and accessibility of the glovebox at the general purpose workstation	50, 58, 73
2.	Procedures and operation of the Lower Body Negative Pressure Device	50, 58
3.	Management, stowage, deployment, and restraint of electric power and data cables	40
4.	Obstacles and facilitators for task procedures and timelines baselined on Earth	40
5.	Advantages and difficulties using a computer touchscreen in microgravity	70
6.	Electronic and paper procedures in microgravity	57
7.	Perceptions and effects of mechanical vibration on task performance in flight	40
8.	The noise environment	40, 50, 57
9.	The lighting environment	57
10.	Crew member translation and equipment manipulation through the tunnel joining the Shuttle middeck and the pressurized SpaceHab or Spacelab	40, 47, 57
11.	Assessment of crew neutral body posture	47, 57
12.	Questionnaire responses	57

Source: Charles F. Sawin, Gerald R. Taylor, and Wanda L. Smith, eds., *Extended Duration Orbiter Medical Project: Final Report 1989–1995*, SP-1999-534 (Houston: NASA-JSC, 1999), 6–1.

as to how intrusive, painful, and public an experiment and its results could be.[49]

This placed constraints on research since there might be just one female, one rookie, one pilot, or one person in some other category generating data that had to be unattributable to any one subject. PIs also had no input as to crew selection. They did not have the option to add more subjects. Only if the PI or management learned years before flight that a member of a crew not yet in training for a mission was opposed to being a test subject could they ask that individual to take themselves out of consideration for that assignment.[50]

Individual scientists and astronauts took different views of that free choice mandate. Some, pointing to the post–World War II Nuremberg

Code, believed the Agency had to uphold the highest standards in medical ethics regardless of the costs scientifically and monetarily.[51] NASA must allow an astronaut to refuse to participate on any grounds and without explanation, even if flight surgeons needed the data to protect future astronauts or if declining meant a science opportunity lost or an experiment ruined. Others felt the good of the many outweighed individual wishes and that functionally NASA was a quasi-military organization, so highly trained astronauts ought not to have the luxury of refusing assignments.[52] One longtime flight surgeon said Apollo astronauts had been simply told they had to do what flight surgeons asked. Astronauts were themselves part of a bigger experiment and knew that when they signed on.[53]

EDOMP, after six years of gathering data, instigated a few new flight rules. NASA made g-suits and exercise mandatory. All shuttle missions flew the Combustion Products Analyzer to track electronics thermodegradation and inorganic volatiles dispersement that astronauts ingested as they breathed.[54] Largely intramural, EDOMP also likely eased the Agency into the more daunting international experiments that would come with Neurolab three years later.

Neurolab (STS-90)

Spacelab was originally a European project. In the late 1960s through the early 1970s, plans had been afoot to challenge US and Soviet space preeminence. Concerned that France's Ariane rocket would cost NASA the launch business it was hoping would pay for STS, NASA Administrator Tom Paine urged the Nixon administration to pressure Europe to pursue new areas of mutual benefit to themselves and the United States if they wished to gain experience and jobs for European workers. After much argument among White House, State Department, and NASA leadership over what sort of cooperative projects NASA would be willing to pay or barter for, the idea that won out was the "sortie can," essentially a free-flyer. The Europeans—getting less, working harder, and spending more money—were unhappy enough to resolve to "pool the better part of their future projects in a multilateral alliance for space," namely ESA. The sortie can did not remain a can for long, but became Spacelab, which was not simply a vessel but a commitment by NASA to launch several of the shuttle payload bay laboratories, reserving use of the first one for itself.[55]

The Spacelab concept was on its way to success with three NASA missions and the all-German D-1 by January 1986 and the *Challenger* disaster. It was five years before a Spacelab mission orbited again, the first dedicated Space Life Sciences (SLS-1) flight in June 1991. Then came roughly three Spacelab missions per year, dedicated to national (Japanese, German) or thematic projects (life sciences, materials processing), a record marred by the January 1994 cancellation of SLS-3, -5, -6, and -7 (including a joint US-French primate flight). In spite of a Space Studies Board (SSB) recommendation that NASA "make every effort to mount at least one Spacelab life sciences flight in the period between Neurolab and the completion of ISS facilities," station assembly flights took precedence and the program ended.[56]

However, Spacelab's life sciences work finished on a high note. The April 1998 SLS-4 mission aboard STS-90 stood out as a model of international collaboration, along a well-defined path and with achievable goals.[57] Presidential Proclamation 6158, which George H. W. Bush had signed eight years earlier in response to House Joint Resolution 174, declared the 1990s to be the "Decade of the Brain" and inspired the mission. The National Institute of Mental Health (NIMH) and Library of Congress (LOC) spearheaded the Decade of the Brain program to raise public (and congressional) awareness of the benefits of research on brain and central nervous system conditions such as Alzheimer's, Huntington's, and Parkinson's diseases, epilepsy, depression, paralysis, stroke, addiction, muscular dystrophy, schizophrenia, and autism.[58] Frank Sulzman at Headquarters, J. Wally Wolf of USRA, and Rodolfo Llinás of the New York University School of Medicine suggested the space agency participate with a dedicated mission. NASA, the NIH, NSF, and DOD partnered with the Canadian Space Agency, Japan, the German Space Agency, France's Centre National d'Etudes Spatiales (CNES), and ESA, publishing an AO in 1993 that brought in 172 proposals. An NIH panel selected 26 experiments for flight.[59]

JSC led the human experimentation effort for Neurolab, while Ames Research Center managed the animal studies. For some experiments, such as anesthetizing rats and then performing microsurgery on them inside the GPWS, PIs explored procedures that might be carried out aboard the ISS. Fulfilling its educational mandates, NASA filmed documentaries, created an interactive website, designated a "Professor in Space," and sponsored a K-12 neurology curriculum developed by the Morehouse School of Medicine in Atlanta. Overall, intentions were to

Table 4.4. Chart of Spacelab missions with life science experiments in a habitable module

Flight Number	Flight Date	Mission Name	Participants
STS-9	Nov. 28–Dec. 8, 1983	Spacelab 1	John W. Young, Brewster H. Shaw, Owen K. Garriott, Robert A. Parker, Ulf Merbold (ESA), Byron K. Lichtenberg
STS-51-B	Apr. 29–May 6, 1985	Spacelab 3	Robert F. Overmyer, Frederick D. Gregory, Don L. Lind, Norman E. Thagard, William E. Thornton, Lodewijk van den Berg, Taylor G. Wang
STS-51-F	July 29–Aug. 6, 1985	Spacelab 2	C. Gordon Fullerton, Roy D. Bridges Jr., F. Story Musgrave, Anthony W. England, Karl G. Henize, Loren W. Acton, John-David F. Bartoe
STS-61-A	Oct. 30–Nov. 6, 1985	Spacelab D-1	Henry W. Hartsfield, Steven R. Nagel, Bonnie J. Dunbar, James F. Buchli, Guion S. Bluford, Reinhard Furrer (Germany), Ernst Messerschmid (Germany), Wubbo Ockels (ESA)
STS-40	June 5–14, 1991	Spacelab Life Sciences-1 (SLS-1)	Bryan D. O'Connor, Sidney M. Gutierrez, James P. Bagian, Tamara E. Jernigan, M. Rhea Seddon, F. Drew Gaffney, Millie Hughes-Fulford
STS-42	Jan. 22–30, 1992	International Microgravity Laboratory-1 (IML-1)	Ronald J. Grabe, Stephen S. Oswald, Norman E. Thagard, David C. Hilmers, William F. Readdy, Roberta L. Bondar (Canada), Ulf Merbold (ESA)
STS-50	June 25–July 5, 1992	US Microgravity Laboratory-1 (USML-1)	Richard N. Richards, Kenneth D. Bowersox, Bonnie J. Dunbar, Ellen S. Baker, Carl J. Meade, Lawrence J. DeLucas, Eugene H. Trinh
STS-47	Sep. 12–20, 1992	Spacelab-J	Robert L. Gibson, Curtis L. Brown Jr., Mark C. Lee, N. Jan Davis, Jay Apt, Mae C. Jemison, Mamoru Mohri (Japan)
STS-55	Apr. 26–May 6, 1993	Spacelab D-2	Steven R. Nagel, Terence T. Henricks, Jerry L. Ross, Charles J. Precourt, Bernard A. Harris Jr., Ulrich Walter (Germany), Hans Schlegel (Germany)

(continued)

Table 4.4—*Continued*

Flight Number	Flight Date	Mission Name	Participants
STS-58	Oct. 18–Nov. 1, 1993	Spacelab Life Sciences-2 (SLS-2)	John E. Blaha, Richard A. Searfoss, M. Rhea Seddon, William S. McArthur, David A. Wolf, Shannon W. Lucid, Martin Fettman
STS-65	July 8–23, 1994	International Microgravity Laboratory-2 (IML-2)	Robert D. Cabana, James D. Halsell, Richard J. Hieb, Carl E. Walz, Leroy Chiao, Donald A. Thomas, Chiaki Mukai (Japan)
STS-71	June 27–July 7, 1995	Spacelab-Mir	Robert L. Gibson, Charles J. Precourt, Ellen S. Baker, Bonnie J. Dunbar, Gregory J. Harbaugh
STS-73	Oct. 20–Nov. 5, 1995	US Microgravity Laboratory-2 (USML-2)	Kenneth D. Bowersox, Kent V. Rominger, Kathryn C. Thornton, Catherine G. Coleman, Michael E. Lopez-Alegria, Fred W. Leslie, Albert Sacco Jr.
STS-78	June 20–July 7 1996	Life and Microgravity Spacelab (LMS)	Terence T. Henricks, Kevin R. Kregel, Susan J. Helms, Richard M. Linnehan, Charles E. Brady Jr., Jean-Jacques Favier (France), Robert Brent Thirsk (Canada)
STS-83	Apr. 4–8, 1997	Microgravity Science Laboratory-1 (MSL-1)	James D. Halsell, Susan L. Still, Janice E. Voss, Donald A. Thomas, Michael L. Gernhardt, Roger Crouch, Greg Linteris
STS-94	July 1–17, 1997	Microgravity Science Laboratory-1R (MSL-1R)	James D. Halsell, Susan L. Still, Janice E. Voss, Donald A. Thomas, Michael L. Gernhardt, Roger Crouch, Greg Linteris
STS-90	Apr. 17–May 3, 1998	Neurolab	Richard A. Searfoss, Scott D. Altman, Richard M. Linnehan, Dafydd Williams (Canada), Kathryn P. Hire, Jay C. Buckey, James A. Pawelczyk

Source: Created by author.

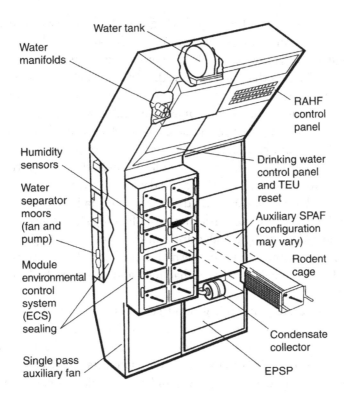

The Research Animal Holding Facility. From Jay C. Buckey Jr. and Jerry L. Homick, eds., *The Neurolab Spacelab Mission, Neuroscience Research in Space. Results from the STS-90, Neurolab Spacelab Mission,* SP-2003-535 (Houston: NASA-JSC, 2003), 296.

make the flight fit the science, not the other way around. Experiments in Earth orbit would be as equivalent as possible to what ground labs did.[60]

Life forms for Neurolab's experiments ranged from insects to snails to fish to small mammals (rats and mice). Rodents were in the animal enclosure module or AEM, which housed groups, and the Research Animal Holding Facility (RAHF), for pairs or mothers with babies. The vestibular function experimental unit held oyster toadfish (*Opsanus tau*), the Closed Equilibrated Biological Aquatic System (CEBAS) contained swordtail fish, and crickets were in the Botany Experiment (BOTEX) incubator with a centrifuge. Astronaut Rick Linnehan, a veterinarian, was on board to oversee their care. Crew carried out in-flight experiments at the GPWS, which provided an enclosed space in which animals could be removed from their containers.[61]

Not everything went as hoped. Of greatest concern was the death of more than half the juvenile rats, 14 and 8 days old, when housing and health-monitoring protocols both proved to be insufficient for that age group. The rats that died were in the RAHF, which had a controlled environmental system and methods for measuring food and water consumption—practical indicators of animal health and vitality—but the crew could not easily see the animals. The AEM, on the other hand, had good visibility, but crew could not measure consumption by individual animals. Astronaut handlers did not pull and check the cages until Flight Day 8 (a two-day delay), which was 10 days after they had been put on board the orbiter. Once the crew discovered they had 38 dead baby rats on their hands (out of 96), with another 19 looking sick, they undertook rescue measures. These included antibiotics, fluids (including dilute Gatorade), and warming, but the crew had to euthanize 12 rats. By the time *Columbia* had landed, another 10 had died and one more had to be put to sleep.

Neurolab astronauts and ground scientists concluded the deaths were due to a combination of environment ("soaking wet" cages), lack of close monitoring (measuring adult food and water consumption did not provide enough information on nursing infants), neglect by new rat mothers, and cage design. The RAHF did not provide sufficient grasping surface that would allow very young rats, whose eyes were not yet open, to remain attached to their mother or find her readily if they became separated. The experience also highlighted the importance of having adequate medical tools on hand. The veterinarian kit proposed initially "had minimal capabilities," and the crew had been only serendipitously able to find room for the fluids and veterinary antibiotics they turned out to so urgently need.[62]

Results offered new understanding of effects of microgravity on the brain and nervous system. For example, by centrifuging Neurolab astronauts on orbit as well as on Earth, the study team realized that earlier theories about perception of movement, direction, and tilt in zero-g were incorrect. Via rat experiments and subsequent necropsies, researchers evaluated individual regions of the brain for their roles in gravity perception and reaction. They learned that the nervous system adapted to space by forming new nerve connections so crew returned with a changed nerve physiology that altered how they responded to tilting and other motions. This would be especially significant on a long flight to Mars, and on Earth in understanding why the elderly fell more

In-flight centrifugation studies were carried out during the Neurolab mission, STS-90. Photo credit: NASA.

often, in rehabilitating stroke victims, and treating balance disorders like Ménière's Disease. Experiments also provided instructive new data on Space Motion Sickness (SMS) and indicated that artificial gravity would lessen the reduction of a specific ocular reflex that helped maintain balance. Tests on rats trained to run a three-dimensional track and various mazes showed that the hippocampus, the region of the brain that controlled navigation, adapted to gravity disturbance in its processing. Visual-motor coordination was good but required more "brain-power" to maintain in LEO. This was important because astronauts became more dependent on visual, rather than inner-ear cues.[63]

Of interest for long-term space colonization were questions of neurological development: would humans conceived, born, and raised to a certain age in weightlessness or fractional gravity have all their body parts and physiological systems? Would they be wired correctly to function on Earth? Would they be able to balance? Would the muscles that controlled the flow of blood to the brain (and prevent fainting) do their job? Would any anomalies be permanent? Lower life forms were uniquely useful as test subjects for asking these questions. Insects and gastropods took up much less space and required fewer consumables. Crickets had "easily accessible external gravity receptors" that regenerated if damaged during development and a nervous system much simpler to understand than that of mammals. Rats on a 16-day mission would grow from infancy to the approximate equivalent of a five-year-old human. Fish had a short enough life span that it was possible to observe several cycles of development.

STS-90 proved that gravity did matter. Crickets and snails compensated for the lack by growing many more and larger otoliths, the calcium carbonate granules in the inner ear that signal body position. Observations (and later necropsies) of rats swimming, walking, and righting themselves indicated that early development optimized an animal for its expected environment, even if it lacked gravity. Cricket and rat gravity sensors developed normally—cricket sensors removed in flight even regrew, albeit with fewer sensory units—but other nervous system components did not. Thus, reflex development and adaptation to one g was reduced in flight rats because they lacked enough links between brain and vestibular structures to balance and coordinate movement well. Even 110 days after landing they had still not developed "normal mature righting behavior." Development of antigravity muscles and structures used in baroreflex (blood pressure regulation) was affected, too, as fewer nerve terminals developed and muscle growth was less. Rats seemed able to adapt if returned to one g before reaching adulthood, but not quickly, and the lower number of nerve fibers appeared to be permanent. Investigators concluded that they could not reliably predict what would happen to humans conceived and raised in microgravity.[64]

One surprise was that orthostatic intolerance on return to Earth could be caused by a decrease in overall blood volume, triggered by the body perceiving a need to produce less blood in microgravity. This might be due to a lack of responsiveness on the part of blood vessels that had gotten out of shape in space.[65] This indicated that OI was treatable, even preventable, by fluid administration during flight.

Sleep deprivation studies included circadian rhythm due to its impact on mood, alertness, and productivity and also the changes in throat shape in microgravity. Did it allow the tongue to "float" and impair breathing, resulting in sleep apnea? The results showed that weightlessness, surprisingly, cured snoring rather than making it worse because

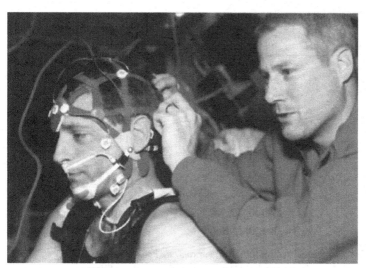

Payload commander Richard M. Linnehan, a veterinarian, being instrumented for circadian rhythm sleep studies on STS-90, Neurolab by payload specialist Jay Buckey. Photo credit: NASA.

the airways remained open. It did induce circadian changes and melatonin failed to bring on sleepiness, hinting that environmental factors such as noise and crowding kept astronauts awake. Sleep researchers observed extra REM (rapid eye movement) after return to Earth and speculated that the body might be busy "relearning" how to function in one g while the individual slept.[66]

Neurolab also pointed out fruitful avenues of study for future ISS missions. For example, crickets that regrew sensory organs with fewer sensory units exhibited no change in gravity-related behavior. Did they have other channels for perceiving gravity, and if so, did they compensate for the missing sensory input? All the rat experimenters noticed a much lower body weight among the flight population. Was that medically significant, or just a sign they needed to create better feeding devices? Optimum technology and research protocols would be crucial for ISS productivity.

To study questions like these, a team from the Massachusetts Institute of Technology (MIT), the Integrated Project Development Laboratory in the Space and Life Sciences Directorate at JSC, and Lockheed Martin Engineering Services in Houston created a prototype for the "higher-performance" Human Research Facility graphics workstation, to be used on the ISS. Investigators Danny A. Riley and Margaret T. T.

Table 4.5. Chart of Spacelab funding history

Year (Fiscal)	Submission	Programmed
1989	80,400/88,600	87,600
1990	98,900/95,600	93,700
1991	130,700/129,300	129,300
1992	150,200/96,000	99,200
1993	122,600/114,459	112,800
1994	139,900/125,500	125,500
1995	92,300/98,600	90,000
1996	97,000/86,700	86,700
1997	62,400/50,300	40,100
1998	14,200/11,900	9,100

Source: Judy A. Rumerman, comp., NASA Historical Data Book, Vol. VII: NASA Launch Systems, Space Transportation/Human Spaceflight, and Space Science, 1989–1998, SP-4012 (Washington, DC: NASA, 2009), Table 3-4, 332.

Wong-Riley of the Medical College of Wisconsin in Milwaukee, with a team of technical assistants from Ames, developed a means for Payload Specialist Jay Buckey and Mission Specialist Dave Williams to carry out painstakingly delicate microsurgery on rats in zero g, administering anesthesia and performing retrograde labeling of individual muscle fibers. Here the surgeon-astronaut placed a small amount of a tracer chemical on a lone fiber. Nerves then transported it to the spot on the spinal cord, controlling that particular fiber for the researcher to correctly observe correspondence between nerve and muscular response.[67] Such work was to ensure that ISS scientists would have productive equipment and specimen-handling techniques.

The Neurolab program began under Richard Truly but finished under Daniel Goldin. A vigorous proponent of life sciences research at NASA, Goldin predicted before the Federation of American Societies for Experimental Biology (FASEB) four months prior to the STS-90 launch that, as a collaboration among federal agencies and foreign partners, Neurolab would "serve as a model for how we plan to conduct research on the International Space Station."[68] Two decades later, nothing like a hoped-for Neurolab 2 had yet to come to pass.[69]

Table 4.6. Chart of space station funding

Year (Fiscal)	Submission	Authorization	Apppropriation	Programmed	Budget Authority (Full Cost)
1989	967,400/900,000	900,000	900,000	900,000	n/a
1990	2,050,200/1,749,623	1,800,000	1,800,000	1,749,623	n/a
1991	2,451,000/1,900,000[a]	2,907,000	1,900,000	1,900,000	n/a
1992	2,028,900/2,028,900	2,028,900	2,029,000	_[b]	n/a
1993	2,250,000/2,122,467	2,100,000	2,100,000	2,162,000	n/a
1994[c]	-/1,937,000	1,900,000	2,100,000	1,939,200	2,106,000
1995	1,889,600/1,889,600	1,889,600	2,100,000	1,889,600	2,112,900
1996	1,833,600/1,863,600	2,121,000	2,144,000[d]	2,143,600	2,143,600
1997	1,802,000/2,148,600	1,840,200	1,800,000	2,148,600	2,148,600
1998	2,121,300/2,501,300	2,121,300	2,351,300	2,331,300	2,121,300

Source: Judy A. Rumerman, comp., *NASA Historical Data Book, Vol. VII: NASA Launch Systems, Space Transportation/Human Spaceflight, and Space Science, 1989–1998,* SP-4012 (Washington, DC: NASA, 2009), Table 3-5, 333.

[a] Congress reduced the FY 1991 funding requested for the space station by $551.0 million. A study to restructure the program was incomplete and did not allow for sufficient definition of requirements to develop detailed estimates.

[b] Program was being restructured and no programmed amount was shown.

[c] Space Station Freedom program was budgeted within the Office of Space Systems Development.

[d] Marcia S. Smith, Congressional Research Service, *Space Stations,* IB85209 and IB93017, 1999 edition, p. CRS-10. Neither the appropriations bill (*Making Appropriations for Fiscal Year 1996 To Make a Further Downpayment Toward a Balanced Budget,* and for Other Purposes, Public Law 104-134, 104th Congress, 1 sess., April 26, 1996), nor conference report H. Rept. 104-537, gives any figure at all for the space station for FY 1996.

A Centrifuge on Station

Early space station plans had called for a centrifuge large enough to hold a recumbent human.[70] The hope was that physiologists would find that a certain number of hours each day spent at one g or some fraction of it was enough to keep astronauts healthy. Maybe station personnel could spend that number of hours each day spinning, even while they slept, then arrive on Mars or back on Earth fit and ready to work. Thus, the first station designs showed a 13-foot (4-meter) diameter centrifuge.[71]

In the fall of 1980, Ames had a design in hand, a Lockheed Missiles and Space Company concept called a Vestibular Research Facility. Intended to ride in the shuttle cargo bay on an unspecified Spacelab mission, it was notable for its flexibility in configuration. The device could manage animal specimens ranging in size from a mouse to a monkey (and including containerized specimens such as fish), spinning them at up to 1.25 g. It came with a gimbal mechanism for three-axis specimen rotation and a chair for accommodating human beings. TRW Inc. had designed the container and SCI Technology, Inc. of Huntsville the computer controls.[72] McDonnell Douglas worked on a design in the early 1980s as well, one for which Ames had asked them to "provide the capability of removing specimen cages without stopping the centrifuge to avoid the effects of stops and restarts on the remaining specimens."[73]

Extramural scientists had been strongly supportive of a centrifuge on station. The Committee on Human Exploration of Space of the National Research Council, in reviewing NASA's 90-day study of George H. W. Bush's Human Exploration Initiative (aka the SEI), had suggested integrating research on the station centrifuge, bed rest studies, and animal experiments to explore effects of long-term weightlessness, the safety of the proposed Mars mission, and even the necessity of artificial gravity.[74] In 1987–88 the NASA Life Sciences Advisory Committee made several recommendations, including one that NASA "build a series of centrifuges that are critical for basic and applied biological and biomedical research and as part of the enabling technology for long duration manned space flight." The Life Sciences Division concurred in August 1988, stating that FYs 1989 and 1990 budgets requested a start on a 5.9-foot (1.8-meter) centrifuge project "as the first step in this series." The Division very definitely intended this to be only one of many such devices, adding that "options regarding larger centrifuges for Space

Ames Research Center's 1.8-meter (5.9-foot) centrifuge in motion, 1987. Photo credit: NASA-ARC.

Station deployment are being examined and planning budgets for these items are being developed."[75]

The following year, 1989, station designers at JSC were wrangling over where to put the centrifuge, unsure of what size instrument NASA was going to select, the 8.2- or 5.9-foot (2.5- or 1.8-meter) diameter. Ames planners showed a 5.9-foot instrument in their publications, suitable for cell, plant, and small animal experiments.[76] The best JSC could do was specify a place then referred to as "Node 2." It was the "only viable option," according to the Space Station Project Office (SSPO) in Houston, given the situating of support racks that astronauts using the centrifuge would need. These were in the US lab, and the other node attached to it, number 4, was for command and control equipment.[77]

Three years later, Agency literature indicated continuing, strong support for a life sciences centrifuge on Freedom, emphasizing the need to keep one in space over a long time period rather than just a two-week mission. Only thus could scientists breed many generations of one organism or test the same biospecimen in differing amounts of gravity

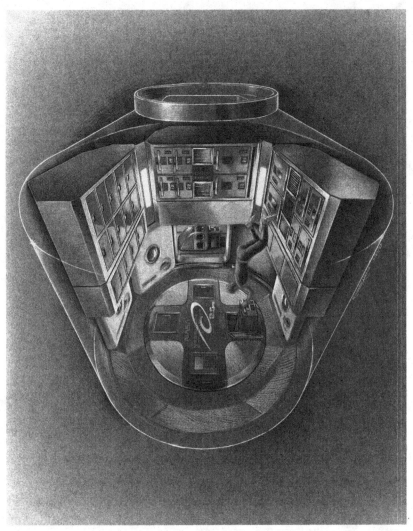

Artist's concept, centrifuge on station, 1989, #AC89-0771-1, by Roger Arno. In folder 56.14.43 "Pictures," shelf 2, cabinet 3, Lawrence Chambers Files, NASA Headquarters.

over a single life cycle and with reliable controls. Scientists needed such tests to develop the protocols for human testing and to replicate the time frame and changing gravity levels of expected missions, right down to the cellular level. Astronauts might be going from the three g's at shuttle liftoff to weightlessness for three days (a Moon shot) or nine months (one-way to Mars), then to fractional gravity (one-sixth or one-third on the surface) for weeks or months, then liftoff again, more weightlessness, a possibly high-g reentry, and one g back on Earth. The 1992 media guidebook on station spoke of an 8.2-foot centrifuge that could go from 0.1 to 2.0 g's.[78]

One school of thought was that a centrifuge rotating a single person was inadequate. Why not have an entire habitable module slowly spinning about an axis, perhaps on a tether attached to a fixed station, instead of installing a heavy device that was difficult enough to operate on the ground, let alone in LEO? Regarding tethers, in a 1991 interview retired Administrator Thomas Paine commented that "If you were starting with a clean piece of paper, that's certainly the direction you'd go." That vision did not last long, however, and centrifuge designs were later cut to 6 or 8 feet (1.82 to 2.44 meters)—big enough for a monkey, but not for an astronaut.[79]

By 1991 NASA had started looking at options like putting the device in its own separate module, to be attached at some unspecified date after the basic station was complete. Apparently, this was either not widely known within the Agency or not acknowledged in public, because in NASA's own monthly newsletter on the station's progress, Freedom's new Chief Scientist, Robert Phillips, gave an interview the following year referring to the centrifuge as though it were on track exactly as planned originally. Per Phillips, it was to be delivered on the next flight after Freedom achieved permanent occupancy. The Public Affairs Office within the Office of Space Systems Development at Headquarters also showed it as being deployed on the eighteenth shuttle flight to the station, the one immediately following the installation of the "Assured Crew Return Vehicle that will allow for the station to be permanently manned." Phillips called the centrifuge "essential for life sciences research," and "the cornerstone of gravitational biology research."[80]

By mid-1991, there was talk at NASA about international partners taking on the task. NASA and the Agenzia Spaziale Italiana (ASI) signed an MOU that December in which the Italians agreed to design and develop two "Mini Pressurized Logistics Modules" for Freedom. The two agencies would "work toward expanding the relationship to include provision of a Mini Laboratory as well." This would be a dedicated life-science facility, "at a minimum capable of accommodating an 8.2 foot (2.5 meter) tilting centrifuge and three international standard payload racks." A decision on building the life sciences Mini Lab was to have been reached "no later than February 1993," with an eye toward an October 1999 launch.[81]

Whatever the initial reason for jobbing out the centrifuge, by 1993 it was money. Even though NASA downgraded the scope of the program significantly, the centrifuge still was not affordable. The White House ordered another scrubbing of Freedom by a new Space Station Design Review Panel, which Vice President Al Gore appointed Charles Vest of MIT to head. It included several members from partner nations. At a June meeting before the group released the report to the president, Goldin brought up the idea of hiring the Russians to build the centrifuge on the basis of their experience with flight hardware.[82] In retrospect, that might have been productive.

The decade and a half between Spacelab-1 and Neurolab was thus a period in which the STS began to mature as a platform on which to conduct life sciences research. It was also a time when the Agency tightened the rules for its own conduct of research and contract scrutiny. It took first steps toward committing to and designing sophisticated tools for the ISS, intended for use by both US and partner agencies and scientists. Although plans for *hundreds* of such shuttle flights would be dashed with the *Challenger* tragedy, in retrospect perhaps that made everyone work more methodically toward getting maximum scientific return on investment.

5

Organizing in
the 1980s-1990s
Ethics, Institutes, and Biological Modeling

Research opportunities actually *in* space were—and are—the most competitive, sought-after, challenging, and rare. The United States performed or facilitated most of its space life sciences research on the ground and at non-NASA sites. Attempting to continually broaden its "customer base"—the scientific community—NASA sometimes created new organizations, attempting to spin up scholarly interest in, capacity for, and expertise in life sciences research.

Medical Ethics and Test Subject Rights

Support for life sciences research required an understanding of and adherence to a new set of ethics that emerged in the 1980s. Animal and human experiments in flight and ground-based testing going on or done earlier by NASA-funded researchers became an open controversy. Public challenges forced rapid, even proactive, response.

For NASA, it started with a summons in 1981 by the House Subcommittee on Investigations and Oversight, Committee on Science and Technology, chaired by then-Representative Al Gore Jr. (D-TN).[1] An article in *Mother Jones* had brought to the Subcommittee's attention human experiments performed at what was originally the Department of Energy's Oak Ridge Institute of Nuclear Studies (ORINS), a southern university consortium that became the ORAU (Oak Ridge Associated Universities) in 1966.[2] It operated in Gore's home state under the auspices of the Atomic Energy Commission (AEC), later the Nuclear Regulatory Commission. From 1964 to 1974 NASA had contributed $2.3 million to a study involving 194 patients (since 1957) on the effects of whole-body radiation on human physiology and its efficacy as a cancer treatment.[3] Such research ultimately would lead to the bone marrow transplant for treatment of childhood leukemia, but NASA participated to study acceptable radiation exposure levels for astronauts in space.

The *Mother Jones* story charged that Oak Ridge researchers had not communicated fully to patients and their guardians the consequences of the treatment—that is, obtained "informed consent"—and that they had withheld better remedies, sacrificing patients' lives for the sake of gathering data. The magazine singled out one experiment that NASA had partially funded, involving the death of a little boy. Two important ethical questions were on the minds of Subcommittee members. First, had patients been offered standard chemotherapy treatment in addition to or instead of radiation, or had their best interests been sacrificed to supply statistics to the prestigious space agency? Second, had an organization that possessed no expertise in childhood leukemia (NASA) dictated or even suggested patient treatment regimes to the medical personnel at Oak Ridge?[4]

The Subcommittee held its hearings in September 1981 and invited representatives of the patients and their families, the Oak Ridge experimenters, NASA, and expert witnesses in cancer treatment.[5] Acting Associate Administrator for Space Science Andrew Stofan explained that NASA had not instigated the connection with ORINS, nor asked anyone to follow specific experimental protocols or treat patients in any particular way. At no time did it have any interest in pediatric data. NASA had not paid the salaries of the physicians and staff attending the sick but only paid for some radiation sources and instruments used in physiologic telemonitoring, a technology which the Agency had pioneered.[6]

Literature searches (i.e., gathering statistics by reviewing patient records) were and still are standard procedure in medical research. NASA had asked Oak Ridge scientists to obtain data on individuals given radiation therapeutically and on nuclear workers exposed accidentally to radiation. The database included information on 3,000 therapeutic cases, contributed by 45 institutions, and 100 accidental exposures. After testimony by all the medical experts, who corroborated NASA's understanding of the work done and explained that radiation was an accepted treatment for cancer, that the young patients chosen had been at the terminal stage of their illness, and that they suffered from forms of cancer with survival chances that were "about nil," former patients and families were given a chance to speak. They testified that they held "high regard for Oak Ridge," had been informed about the treatments, and that they "knew NASA would be using data of interest for the Agency for the health of astronauts, and that they didn't mind at all."[7]

The congresspersons were satisfied, but as counsel to the subcommittee Ray Brill said in a prehearing meeting with NASA legislative affairs personnel, the Agency was "clearly 'tarred with the AEC brush.'" During the meeting, Brill repeatedly referred to the Oak Ridge treatments as "the NASA experiment."[8] This exemplified the connection many layperson made between the Agency and government-sponsored human and animal experimentation.[9]

It is somewhat surprising that the DOE did not ask NASA to fund more such projects. NASA could have underwritten work on bone calcium metabolism, radiation poisoning, fallout (after breakup of nuclear-powered satellites), and physiological damage from space radiation as part of its mission. Researchers seeking funds might have overlooked the Agency because they did not recognize the need NASA had for radiation data, or perhaps because they perceived it as an engineering organization only, rather than one interested in life science (see Chapter 6).[10]

Using animals as test subjects was a long-standing practice, justifiable for lessening the need for intrusive experiments on human beings. Specific creatures were selected because, as "model organisms," some physiological system they possessed closely replicated the same system within humans and so might be expected to produce data applicable to people. Also, if scientists had worked with a particular species for a long period of time, they had likely accumulated enough data to use as a solid baseline.

Animal rights organizations had existed in North America and Western Europe since the nineteenth century, but in the 1980s new groups with a variety of agendas came into being. The best known, People for the Ethical Treatment of Animals (PETA), formed in 1980. Around the same time, the Animal Liberation Front (ALF) began surreptitiously entering labs, freeing caged animals, taking records, and destroying equipment. NASA and USAF chimp facilities were targets.[11] On the other ideological side, groups like the States United for Biomedical Research, whose members were executive directors of biomedical research associations, the Foundation for Biomedical Research, Americans for Medical Progress, and the National Animal Interest Alliance organized themselves to fight the tactics of ALF and PETA.[12] Somewhere in the middle, the Scientists Center for Animal Welfare formed in the mid-1980s to self-regulate the community of lab animal users and foster debate on ethics issues.

Every institution that used lab animals came under scrutiny, even those that, like NASA, held a favorable place in the public eye. Headquarters' policy was to "generally . . . not respond" to activists with an "extreme" cause, such as United Action for Animals, Inc., which objected to astronaut "surrogates" (actually payload specialists) doing the work of Earthbound PIs. As the decade wore on, however, NASA's Public Affairs Office (PAO) and Office of the General Counsel had to respond to more and more pointed questions.[13] The White House and Congress also received queries and complaints from constituents and passed those along to the space agency. Some, such as a request to stop breaking cats' bones (to test their ability to heal in simulated weightlessness), were a distortion of fact or a response to outdated information. In January 1985, the Headquarters Legislative Affairs Office hosted a presentation by the Ames PAO to prepare for queries from congressmen receiving constituent inquiries over animal testing aboard two scheduled Spacelab missions, given "the vigilance and tenacity of animal rights groups and anti-vivisection organizations."[14]

Now and then, opposition arguments became strong enough to affect federal policy on a broader level. Manager of Space Biology Thora Halstead attended a September 1984 meeting called by the staff of Representative Tom Lantos (D-CA) regarding N.R. 5964, a proposal that federal agencies give their unused test animals to humane organizations for, it was hoped, adoption. Representatives of the DOD, NIH, US Department of Agriculture (USDA), and House Subcommittee on Government Activities and Transportation also attended, and individuals familiar with both lab animals and the researchers who used them made several points: (1) 95 percent of all government lab mammals were rodents and therefore unlikely to be adopted at local shelters; (2) due to the high cost of procuring and maintaining lab animals, the idea that there were "'excess' mammals . . . lying around" was not true; (3) most had to be necropsied as part of the experiment or destroyed for toxicity reasons; (4) most lab animals were not the property of the federal government but of contractors; and (5) if humane societies had to destroy millions of unwanted family pets each year, how could they handle more animals?[15]

One outcome was a 1985 amendment to the Animal Welfare Act of 1966, originally written to prohibit the use of stolen animals and to regulate the handling, treatment, and transportation of creatures used in laboratories, zoos, circuses, and exhibitions and by animal breeders. The USDA could issue cease-and-desist orders and fines, and confiscate and destroy suffering animals—although not rodents or cold-blooded animals. The amendment did not mandate anesthetics to reduce suffering but did add exercise requirements and limited animals to one major surgical operation (with a loophole for "scientific necessity"). It urged labs to consider the psychological well-being of nonhuman primates and look at minimizing or eliminating the use of animal subjects. The law now mandated the formation of Institutional Animal Care and Use Committees (IACUCs). NASA established one at each Center and charged them with evaluating each experiment proposal for compliance with federal regulations.[16]

The call for scrutiny was not without merit. As an example, to better understand and thereby prevent or control the debilitating nausea that affected many astronauts, NASA had given Victor J. Wilson, a highly regarded research neurophysiologist at Rockefeller University, funds for a series of experiments on feline nervous and vestibular systems.[17] One grant, NSG-2380, sponsored by the Ames Biomedical Research Division, covered the period 1979 to 1987. In these experiments, Wilson and coexperimenters would decerebrate anesthetized cats (sever their spinal cords just below the brain), fix them in a frame, then stimulate various nerves and observe the physiologic response. They destroyed some animals' inner ear mechanisms during the experiment to mimic the sensation of zero gravity. Other labyrinthine bones were destroyed earlier, while still others left "intact." Although Wilson wrote in his final report to the space agency that the NASA grant had funded "three projects," he listed 37 publications written using the results of this research and a sampling of just 5 of the articles tallies 49 decerebrated cats.[18]

Decades later, database searches show that entities worldwide have continued to decerebrate animals for medical scientific research. What actually ended Wilson's work was not the number of cats killed but the query: Did the animals suffer? Rockefeller University announced that the experiments had ended in 1997, and as of January 2007, Wilson's page on the university website said only that the experiments ended when the professor emeritus retired. There was no mention of the 18-month campaign by In Defense of Animals (IDA), which pointedly singled out NASA's backing. IDA pushed for cancellation on the basis that Wilson did not monitor feline brain waves for pain.[19]

In the 1990s, it became possible to obtain physiological data at the molecular level using the new tools of genomics. For example, to understand how bone demineralization affected the buildup of calcium in the kidneys, researchers could use an animal that had kidneys and bones that functioned like those of humans *and* for which the genome had been sequenced. The idea would be to get more and better data using fewer or even no test animals.[20] NASA found itself rethinking previous strategies for conducting many aspects of its life science research over the course of the decade. Life sciences leadership accepted the challenge of refashioning their approach and moved forward.

Nongovernmental Organizations

The University Space Research Association (USRA)

NASA had long worked one-on-one with universities, but in early decades had also used the research institute model. In 1968, with unknown quantities of Moon rocks expected at JSC, Administrator James Webb asked the National Academy of Sciences to help organize their access by academic researchers. The result was the Lunar Science Institute. In December 1969, a group of universities took over management and incorporated the following March as the Universities Space Research Association (USRA), located adjacent to JSC.[21] The association was selective; participating academics had to be recognized experts in a specific USRA discipline and distinguished faculty at their home institutions, which had to be PhD-granting universities.[22] Membership grew from 49 to 97 by 2005, including institutions in England, Canada, Israel, and Germany.[23]

USRA was to be "a not-for-profit corporation that provides universities and other organizations the means to cooperate in acquiring knowledge about space science and technology . . . [and] . . . to operate laboratories and other facilities under contracts, grants, and cooperative agreements, mainly with the federal government, for research, development, and education."[24] Standing panels of scientists called Science

Councils provided guidance to other agencies under contract, including peer review, but also served as a scientific board of directors for USRA itself and individual programs.[25]

Life sciences played a modest but significant role at USRA through the auspices of its Division for Space Life Sciences (DSLS). The Division provided administrative support for a postdoctoral program in space medicine that JSC and the UTMB in nearby Galveston offered jointly. One component was the Aerospace Medicine Grand Rounds, alternating between UTMB and USRA's Center for Advanced Space Studies (CASS) sites, and covering topics as diverse as mapping Mars for future human exploration, reproduction and space flight, wintering-over in the Antarctic, and the science-fiction universe of *Star Trek*. It provided science support to NASA during the Extended Duration Orbiter study (1989–1995) and Neurolab (STS-90). Its International Programs Office provided translation and interpretation services as well as meetings and workshops on interacting and negotiating with the Agency's Russian counterparts during the Shuttle-Mir program of the 1990s. USRA contracted with investigators at other institutions to carry out diverse studies, including artificial gravity, telemedicine, and changes brought on by bed rest. DSLS hired a staff of scientists (who worked primarily in JSC labs in support of JSC programs), arranged and facilitated seminars in the United States and abroad, and ran programs for undergraduate and graduate students of science and medicine whom NASA wished to coax into the nation's science pipeline.[26]

USRA founded the Exploration Science Institute in 1993, during the active years of Planetary Surface Operations planning at JSC, in anticipation of an imminent SEI-funded return to the Moon and a planned crewed Mars missions. Its purpose was to serve as a repository of NASA information and to disseminate data more widely. Both the DSLS and the Institute were located at the Center, established in 1991 adjacent to the University of Houston–Clear Lake. An USRA Division of Microgravity Sciences and Applications at Huntsville had expended some of its resources on protein crystallization since its founding in the early 1970s and managed a visiting scientist program in microgravity sciences at MSFC, provided science guidance to Marshall employees, and collaborated with University of Alabama researchers.[27]

The Aerospace Medicine Grand Rounds were under way by 2000 as a monthly seminar originating at USRA, teleconferenced to sites as diverse as the Gagarin Cosmonaut Training Center near Moscow and the

National University of Colombia in Bogotá. Topics that year included g-induced vestibular abnormalities; medical care at large public events (such as a shuttle launch); a Mars-mission medical operation simulation in a crater on Devon Island, Nunavut, Canada; and managing fatigue in flight operations. The UTMB residency in aerospace medicine had been augmented in 2002 with a brown bag lecture series, and in 2003 the DSLS began to offer the Bioastronautics and Fundamental Space Biology Postdoctoral Research Program. The Division would also coordinate the NASA Space Radiation Summer School, begun in 2004 at Brookhaven National Laboratory (BNL) as an invitational pilot program sponsored by USRA, NASA's Space Radiation Research Program, Brookhaven, the DOE, Loma Linda University Medical Center, and Lawrence Berkeley National Laboratory (LBNL). The next year, 15 students, including some from Italy and the United Kingdom, participated in the first competition and the program offered 23 research opportunities, each for two years, at JSC, KSC, and Ames. Senior scientists there were to serve as mentors. The DSLS also made available part-time jobs at JSC for local college and university students. In 2005, DSLS scientists saw 26 peer-reviewed journal articles published, along with two book chapters, four NASA technical memoranda, a technical brief, and an activity book for students. DSLS staff also presented 98 talks, seminars, posters, and abstracts at national and international scientific meetings. The administrative unit coordinated 102 workshops, seminars, and meetings, including the Bioastronautics Investigators' Workshop in Galveston and the 16th Annual NASA Space Radiation Investigators' Workshop in New York.[28]

USRA's material contribution to the space life sciences remained tangential, however, and began to decline, perhaps due to its leaders' interests—none of its presidents were life scientists—and its original charter: providing access to Moon rocks. Psychology and behavioral science, for example, were disciplines USRA did not underwrite during its 20-plus years fostering medical research. By 2000, its microgravity sciences group no longer listed protein crystallization among its activities.[29] The Exploration Science Institute disappeared as a separate entity. An Atlanta-based virtual institute called the NASA Institute for Advanced Concepts (NIAC) had gotten off to a promising start in February 1998 under the sponsorship of Goddard and with the goal of fostering "an atmosphere of creative examination of seemingly impossible aerospace missions and of audacious, but credible visions to extend the limits of technical achievement."[30] However visionary the thinking and in spite of receiving recognition for its work in 2003 and 2004, NIAC would close in August 2007, a stated casualty of the Vision for Space Exploration.[31]

The National Space Biomedicine Research Institute (NSBRI)

Planning for what would become the National Space Biomedical Research Institute (NSBRI)—ultimately replaced by the 2005 JSC Human Research Program and the 2016 Translational Research Institute—began around the fall of 1992, under Carolyn Huntoon, then Director of the Space and Life Sciences Directorate at JSC.[32] Huntoon wanted to lure life sciences researchers away from the much better funded National Institutes of Health. NASA "could not get a consistent group of high-level [researchers] to pay attention" to its biomedical research needs because the NIH was where investigators, "particularly the top

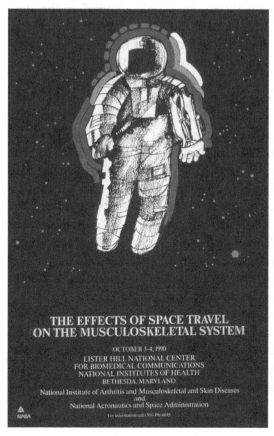

THE EFFECTS OF SPACE TRAVEL
ON THE MUSCULOSKELETAL SYSTEM

OCTOBER 3-4, 1990

LISTER HILL NATIONAL CENTER
FOR BIOMEDICAL COMMUNICATIONS
NATIONAL INSTITUTES OF HEALTH
BETHESDA, MARYLAND

National Institute of Arthritis and Musculoskeletal and Skin Diseases
and
National Aeronautics and Space Administration

For information call (301) 496-6045

NASA had a competitive yet collaborative relationship in recruiting researchers and sponsoring projects, as this conference poster implies. *The Effects of Space Travel on the Musculoskeletal System.* Bethesda, MD: National Institute of Arthritis and Musculoskeletal and Skin Diseases, 1990.

ones," focused their efforts in proposing new research and applying for grants, she explained to a planning group at JSC. To get highly regarded scientists to turn their minds to space, NASA would seek input from "selected academic institutions," then nonacademic bodies and industry, on how best to organize and manage the new institute and fund top notch biomedical research. JSC was "rather isolated, in a lot of ways," Huntoon believed, and this was an opportunity "to reach out and get the best people in the country thinking about some of our problems."[33]

NASA was open to either a brick-and-mortar or a virtual institute, and the financial and oversight relationship between an institute and the Agency had yet to be determined. Dan Goldin gave the concept his blessing, however, as a model of external institution involvement that would increase visibility and the perception of NASA as worthy of federal funds for higher level life sciences research.[34] By the spring of 1993, there was talk Agency-wide of NASA Technical Research Institutes (NTRI), and Headquarters had sent out a draft proposal on the idea. The committee tasked with pursuing the concept chose life sciences as the first NTRI simply because it had the most trouble attracting researchers. The JSC connection remained firm, as evidenced by the newly proposed name, JSC Institute for Biomedicine and Biotechnology, and a suggestion that 10 percent of its activities employ that Center's labs, equipment, and personnel. This physical connection would further link research to the long-term biomedical needs of humans in space.[35]

Houston-area proponents of a JSC institute, including Science Institute Planning team member John A. Rummel and Headquarters Life Sciences Division Senior Program Scientist Ron White, set out to steer potential institute funding toward issues of direct import to crew health and safety and to keep it under local supervision. If JSC itself didn't control the money and a scientist somewhere found anything that would hinder plans related to human space flight, Headquarters might shut down the whole program, and therefore the best strategy, they reasoned, was to keep Headquarters out of the loop.[36]

In early 1993, a draft proposal stated that there was to be at least one participating university, physically near the Center, that would be responsible for the institute. It would be considered an extension of that campus and each member university would "develop reciprocity agreements for approving course materials contributing toward award of graduate degrees." In addition, Headquarters wanted the promise of significant involvement of minorities and women in science and engineering. The JSC group included those points and added that institutes must be willing "to incorporate any available national source for accomplishing goals of the NTRI," to demonstrate "innovative methods for developing dual-path technologies," and to "address identified critical U.S. technologies." Funding would come primarily from the consortium itself for the five years of the initial contract, with JSC contributing 10 percent of the budget.[37]

By June, JSC had gained permission to draft an announcement looking for proposals by "consortia of universities and other nonprofit entities, including Historically Black Colleges and Universities and minority educational institutions" to establish the Institute for Biomedicine and Biotechnology. The proposal was also to be evaluated in part for the institution's "strong representation of ethnic minorities and women in science and engineering."[38]

Over the course of 1993, the draft also added economic components, such as references to national competitiveness, teaming with "commercial entities," and encouraging spinoffs and the transfer of consortium-developed technology to industry. All NTRIs were to be "customer-oriented with rapid response to the private sector." An institute's mission was to "bring the combined expertise and experience of its . . . members to bear on scientific, technical, and management issues central to improving the American economy and life on Earth in the areas of biomedicine and biotechnology."[39]

In late 1994, Republicans delivered a political blow to the Democrats in the White House by gaining control of Congress and claiming a mandate from the people to put the country on a more conservative track by means of less taxation and a smaller federal government. President Bill Clinton hoped to seize back the economic initiative, so in January 1995 he proposed a middle-class tax cut, a balanced federal budget, and less government spending. He asked all agencies to find items in their FY 1996 budget requests they could do without for the five years projected forward. NASA's share would be $5 billion, Clinton told Goldin, paid out in cuts of 3, 5, 7, and 9 percent of the baseline FY 1996 budget from 1997 to 2000. This was in addition to earlier budget tightening that had threatened to eliminate the Spacelab program in late 1995, and a 31 percent cut Clinton had more recently told the Agency to make. Adding to the pressure, an executive-congressional summit agreement to save the space station had placed ISS money out of bounds and no cuts could be made there.[40]

Goldin turned in the new budget request in February 1995 without specifying precisely where NASA would make those cuts. Per guidelines of Vice President Gore's National Performance Review, he had instituted the "zero base review" concept the previous September to scrutinize every budgeted dollar. Goldin formed a "ZBR team" from two groups: the Center Assessment Review Team and a cross-cutting team comprised of the Associate Administrators of Space Science, Mission to Planet Earth, and Life and Microgravity Science and Applications. Chief Scientist France Cordova would lead the ZBR team, with a goal of identifying items to cut. Every NASA entity would have to justify its existence, emphasizing technology and science over other functions in order to save entire programs from being terminated.[41] Goldin told his team to attack the problem with gusto. "Achieve OMB directed . . . reductions and more where it makes sense," he wrote in November. "Down size NASA installations but increase overall NASA effectiveness—do the right things the right way."[42] By May 1995, the ZBR team had found $4–5 billion in savings, but at the cost of 2,606 civil service jobs and dependent upon cutting another 4,000. At that rate, NASA employment would slip to 17,500 by calendar year 2000—the lowest level since 1961—and the Agency would have to eliminate 25,000 contractor jobs. The explanation for the firings was that NASA had been staffed in anticipation of important new programs and of a $20 billion budget by 2000. It was obvious now that these would not materialize and that management should trim the Agency to match grimmer expectations.[43]

The ZBR team also recommended that the science done at each Center be more tightly focused, to very explicitly contribute to the Center's role. KSC would phase out its CELSS work, for example, and send the water filtration component to JSC. Ames should have no life support role and no animal research but be "institutized" with an emphasis on astrobiology, so the centrifuge could go to JSC and the space station program. GAS experiments could move from Goddard to KSC, since they flew on the shuttle.[44] Additionally, programs would have to fund their own infrastructure, including the cost of civil-servant research scientists and their overhead. In theory, this would better enable more accurate comparison of costs against nongovernmental research organizations.[45] A closer alliance with academia might allow science to be improved at no cost, assuming NASA could piggyback onto university projects that other government agencies were funding. Looming over all this was the threat by Congress to cut the budget even more deeply, by

$3.4 billion over five years (as proposed by the Senate) or by $5.8 billion *after* a $700 million cut to the FY 1996 baseline (the House plan). Some in NASA foresaw Centers closing, whole programs ending, possibly the entire Agency going out of business.[46]

This drastic downsizing actually dovetailed with the institute concept for NASA. The ZBR team had pursued the idea of joining forces with academia, in line with Goldin's concept of NASA being primarily in the R&D business rather than in operations.[47] A Science Institute Planning Committee under Deputy Associate Administrator for Space Science Alphonso Diaz conducted a review of how other scientific institutes were organized and operated. It put suggestions culled from these models on the desks of Cordova and Goldin. The Science Institute Plan called for *eleven* such bodies, and that universities, nonprofits, and consortia would operate under competitive contracts or cooperative agreements.[48]

NASA negotiated for three institutes under the auspices of USRA: the National Center for Microgravity Research on Fluids and Combustion at Lewis Research Center, the Institute for Global Change Research and Education at Marshall for climate studies, and the Research Institute for Advanced Computer Science at Ames.[49] In life sciences, Ames would establish an astrobiology institute, and JSC would become a Center of Excellence for Human Operations in Space.[50] In July 1995, Goldin appointed Diaz to head a 90-day effort to develop plans for the two "pathfinder" institutes, one at JSC in biomedicine and the second at Ames, focusing on the origins and evolution of life in relation to space science.[51] Less than a month later, Goldin appointed Carolyn Huntoon, by then JSC director, to lead plans there.[52]

The Diaz committee paid visits to the Space Telescope Science Institute (STScI) in Baltimore—where Goddard Space Flight Center, in a contract with the consortium AURA (Association of Universities for Research in Astronomy, Inc.), managed the Hubble Space Telescope (HST) project—and to the Goddard Institute for Space Science, a division of the Maryland NASA Center working under agreement with Columbia University. Two days later, the committee was off to Langley, and the following week to the Search for Extraterrestrial Intelligence (SETI) project at Ames and the LBNL, which partnered with the University of California. Members then headed south to Scripps Institute for Oceanography, part of the University of California–San Diego. North of that was the Infrared Processing and Analysis Center on the campus of the California Institute of Technology (Caltech), linked to NASA's Jet Propulsion Laboratory (JPL). Over the next 11 days, the committee members toured Rice University's Center for Research on Parallel Computation and the Lunar and Planetary Institute (LPI), both in Houston; the National Center for Atmospheric Research in Boulder, Colorado, overseen by the nonprofit University Corporation for Atmospheric Research; the Smithsonian Astrophysical Observatory, allied with Harvard; and finally to Lewis (later the Glenn Research Center), which operated two institutes cooperatively with Case Western Reserve University, one for space power and on-board propulsion, the other for microgravity research.[53]

The visitors asked how each was funded, about the academia–government relationship, what sort of peer review process they used, who wrote proposals, how each managed conflicts of interest, how they conducted internal audits, what points of tension existed between the institute and NASA or the managing organization, how adequate funding was, how they handled academic tenure, what international activities they engaged in, and more. The institutes responded with detailed information and included personal concerns, worries, and hopes.[54]

The Committee also held meetings with the Centers at Headquarters to discuss organization, including funding and lines of authority. From the records, it is clear that JSC leaders, at least, thought that the ultimate decisions must rest with the Centers. During one meeting, Diaz raised the possibility of a single NASA entity unaffiliated with any Center taking charge of "procuring" the new institutes. JSC's John A. Rummel came back at the next meeting with an answer: "No." The people in Houston did not see that approach "as a successful model" and did "not believe" that "JSC requirements are well understood and represented" in such a process. Diaz refused to shut the door on the matter, though. One JSC request was for the "binding functions" of the institute, meaning the actual project or mission scientists and engineers, to be NASA employees. A committee member suggested stipulating that NASA reexamine such a structure in five years, and Diaz took that a step further, stating that the initial call for proposals must say that a transfer of control *would* take place at the five-year mark, not *might*. Rummel refused to bend on that matter and replied that JSC management would have to consider the issue further.[55]

Diaz also told Rummel that he "was to come back to the next meeting" (11 days later) "with $50M for the life sciences institute." JSC did not have that kind of money, Rummel responded, but Diaz "said to 'work it.'" To Rummel, that meant shifting his scientists and engineers to the institute, reprioritizing high-profile and expected high-return life science missions such as Neurolab and Shuttle-Mir, and reducing their funding correspondingly. Another idea would be to make the institute itself bigger, "to include ARC activities."[56]

Pitting Center against Center, however, was not likely what Head-quarters had envisioned. Eventually, the Science Institute Planning Committee gave its findings to the administrator and a 90-day discussion period ensued on October 17 among the "key stakeholder groups," including all the advisory committees, scientists, Congress, the White House, and industry. A formal comment period commenced December 1 and ran through January 5, 1996, inviting review and remarks by the science community, industry, NASA employees, and the general public.[57] The final report noted that the goals included increasing NASA's contribution to science and getting more outsiders to participate, but not cutting the Agency's work force. NASA would also be transitioning to a new accounting system, with full cost accounting that more explicitly revealed the actual cost to the Agency of its work. Diaz's team incorporated explicit instructions for the institutes to retain intellectual property rights in order to promote dissemination of NASA science and technology to the national economy, yet be able to control profits from data or inventions it created.[58]

By the end of January 1996, planners at JSC were awaiting Head-quarters approval of their "acquisition process." The Center was to form an institute, via a cooperative agreement, for four five-year intervals, with a performance review by senior NASA management and outside consultants in the third year of each interval. The work of the institute would be in cardiovascular and pulmonary physiology, endocrinology, immunology, general metabolism, neurosensory physiology, behavior and performance, and human factors, with "access to in-flight crew time" to carry out its research.[59] In spite of earlier assurances that the institute concept was not intended as a money-saver, a few new twists were added to the proposal. First was the statement that JSC intended to "minimize/eliminate new brick and mortar," and create an "Institute without walls"—that is, spend little or no money on new buildings. The second was a move to "core funding," which would be set at an initial level (adjusted for inflation) for five years, then *decrease* to 60 percent of the five-year inflated level by year 10.[60] The following month, NASA

Headquarters issued the final plan for the acquisition, structure, and operation of science institutes in general. It called for the eventual development of all 11 institutes proposed by the ZBR, but declared that just three were ready to commence: the Biomedical Research, Astrobiology, and Microgravity Institutes.[61]

The legal format of the affiliation would be a cooperative agreement, unless the peculiarities of a particular situation required a contract, in which case it would be termed a "Sponsoring Agreement." The bidding should be done competitively "whenever it appears appropriate to do so," and instead of asking certain entities to bid, the plan encouraged broader participation, with Centers holding workshops to explain the plan's potential and giving attendees a tour of the facilities.[62] USRA, Texas Medical Center, MIT, University of Texas Medical Branch at Galveston, University of Houston, and Science Applications International Corporation were already expressing interest in bidding.[63]

Daniel Goldin put his own stamp on it, too. In the mid-1990s he had introduced the Enterprise concept, an industry term intended to bring an entrepreneurial spirit to NASA. An Enterprise might be defined as a functional area charged with making a specific component of NASA's overall mission happen. Initially there were four, each under an associate administrator. Each Enterprise was responsible for strategic planning, policy development, budgeting, resources allocation, program direction, advocacy, oversight of their specific area, and directing approved international activities. In regard to the proposed science institutes, it would be these Headquarters-based Enterprise Offices—not Centers, not programs—that chartered, funded, delineated research requirements, decided the role an institute might play in another Center's work, and conducted periodic performance reviews. If in close physical proximity, an Enterprise Office might delegate administration of the contract to a Center, and certainly Centers were to give institutes access to facilities and to employee expertise, albeit on a "full cost recovery basis." Since the NSBRI and other institutes would be representing the outside science community to NASA management in return, the governing Enterprise might authorize it to manage an extramural scientific program on the Agency's behalf, possibly on a subcontract basis.[64]

Some institute funds would come from the Enterprise to support the overhead costs of senior staff, but only from the existing appropriation, in line with President Clinton's goal of a balanced budget. NASA also would fund the cost for any services or hardware bought for its own use.

Institute PIs could bid on NASA Research Announcements (NRAs) like any other scientist. However, institutes were "expected to become partially self-sufficient." They would be "encouraged to seek funding from a variety of sources and to market their capability in ways that are appropriate to their primary NASA mission." The private sector could be a source of funding and expertise. Outside income might also come from selling or licensing an institute's intellectual property. Scientists' home organizations would continue to pay their costs. The Institute and its directors would have authority to decide whether any non-NASA work fell within the scope of their contract or conflicted with NASA's mission.[65]

The planning team took care to depict these institutes as essential to the "budgetarily untouchable" space station. In the preface, Diaz wrote of the effort required to keep the personnel pipeline full of new scientists and engineers from academia, future leaders to "organize the efforts of the community at large and bind together the engineers and scientists who produce the results that the world has come to expect from NASA." He linked the institutes with ISS success by adding, "NASA's ability to replenish and revitalize these [science and engineering] capabilities has been adversely affected by declining budgets and workforce constraints, which threaten to stifle the in-house technical capability that NASA has depended upon in the past to assure scientific success. The threat is especially severe in the newer disciplines of microgravity and space life science, which must be nurtured and developed for most effective use of the International Space Station."[66]

To attract bidders and potential partners, JSC planners used the figure of $17.2 million for year 1 (FY 1997) as a straw man. As that began to appear less achievable, they instead looked at ways to move money around. There were different *kinds* of money, for example: core money (for senior staff support), program money, and project money. The latter two might be encumbered—meaning already pledged to some other entity—or unencumbered. In late February, a JSC team came up with $47 million in FY 1997 by combining Ground and Flight Research ($28 million) and shuttle and ISS Utilization ($19 million). This imaginary budget also grew by more than 10 percent the first year, nearly 50 percent the second, and 20 percent the third, with a dip during the final contract year that still would give the institute $94 million for FY 2001. Regrouping and planning separately, the Headquarters and JSC groups struggled to reconcile Washington's supposition that $30–90 million per year could be found in Texas with Houston's attempt to locate even

$8 million and assuming the difference would be "new HQ money." In early April, Headquarters and JSC discussed offering $27.2 million for eight months of FY 1997, increasing to $93.9 million by FY 2001. A few days later this plan, with "stable core funding for initial 4-year period (decreased to 60% over next 5 years)," was in play.[67]

On May 1, the JSC Contracting Office posted a Rummel-authored draft Cooperative Agreement Notice (CAN) on the Internet and gave copies to attendees at a summary briefing at JSC, making no mention of money. There was also a tour of JSC mid-month as an enticement, and informational material was mailed to 6,000 in the biomedical research community. In-house planners were using the figure of $10 million per year, still with four annual cuts of 10 percent beginning in year 5 until the budget was 60 percent of the original. JSC would allow, even encourage, the institute to bid on NRAs (and so earn more than the budgeted support funds), but limit it to 25 percent of that total pool. However, JSC already had a project to bid on, a $400,000 contract for the Life Sciences Data Archive (LSDA).[68]

NASA issued the final CAN June 10, telling potential bidders to assume a $10 million annual budget for years 1–4, cut to 90 percent of that in year 5, 80 percent in years 6–7, and so on until years 10–20, when their core funds would be 60 percent of the first year's. To sweeten the pot, the Agency stated that JSC scientists were currently engaged in $2 million worth of peer-reviewed research annually. As soon as the institute was established, this and related funding would become part of the institute, as would the $400,000 LSDA contract. Regarding Research Announcement bidding, a chart at the end of the CAN using the generous figures from the March Headquarters plan said that the NSBRI could have 25 percent of nearly $160 million over the next five years. In the undefined long term (that offerors probably expected to be three to five years), NASA "envisioned" giving it even more functions. "Considered for inclusion" would be management of ISS biomedical research, and it was the "intent" to give the NSBRI "the overall NASA biomedical research and analysis (R&A) program," seemingly implying each Center's biomedical work. These were irresistible carrots. Even though no precise dollar figure was available, the challenge and prestige would be priceless.[69]

It took until March 14, 1997, for NASA to announce the winner: Baylor College of Medicine in Houston. Its strategy had been to form a consortium of seven schools: Baylor, Harvard Medical School, Johns

National Space Biomedical Research Institute
Distribution of Funded Research and Education Projects
and External Advisory Council Representation

Map showing the geographic influence of the NSBRI circa 2004. National Space Biomedical Research Institute image.

Hopkins University's Applied Physics Laboratory (APL), MIT, Morehouse School of Medicine, Rice University, and Texas A&M University.[70] Baylor had chosen these schools based on their reputation, interest, and expertise in specific fields needing study. It brought senior management from each school together and told them they were organizing their proposal around a distributed management approach, with each school heading a research topic. Baylor asked each to invite their own expert in the discipline for which the school would be responsible to be team leader. Then he or she would nominate other outside scholars, to form teams of six to eight. Teams would be allowed to draft a budget for $1 million annually, and the remaining money would go toward education, outreach, administration, and other tasks.[71]

The Cooperative Agreement Management Plan (CAMP) signed on May 29, 1997, listed eight research areas: human performance factors, sleep, and chronobiology; neurovestibular adaptation; cardiovascular alterations; muscle alterations and atrophy; radiation effects; DNA damage and repair, bone demineralization/calcium metabolism; hematology, immunology, and infection; and advanced technology. The NSBRI incorporated as a not-for-profit with Baylor as its only member, presumably therefore with full authority to acquire and dismiss other consortium members as it saw fit.[72]

The document repeatedly emphasized the extent to which this was a pilot project, with expectations of flexibility, rapid response, and full accountability. "An essential aspect" of the agreement was that the "significant assets" of both NASA and the consortium "be synergistically blended to become significantly more productive than either working alone." The NSBRI had authority to find "top level guidelines and policies to enable this interaction without creating unnecessary constraints." There would be a learning curve for both parties, and there was "considerable interest both within and outside of NASA for establishing metrics to determine interim progress." NASA would "qualitatively assess" the NSBRI during its first three years on the basis of the research plans' development, implementation, and results (35 percent of total score); ability to maintain quality key personnel (30 percent); core management approach to cost and schedules (15 percent); effectiveness in applying results to commercialization, education, and public outreach (10 percent); and diversity in both participation and outreach (10 percent). The NSBRI would submit an annual status report. At the end of the third year NASA would conduct its own formal review per procedures the Chief Scientist would set nine months prior, tentatively September 30, 2000.[73]

The team leaders of the eight research areas were required to communicate regularly and have monthly formal meetings, with minutes provided to the NSBRI director and to the Contracting Officer's Technical Representative (COTR), Charles F. Sawin of NASA. The space agency would host a yearly symposium for biomedical research presentations and jointly with the NSBRI a workshop for participating PIs to "review, analyze, coordinate, and integrate research results and progress within and across all disciplines and . . . [to] update future research plans, as required."[74]

The Agency was sensitive to the NSBRI seeming to be "just another support contractor" if it were authorized to manage the JSC peer-reviewed intramural research process in the fullest sense of the term. Thus the CAMP stated that the institute would not make decisions regarding the science content or the implementation of research projects. JSC PIs would maintain full control. It went on to describe in some detail a process by which the NSBRI would essentially look over the shoulders of JSC PIs, implying some amount of imposition of NSBRI's methodologies, visiting scientists, and postdoctoral candidates on JSC. Operational crew medicine was still sacred and off-limits. However, NASA would provide NSBRI management with a medical debrief after each mission. In addition, NASA would adhere to crew confidentiality and data privacy rights in a to-be-determined manner, and so would not share a planned crew health database without restriction.[75] NASA added to the contract wording that allowed fully funded "additional specific work elements" to be added to the core money, not to exceed $4 million per year. The NSBRI could also submit unsolicited project plans. There was even a statement that there were "no limitations to the level of core funding that NASA may apply over and above the original core funding specified in the CAN."[76]

Perhaps emboldened by initial success attracting collaborators, NSBRI Chairman and CEO Bobby Alford began to push for a huge expansion plan for the NSBRI less than a year after signing the CAMP.[77] Alford asked Goldin for his "perspective of the research priorities addressing the biomedical activities." In a July 9, 1998, reply Goldin stated his opinion that the priority would be radiation protection, human factors (including man-machine interface and psychology), "smart" medical care systems, and fitness.[78] Four months later, Alford had taken Goldin's letter and created "Vision 2005: A Comprehensive National Program to Pave the Way to Go Beyond Earth Orbit." This he sent to his board, Ron White, NSBRI director Dr. Laurence Young, and "appropriate NASA personnel," including procurement officers, in a pitch to take his budget from a $10.2 million NASA stake to $63 million in FY 2000, of which $48 million would be Agency money. He stated that the NSBRI had "already formed 8 research teams, incorporating 41 projects and 133 investigators from 25 institutions, including NASA, to begin to carry out our research mission to develop countermeasures against the effects of long duration weightlessness and space radiation. But the current funding level . . . forces us to select only a few programs for study from among the many critical areas that require further research and severely restricts the scope of the NSBRI's programs. In addition, large numbers of interested and potentially valuable collaborators cannot be included in the Institute's existing research efforts."[79]

Referring to the NSBRI's "alignment" with goals in NASA's Strategic Plan and claiming "success" for the institute's first year and a half of work, Alford wrote that given the "long lead time required to go from biomedical research to proven space countermeasures," this was an excellent time to "enhance and augment the NSBRI." (This request was a response to Goldin's letter, he added.) Additional money could get "three times the number" of intramural investigators involved in life sciences research and triple the number of extramural researchers, double the number of consortiums, and add four new teams. Achieving this, however, required NASA to increase its funding *eightfold*. Alford proposed a $117 million annual budget by FY 2003, including private, federal, and foreign money, up from $14.853 million in FY 1999.[80]

The budget summary Alford and the board offered for FYs 1999–2003 showed NASA contributing $77 million of the requested total for FY 2003 (vs. $10 million in 1999). The total foreign, non-NASA, and private funding for FY 1999 was only $4.698 million, but Alford was predicting that it would be $40.150 million four years later. The only concrete evidence the report offered for non-NASA aid was a statement that an "agreement of affiliation" with the German Space Agency would "enabl[e] German-funded projects to become part of the . . . program," and that "partnerships" had been "developed" with the NSF and the NIH's National Institute on Deafness and Other Communications Disorders. That research would be jointly funded, and NSBRI was paying for its share out of unspecified private funds.[81]

Alford may not have known that by then some of NASA's own senior management were accusing shuttle upgrades and the ISS of soaking up funding intended for other departments, especially the Human Exploration and Development of Space (HEDS) Enterprise, which included some life sciences.[82] The NSBRI's very existence was then being questioned at Headquarters. Internal rearranging of funds and congressional imposition of unfunded earmarks had left Code U, the NASA entity responsible for the NSBRI, with $273.1 million for that year, or $22 million short. OLMSA was debating how much to cut university research budgets, how many new PIs *not* to fund, and to what extent NSBRI funds might be reduced. NASA's Capital Investment Council, the advisory group on facilities, staffing, and other big expenses, was talking about making deep cuts. The Life and Microgravity Sciences and Applications Advisory Committee recommended waiting until the year 3 performance review to consider increases for the NSBRI. It also hinted at second thoughts, specifically a worry that "consolidation of management responsibility for fundamental research to a single consortium may unfairly diminish opportunities for funding meritorious research by non-NSBRI connected institutions."[83]

Those concerns proved to have some merit. The NSBRI self-review of December 31, 1998, cited a "stable, high caliber management unit" and said the institute "had successfully implemented its research plan and efforts to date are very good." External relationships were "exceptionally successful," in spite of admitted weakness in attracting the best proposals. Just 58 came through the door, with the result that only those rated as "weak" were rejected outright. There hadn't been enough proposals in immunology, several projects were "too fundamental or basic in nature," and there were only four female PIs among the 133 investigators from over 25 institutions working on 34 projects.[84]

NASA's formal response to the plea for an increase in FYs 1999–2003 funding came four months later in a memo from the JSC contracting officer for NSBRI. She reminded Institute management that the CAMP allowed for "unsolicited project plans" but that NSBRI would have to "submit a . . . proposal . . . per the enclosed technical requirements description." This bought Headquarters another four months to think about things.[85]

Ultimately, the exchange set the tone for the fiscal relationship between the NSBRI and NASA management for the next half-decade. The institute and its advocates would report successes and ask for hugely increased budgets or additional facilities, arguing that from the beginning it had been undercapitalized. NASA administration would respond with promises of increased authority, visibility, and funds. Funding proved elusive for all parties and the responsibility for that lay everywhere: in the wrangling between Headquarters and JSC over authority for ISS research selection; in a mixed year 3 performance review, specifically concluding that the review committee had "no basis to endorse a large increase in funding"; in post-9/11 budget cuts; and in OMB concerns that having an NSBRI stifled other research. After increasing to $25.5 million in NASA funding by FY 2001, the institute found itself back at square one in early 2002 with a FY 2003 budget of $10 million per year projected out to FY 2007.[86] At that point "specific OMB direction" was that "a significant portion of planned funding for the NSBRI was put in a 'placeholder wedge' for highest priority OBPR activities." The Office of Biological and Physical Research was to work with JSC and the NSBRI on the funding issues.[87] The NSBRI and Headquarters achieved something of a truce, finally, in June 2003, with a Revised Strategic Plan for the institute that achieved "budget stability" of $30 million per year for FYs 2003–2007 by adding the new Bioastronautics Initiative, including its Critical Path Roadmap.[88]

Bobby Alford expressed the institute's collective frustration, anger, and embarrassment in front of the life sciences community in a February 2002 letter to JSC Acting Director Roy Estess. Ninety-four grants had already been selected for FY 2001 funding, most for three years, but in January 2001 (nearly a year after NASA had submitted its budget) the institute was hit with a cut of 9 percent. Now, the FY 2002 budget was being submitted with 13 percent less money for that year and 23 percent less for FY 2003. "This causes shock, discouragement, and lack of confidence within the biomedical space life sciences community so far as NASA's commitment, reliability, credibility, and intentions are concerned," Alford wrote. "The aim had been that the NASA/NSBRI FY03 budget would be as much as $37M which was a goal consistent with the pledge of the NASA Administrator to bring . . . support eventually to an annual base of $48M." In fulfillment of its research mandate, the NSBRI had released a Research Announcement on October 31, 2001, but given the cuts would be "unable to select any of these [50] proposals for funding." A "failure to fund worthy proposals . . . [would] be demoralizing to the space life sciences community and will undoubtedly lead to anger from some. It surely was misleading for NASA to agree to this simultaneous release, in the same mail out package, inviting NASA/NSBRI proposals when there was no intention to provide for additional funding."[89] Alford did not name names, but the promises had been made (and apparently broken) by, or at least under, Dan Goldin, and likely dated back to major ISS research cuts made in response to the decision to redefine a "complete" ISS as "U.S. Core Complete," limited in size. Goldin had removed George Abbey as JSC Director due to massive station cost overruns and questions of systemic budgetary mismanagement just seven months earlier.[90] In October Goldin would remove himself from the space agency, one factor being those very same issues.[91]

The Mathematics of Risk Assessment

An engineering agency first and foremost, in 1972 NASA's JSC had funded a General Electric attempt to develop an integrated mathematical model of the human body in preparation for data expected from Skylab. As part of that, GE sent Ron White, then a math professor at the University of Southwestern Louisiana, on a summer faculty fellowship at JSC to the University of Mississippi Medical Center to study under the renowned physiologist Arthur Guyton, sometimes referred to as the "father of systems biology." NASA hoped to use this approach to integrate data coming from multiple physiological tests well enough to understand the process of adaptation to the space environment and also get a clearer picture of what was going on in the biological processes not studied directly. By the mid-1970s, White recalled, that work was mature enough to enable the development of hypotheses that would later be used in Spacelab.[92]

In 1997, JSC and the Life Sciences Directorate commenced what was then called the Critical Path Roadmap, related activities having started in the early 1990s.[93] The Roadmap began as an effort to identify all the risks to astronaut health in different scenarios, during and after space flight, then to prioritize them—in other words, deciding which risks needed to be addressed in a specific sequence, which might take longer to resolve than others, and which had to be handled first because they involved situations more near term. It also addressed some fundamental questions, such as biochemical and cellular changes in bone chemistry in relation to fracture healing, bone resorption and renal stone formation, and the effect of radiation damage to the central nervous system.[94] In an attempt to be as all-encompassing as possible, the team (about a dozen experts at JSC and Headquarters) started with the most difficult human space flight mission NASA had planned for the next few decades: a trip to Mars. If they studied a 30-month Mars voyage, they reasoned, they would cover every other space flight scenario as well.[95]

Step one was a literature search. Experts in each discipline looked at relevant reports made by entities such as the NAS, the NRC, National Academy of Engineering, National Council on Radiation Protection and Measurements, and even the Food and Drug Administration. Mission reports such as *Biomedical Results of Skylab, Crew Health in the Apollo-Soyuz Test Project, Biomedical Results of Apollo,* and *Scientific*

Results of the German Spacelab Mission D-2 (DLR) contained useful flight data and analysis, as did technical memoranda from Shuttle-Mir missions and internal committee reports. Also helpful were conference proceedings, symposium papers, and textbooks written by NASA scientists. Reviewers also found publications in peer-reviewed journals from the United States, Hungary, Switzerland, Germany, the Netherlands, the United Kingdom, Canada, and Russia.[96] Using this data, the group identified 150 distinct risks to human health.[97]

NASA was approaching this task of Roadmap creation with almost 40 years of mathematically defining, predicting, and reducing risk related to space flight. In fact, counting NASA's previous incarnation as NACA, it could claim experience dating back to 1915. Even with machines, predicting the probability of something happening was itself risky, but nevertheless had to be done and the results, no matter how inconvenient, could not be ignored. *Challenger* proved that *qualitative* rather than *quantitative* assessments of risk and nonrigorous application of risk theory (e.g., assuming that if an accident has not happened yet it is unlikely to happen in the future) could have disastrous results.[98]

In life sciences and medical safety there were an almost infinite number of variables that had to be taken into account in doing probabilistic risk assessment—defined as a combination of likelihood and level of damage or loss—because of the unpredictable nature of any living organism's behavior.[99] Posing questions like the possibility of animal diseases being transmitted to astronauts, radiation causing cancerous mutations of human cells, genetic susceptibility to renal stones, or the likelihood of an astronaut adhering to a countermeasures program made life sciences risk assessment extremely difficult. Exacerbating this at NASA was the tendency of its engineers to ask the medical personnel what the "most important" risk was. Risk, defined as "the conditional probability of an adverse event from exposure to the space flight environment," differed by scenario, mission, equipment, and even by gender and age of the astronaut.[100] To some life scientists it was a catch-22; they felt they couldn't ask a question until they knew what the answer was supposed to be.[101]

Risk assessment can be further confounded when a specific negative event has never happened. NASA hired an actuary to calculate the risk of an astronaut breaking a leg. Since no one had ever broken a leg in flight or on the Moon, there was zero data. However, it seemed not out of the question for a Mars mission, when astronauts' bones would have

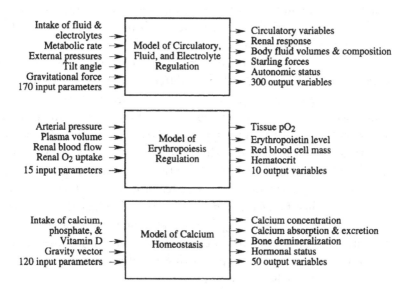

Fig. 1 Three models developed by NASA for simulating long-term physiological responses. The numbers of input parameters and output variables indicate the relative complexity of the models. pO_2, partial pressure of oxygen.

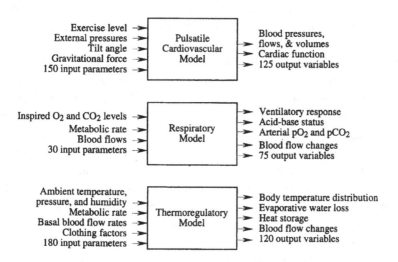

Fig. 2 Three models developed by NASA for simulating short-term physiological responses. The numbers of input parameters and output variables indicate the relative complexity of the models. pO_2 and pCO_2, partial pressures of oxygen and carbon dioxide, respectively.

Charts of models for simulating physiological responses. From R. Srini Srinivasan, Joel I. Leonard, and Ronald J. White, "Mathematical Modeling of Physiological States," in *Space Biology and Medicine*, Vol. III, *Humans in Spaceflight, Book 2*, edited by Nicogossian et al. (Reston, VA: AAIA, 1996), 562.

been weakened due to months in a weightless condition. However, the state of actuarial science in 1997 was such that the expert could not come up a risk assessment for that scenario.[102]

More data existed on renal stones, and in both the US and Soviet/Russian space program, although differences in experimental techniques made comparison difficult.[103] The United States, Europe, and Japan had done a great many studies on radiation and human health and the scientists involved were typically physicists, so the methodology and results had been more mathematical. Knowledge grew, particularly after the Chernobyl accident and as both space powers conducted radiation experiments in orbit. Space research brought about new challenges as scientists had to deal with different types of radioactive particles of different sizes, speeds, energy levels, and frequencies of occurrence. By the late 1990s, it seemed as though instead of answering questions, radiation health research was perplexing scientists and forcing them to rethink their models and suppositions.[104]

On the ground, mathematical modeling was used as a noninvasive investigational tool, using shuttle DSO data as it became available along with the work of other federal agencies. Applications included a better understanding of calcium homeostasis and the mechanics of protein crystallization in microgravity; duplicating the changes caused by bed rest; and studying thermoregulation (for better space suit and environmental control systems design), dispersal of pharmaceuticals through the body, formation of nitrogen bubbles in the blood during EVAs, circadian rhythm, and the heart's pumping action in microgravity. Such modeling was useful not only because plausible hypotheses could be tested before embarking on an expensive, lengthy clinical test, but it also forced a PI to ask hard questions when things didn't turn out as expected. Just creating the model compelled the investigator to think systematically. Along the way, he or she learned something about each element of the system, the processes within these elements, and their connectedness. New relationships and gaps in knowledge could be discovered as the modeling process raised critical questions long before it made any predictions.[105]

This ground research was meant to be a preparation for a robust space station program that included co-orbiting unmanned production facilities in the very near future and with lunar and Mars bases in the not-too-distant future. In hindsight, one can discern strong links with the Agency's NACA roots as a research organization that pushed the envelope for the nation's aircraft manufacturers. NASA was leading the way with mathematical modeling and believed it was fanning the flame of applied research for its life sciences needs by corralling talent into biomedical think tanks. The enormity of building and managing a space station that could do everything for literally everyone, however, would blur even the Agency's own image of itself as the squared-away, go-to place for data, analysis, and nearly fail-safe engineering.

6

Radiation and the Science of Risk Reduction

Fortunately for NASA's life sciences budget, radiation was a danger no one knew much about but everyone wanted to understand. The military and numerous civilian agencies, domestic and foreign, were spending money on research and willing to share the results. This helped the space agency make strides in understanding and mitigating the effects of radiation and in planning for protection.

Early Space Radiation Research

A post-Apollo report called the radiation effects of the 1961–1963 Mercury flights, all but one of which had lasted less than 10 hours, "medically insignificant." It stated that Project Gemini proved humans "could tolerate exposure to the space environment quite well," including radiation, on the basis of simple onboard film packs indicating only the *presence* of radiation and its spectrum.[1] The final two Mercury flights had been equipped with dosimeters and sensors on the capsules and pilots, and a hand-held Geiger counter was added after US atmospheric nuclear tests in July 1962 had created a temporary radiation zone below the Van Allen belts. Some Gemini flights flew with onboard radiation-measuring devices, but over the course of 16 Mercury and Gemini missions, scientists sent just three biosamples for radiation study. Two were human blood and one bread mold, exposed to a calibrated radiation source rather than to ambient space radiation.[2]

Apollo missions used both personal and capsule dosimeters, and flew microbes on one mission, mice on another, and the European Biostack containing plant and animal tissue on four of the 11 flights. The only human tissue studies were before-and-after eye exams to look for damage, including radiation burns.[3] NASA expected capsule shielding and new UV-resistant Lexan EVA suit visors to hold exposure within limits prescribed for laboratory and reactor workers on Earth. Apollo astronaut exposures were indeed no higher than 25 percent of the 5-rem annual limit and created no "operational problems," meaning no in-flight health incidents.[4]

However, the Biostack experiment on Apollos 16 and 17 showed HZE particles, a type of space radiation not studied biologically until the 1950s, to be very disruptive to developing plant and animal cells.[5] Slides of tissue samples showed the particle had tunneled completely through, like a microscopic bullet. Gestational abnormalities in some species reached 48 percent; mortality rates were also extremely high among eggs and larvae hit by particles. On extended crewed flights researchers decided that "the prime candidate" for HZE damage would be "the central nervous system, which consists of highly differentiated nonreplaceable cells." However, the odds of enough such particles hitting the same part of the brain were so small they concluded that "the HZE-particle radiation environment poses no major threat to manned space activities that may be undertaken in the foreseeable future."[6]

Experiments on the same missions showed that light flashes seen by every astronaut but one since Apollo 11, in numbers as high as four dozen, were an effect of charged cosmic particles crossing the retina.[7] Necropsies of five pocket mice flown in a canister on Apollo 17 revealed that the tiny animals averaged 10 cosmic particles penetrating their heads. Investigators could not determine conclusively if they had caused actual damage.[8]

During the 1973–1974 Skylab missions NASA relied for protection on predicting solar proton events (SPE), planning orbits in relation to known radiation fields, shielding vehicles with radioresistant materials, and signaling an alarm on a device that monitored the x-ray output of the sun. A few unplanned events—four French atomic tests and the escape of some radioactive material inside Skylab—were fortuitously situated such that they posed no real risk.[9] Investigators performed three distinct bioradiation experiments: SKYRAD (dosimetry) done on all three missions, Cytogenetic Studies of Blood (chromosomal abnormalities) on Skylabs 2 and 3, and Visual Light Flash Phenomena (HZE particles and the retina) on Skylab 4 during two passes through the South

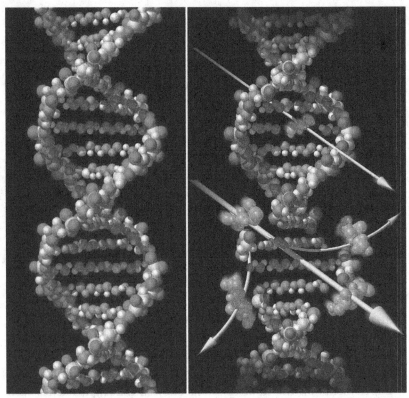

Artist's rendering showing path of damage of HZE particle. Photo credit: NASA-JSC.

Atlantic Anomaly (SAA), a naturally occurring high-radiation zone. The chromosomal structure study was carried out on four astronauts for short periods of time before and after the missions, launch minus 53 days (L-53) to return plus 18 days (R+18) and L-20 to R+20 days.[10] It was not until the Longitudinal Study of Astronaut Health (LSAH) in 1992 that NASA began collecting and archiving data needed to compare the health and morbidity of astronauts to that of Earthbound individuals over their lifetimes.[11]

During the 1970s and 1980s, the Soviets invited Ames researchers to conduct some dosimetry tests and examine effects of radiation on the eyes of rats flown on their Cosmos 782, 936, 1129, 1887, and 2044 missions.[12] European and Japanese scientists in the 1980s and early 1990s flew shuttle experiments dealing with radiation, primarily dosimetry. STS-41G in 1984 carried a Hungarian thermoluminescent dosimeter developed by the Central Research Institute for Physics in Budapest as a test for a new cosmic radiation detection system.[13] Spacelab 1 (STS-9) in 1983 included several ESA space plasma studies, including a German experiment to measure heavy cosmic-ray isotopes. During the D-1 mission (STS-61A) in 1985 the Deutsches Zentrum für Luft- und Raumfahrt (DLR) had a similar study titled "Dosimetric Mapping Inside Biorack."[14]

Results from D-1 of a *Carausius morosus* (stick insect) study showed that larvae hatched from any egg penetrated by heavy ions had "shorter life spans and an unusually high rate of deformities." The German investigators believed they had been able to differentiate between the effects of microgravity and space radiation thanks to an onboard 1-g centrifuge. They concluded that a combination of the two actually had a synergistic effect, increasing the number of mutations. They repeated this experiment on IML-1 as part of the DLR's newest Biostack, exposing bacteria and fungus spores, thale cress seeds, and shrimp eggs to space radiation. The National Space Development Agency of Japan (NASDA) flew something similar: the Radiation Monitoring Container Device, with a sandwich of cosmic ray detectors, maize seeds, bacteria spores, and shrimp eggs to produce a three-dimensional map of the trajectories of the radiation particles.[15]

Outside Agencies and Radiation Research

Many experiments that provided data for NASA radiation-biology researchers were theoretical and materials studies by other entities: the US Air Force, Department of Defense, or Department of Energy; foreign space agencies; or universities. They were usually looking at the effects of radiation on semiconductors in military and civilian satellites and their prevention. NASA was able to efficiently piggyback onto their research.

The DOD flew three radiation experiments aboard the shuttle: Radiation Monitoring Equipment (RME-I, -II, and -III), the Cosmic Radiation Effects and Activation Monitor (CREAM), and the Shuttle Activation Monitor (SAM), directly sponsoring the first two, and the Air Force Space Systems Division sponsored SAM. Research on "hardening" avionics, focusing on the placement, type, and strength of radiation shielding, generated data useful for space vehicle designers incorporating radiation protection. On the ground, a combined USAFSAM/NASA experiment from 1963 to 1969, which the Air Force continued for over two decades, tracked the long-term health of heavily irradiated rhesus monkeys.[16] Also ground-based, in 1981 the Naval Research Laboratory created CREME (Cosmic Ray Effects on Micro-Electronics), a computer model used by Langley Research Center and later JSC.[17]

Early in the Apollo design phase NASA had tasked Langley engineers with radiation shielding and that component of vehicle design remained at LaRC.[18] Much of the data gathered was used to develop computer programs to predict the type and intensity of radiation at various points and times in Earth orbit (and beyond). In the late 1980s and early 1990s Langley engineers created the BRYNTRN (BaRYoN TRaNsport) and HZETRN (HZE TRaNsport) codes—models—to calculate radiation transport, or movement *into* an object irradiated. NASA already had a suitable object. In 1969 Martin Marietta had created a Computerized Anatomical Man (CAM) manikin for the USAF Weapons Lab at Kirtland AFB and what was then the Manned Spacecraft Center (MSC) in Houston. McDonnell Douglas developed for JSC a new, radiosensitized model in 1973 with 1,100 surfaces and 2,450 regions, and included bone marrow, the gastrointestinal tract, skin, joints, and testes. Beginning in the 1970s and continuing for decades NASA and others flew such manikins to gather radiation data. BRYNTRN and HZETRN codes were used to analyze data.[19]

The manikins were sometimes just a head or head and torso. After a significant number of the USAF-NASA monkeys had by the mid-1980s developed brain cancer, researchers thought it would be useful to obtain radiation-level data on the human head in the actual environment of space to compare and see if the animals could be used as a model for predicting astronaut cancer susceptibility.[20] Thus came about DSO 469 (In-Flight Radiation Dose Distribution), the Phantom Head experiment. The Phantom was a human skull embedded in material of radiosensitivity similar to that of human flesh and constructed in sections so that dosimeters could be sandwiched inside at various depths. It rode in a compartment inside the shuttle, close to the aluminum skin, on nine flights from 1989–1992: STS-28, -36, -31, -37, -40, -48, -44, and -42. Three missions, STS-28, -36, and -31, included the missions of the highest and lowest altitudes to date, so consequently the Phantom Head traveled through different types and peak levels of radiation. Data from those three missions were sufficient to disprove the monkey-head model as a valid predictor of humans developing brain cancer. By the end of 1993, when a Phantom Head had flown 11 times, researchers concluded that "(1) neutrons do not pose a significant additional radiation hazard in space; (2) the currently used model for describing galactic cosmic radiation is deficient . . . ; (3) active, time resolved dosimetry can be performed on the Shuttle; and (4) no absorbed dose hot spots occurred."[21]

The civilian DOE funded in part the BEVALAC, combining the Bevatron (Billions of Electron Volts Synchrotron), a diverting tube, and SuperHILAC (Heavy Ion Linear ACcelerator) at Lawrence Berkeley National Laboratory, the only heavy-ion particle generator in the United States and used by NASA.[22] The Department of Commerce's National Oceanographic and Atmospheric Administration (NOAA) was a key contributor with satellites that monitored solar activity and upper atmosphere properties. These provided data to create computer codes of worst-case scenarios, such as solar flares.[23]

NASA's radiation research also benefitted by input from advisory bodies like the National Academy of Sciences and the White House Office of Science and Technology Policy (OSTP). The latter had organized an 18-agency Committee on Interagency Radiation Research and Policy Coordination (CIRRPC) in 1984 and NASA participated in addressing issues such as standards and levels of "acceptable risk." The CIRRPC also looked for gaps in radiation research, and its Subpanel on High-LET (Linear Energy Transfer) Research oversaw an international task group that identified and prioritized radiation research. The committee's recommendations included more attention to the "biological effectiveness" of neutrons and more work in dosimetry, modeling, and genetics.[24] In late 1987 and early 1988, the CIRRPC's Task Group on Biological Effectiveness of Neutrons Research and Policy Coordination lobbied Congress to lock into the FY 1989 budget a national high-LET radiation research program benefitting NASA.[25]

The Academy's National Research Council (NRC) wrote a 1987 report strongly recommending that the space agency continue and even strengthen a four-year-old joint DOD, DOE, and NASA advanced propulsion program then called SP-100. The goal was to develop technology for nuclear power systems for lunar bases, orbital stations, and interplanetary vehicles. Though radiation levels would increase inside spacecraft, moving from chemical propulsion to nuclear would cut vehicle weight by two-thirds.[26] A lighter vessel would make the trip faster, reducing exposure time. NASA stepped up that effort in FY 1986, naming it the Advanced Technology Program, and in FY 1988 folded the work into a planned $900 million, seven-year Civil Space Technology

Initiative (CSTI) program. The Office of Aeronautics and Space Technology managed it, and NASA Lewis did most of the engineering. The SP-100 became a $65 million component of CSTI named High Capacity Power. Research included study of "environmental interactions," particularly radiation. The work to be done included developing new computer models and by the mid-1990s, a demonstration reactor. The idea was to bridge to Project Pathfinder, a study of propulsion technologies for Mars, Jupiter, and similar exploration missions.[27]

NASA chose Lewis because it had done bioradiation studies for the USAF and DOE in the 1950s and 1960s at its nearby Plum Brook facility, one of the most powerful test reactors in the world. Plum Brook had aided the Air Force study of a nuclear bomber and a Los Alamos National Laboratory idea for nuclear rockets, then NERVA (Nuclear Engine for Rocket Vehicle Applications) and SNAP (Systems Nuclear Auxiliary Power Program) in the 1960s and 1970s. Plum Brook had also been part of a National Cancer Institute study on epithermal neutrons to treat glioblastoma, a brain cancer. Employees had carefully mothballed the facility in 1973, hoping NASA would reactivate it in the 1980s for Mars excursion studies.[28]

Lewis had also funded a grant to Texas A&M University to assess the safety of Freedom occupants exposed to onboard radiation power sources, atomic-powered co-orbiting platforms, and two different reactors under consideration for Moon and Mars missions.[29] The space agency created the first NSCORT for radiation biology at Colorado State University (CSU), partnering with the LBNL in early 1992. NASA would renew the CSU contract in 1997.[30]

Biological and pharmacological countermeasures to radiation poisoning were objects of study by universities, private industry, and US military medical research units. From treating Hiroshima, Nagasaki, and then Chernobyl victims in 1986 doctors had learned it was possible to stave off, reduce, or possibly prevent the ravages of radiation poisoning. The Armed Forces Radiobiology Research Institute (AFRRI) in Bethesda, Maryland examined various pharmaceuticals, including selenium and a polysaccharide, Glucan. Standard protocol had been to administer aminothiols (amino acid-sulfhydril organic compounds), which could aid in protecting blood-producing cells from radiation death, but also had toxic side effects. The trick was finding a dose and mix that would not make the patient sicker. Glucan held the promise of allowing a reduced dose of aminothiols. Military researchers tried all of these as preventives or countermeasures to be given immediately before or after exposure, and in all instances got promising, although mixed, results.[31]

Increasing success with cancer patients made autologous marrow or blood transplants—postradiation transfusion of one's own stored blood or marrow—look like an option to allow an astronaut time to regrow healthy blood cells. A Harvard-affiliated team stored both control and test samples of peripheral blood mononuclear cells on the ground and aboard STS-61-C in January 1986 at 39.2°F (4°C), to study the plastic storage medium. Like the ground controls, about 90 percent of the flown cells survived seven days of storage in usable shape.[32] In another study BioTime, a California firm that made blood substitutes, particularly for use in cryosurgery, evaluated rodents, dogs, and baboons to explore the question of whether "cold maintenance" could be an option for "compromised space-borne astronauts."[33]

NASA had not always been involved in experiment selection and design with such outside studies. Some of the data was old; it was useful to correlate and project solar activity cycles but not helpful when dosimetry devices became obsolete or in the case of a radiation field that moved, such as the SAA.[34] Some information—on cancer induction among humans from HZE particles or the influence of neutron particles, for example—did not exist at the outset.[35] The NOAA-USAF satellite expected to take x-ray images of the sun and look for solar storms would not happen until 2003.[36] As a result, an eclectic mix underpinned computer models and the resulting projections of risk to humans and of shielding effectiveness. JSC scientists admitted that the newer CAM and BRYNTRN demonstrated that their previous best guesses had underestimated radiation levels many times over. As knowledge had grown, biophysicists had discovered other potentially hazardous particles and particle behavior, too.[37] The knowledge was not well integrated anywhere, and at NASA was at such a theoretical, nonapplied level that a 1996 joint US-Russian space medicine textbook could call radiation "the area of greatest uncertainty for developing risk assessments," adding that "data on cancer induction in humans by heavy ions are virtually nonexistent, and research is needed to determine the validity of extrapolating such data from animal subjects to humans." Entire physiological structures, such as the brainstem, remained to be studied.[38]

Actions and Reactions in the 1990s

In April 1989 the Life Sciences Division, then part of the OSSA, published a "Strategic Implementation Plan," describing scientific and technical goals and the means by which it intended to achieve them. The division saw its role as *facilitating* the conduct of critical life sciences experiments" (emphasis added), but it laid out an ambitious plan for more actual research on its own part, listing five necessary tasks. Three had some application to radiation and indicated a more active role for NASA's biomedical personnel: "Accumulating state-of-the-art instrumentation . . . flying an augmented series of Spacelab missions . . . [and] using a series of autonomous bioplatforms to study radiation and variable-gravity effects." The Plan advocated for LifeSat, a planned biosatellite series, but hoped that a reestablished biosatellite program would result in "precursors to a human-rated variable gravity facility." The division was already planning "Phase A studies . . . to be initiated in the mid 1990s for bioplatforms capable of generating artificial gravity as follow-ons to LifeSat for long-duration unmanned experimentation." The Plan's creators expected to carry out the studies "in conjunction with the Project Pathfinder in the Office of Aeronautics and Space Technology (OAST) and with the Office of Exploration." The 1989 document contained optimistic charts showing anticipated radiation data plotted on a timeline, including all of the STS missions from 1989 to mid-1997, followed by Freedom the same fiscal year, and punctuated with two Cosmos data points in 1989 and 1992. It showed LifeSat data beginning to flow in 1994 with missions at least once a year until 2000. Meanwhile, Project Pathfinder was also to work on shielding technology, 1989 to 1994. The Advanced Bioplatforms (free-flyers, LifeSat, etc.) would begin contributing results from one or two flights a year near the end of the decade.[39]

Instead, President George H. W. Bush announced his Space Exploration Initiative (SEI) just three months after publication of the Strategic Implementation Plan.[40] SEI looked well beyond station to lunar bases and human exploration of Mars. That the average NASA employee at a far-off Center had been a "true believer" in SEI can be seen in the number of drawings, calculations, meetings, and experiments that the Initiative generated and which were carefully archived at the JSC Collection at the University of Houston–Clear Lake. Much time and effort went into lunar and Mars surface habitats in particular. In Washington,

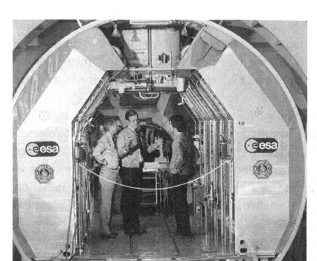

Vice President George H. W. Bush tours the new Spacelab with NASA astronaut Owen Garriott and ESA astronaut Ulf Merbold. Photo credit: NASA.

DC, however, the net result of SEI had been study groups. The first was the Agency's own 90-day study, which reportedly met with "widespread disappointment." Then, former astronaut Thomas Stafford led at the request of the new National Space Council (NSC), chaired by Vice President Dan Quayle, the Synthesis Group, which published its findings in May 1991. It organized the work of manned exploration into four "architectures" but gave only brief mention to the station Bush and Reagan wanted, or even the shuttle.[41]

Reportedly to light a fire under NASA—and as an alternative to ordering an outside top-down probe—the White House, via the NSC, instigated the Advisory Committee on the Future of the US Space Program (AC-FUSSP) under Martin Marietta CEO Norman Augustine in mid-1990.[42] Bush requested Augustine and his group to examine what needed to be done to make SEI happen.[43] The Augustine Committee, which included former Administrator and ardent space exploration enthusiast Thomas Paine, was wholeheartedly committed to NASA's future, even citing life sciences research and crewed operations experience as "the" justification for the station's existence over generic microgravity research. "Hence, life sciences becomes the core mission," Augustine wrote to members in December 1990 as the AC-FUSSP was about to issue its report.[44]

In its executive summary, the report of the Augustine Committee listed their chief concerns, one related to radiation. "The material foundation of any major space project is its 'technological base.' It is this base that produces the key building blocks, or 'enablers,' that make major missions possible—new materials, electronics, engines and the like. The technology base of NASA has now been starved for well over a decade and must be rebuilt if a sound underpinning is to be regained for future space missions."[45]

NASA considered radiation research to be among its "high visibility" projects and science-oriented planners interpreted the just-released Augustine Report as stating that among "program goals . . . science is highest priority." Outranking the development of new launchers in the report, too, was an "expanded technology" program, and radiation sat in both categories, science and expanded technology. SEI planners at Headquarters in December 1990 identified the minimum budget it would need to "make progress" on an "integrated SEI effort" over the next five years and began to work with the OMB and Congress to make at least the FY 1992 budget workable and to reprogram some FY 1991 money.[46]

The Life Sciences Division had set about specifying a plan of action for the radiation component of its 1989 Strategic Implementation Plan. NASA published the document in 1991, two months after Congress cancelled LifeSat funding, so gone were dreams of artificial-g bioplatforms.[47] Maybe to avoid the same fate, the new plan spread radiation research over more congressional districts by formally directing that NASA coordinate with multiple organizations, including external advisory groups. The Agency regrouped to bring control of the science to units with operational astronaut-support responsibilities, yet still outsource the research to the DOE, the DOD, universities, and international partners. The Chief of the Life Support Branch (Code SBM) of the Life Sciences Division would appoint a manager for this new Space Radiation Health Program. In writing, the plan described the program as "closely linked" with Operational Medicine. An organizational chart showed JSC handling "medical and operational dosimetry and research" functions, with other activities at Langley, JPL, and outside agencies. Langley would continue its studies of particle transport and development of computer models. JPL was to work on high-LET radiobiology.[48]

Overall, NASA would continue carrying out crucial ground research, using others' money as much as possible. The Life Sciences Division announced that the new Radiation Biology Initiative would still center on the BEVALAC, even though it expected DOE funding for the device to end in 1993.[49] Beam research was vital because researchers could isolate specific levels of energy and flux, more easily perform ground-level work like experiment design, and better understand the fundamental effects of particle radiation on tissue. The "basic core program research" would be under contract, in particular at the CSU radiation NSCORT, partnered with a national lab. The program would formally include the DOD via the AFRRI, and the solar-flare monitoring capability of NOAA in Colorado. NASA had hopes for a Booster Synchrotron at Brookhaven National Lab to continue accelerator work. The authors stated that the new plan was "developed within the context of" the 1989 document, but also at "the recommendations of internal and external advisory committees."[50]

In early 1992, the Aerospace Medicine Advisory Committee (AMAC) met to create for the NASA Advisory Committee another plan for vigorous radiation research. Few radiation biology experiments had made early shuttle manifests and little European Biostack data had been published by the time the Committee submitted its Strategic Considerations report.[51] More than a decade of radiation data-gathering opportunity was largely unexploited by NASA because its primary focus was Space Motion Sickness (SMS), also known as Space Adaptation Syndrome, likely in expectation of regularly flying large numbers of noncareer payload specialists. The malady was unpredictable but prevalent, and the impact on productivity would be especially severe for researchers making the one flight of their career.[52]

AMAC delivered a two-volume document on its perceptions of research and technology needs in support of the SEI to the NASA Advisory Council. It listed, by priority, critical questions, current readiness in those areas, and possible integration with research in other NASA divisions. The three primary challenges it saw as being radiation, microgravity, and life support. Included in these were a need for countermeasures and operational medical care. The report stated that NASA medical personnel needed direct, ongoing access to space, and the ability to maintain and test human subjects there. Accelerators, which produced pure, focused, unidirectional streams of a single type of particle, were not lifelike simulations of space, where different types of particles struck at varying, even random, concentrations and rates, from any direction, and through differing amounts of shielding. The positive side to lab radiation, however, was that it was under continuous, diverse, and intense study using other federal dollars (or foreign funds), and data were generally in the open literature. Unlike microgravity's physiological effects, all of radiation's issues did not *have* to be studied *in* space, so the AMAC called for even more cooperative research at federal facilities, particularly the Fermi National Accelerator Laboratory and Brookhaven. The committee had also delineated the science "deliverables," complete with milestones and resources required, and differentiated between a "constrained" versus "robust" timetable.[53]

The NASA Advisory Committee "fully endorsed" their findings and turned them over to Goldin in 1992 with the recommendation that he appoint "a single focus of responsibility and accountability, within the NASA top management, for carrying out all agency life sciences/ life support activities." AMAC still recommended experimenting with an unspecified number of "Radiation Health Free Flyers," small and most likely not human-tended.[54] Examples did exist of free-flyer projects making it through the pipeline, especially in radiation physics. The joint NASA-DOD SPACERAD free-flyer had been successfully deployed amidst the Van Allen belts in 1990 and sent back data very useful to life scientists. During a period of 19 orbits the electron fluxes in the outer belt had shown a "much lower" average radiation dose rate than the geosynchronous orbit model had predicted. If that held true under other geomagnetic conditions, NASA's model for electrons would need to be "substantially revised"—downward. That would mean safer working and living conditions overall and significant weight savings, given less shielding.[55] The Space Sciences and Applications Advisory Committee (SSAAC) had urged physicists and engineers at Goddard and the LBNL to study a free-flyer version of NASA's Particle Astrophysics Magnet Facility (ASTROMAG), rather than rely on station deployment. LBNL found by 1991 that a less expensive ELV could place a smaller experimental package into an orbit transecting Earth's geomagnetic lines.[56] Thus, the AMAC put flyers into its recommendations as an Enabling Research and Technology Development Milestone. The hope was that free-flyers would operate over a 10-year period from about 1998 to 2008.[57]

Real-time telemetered data was vital, but the AMAC deemed it especially urgent to deploy *recoverable* satellites with biosamples for collecting HZE-based radiation data, including multigenerational data on different species that would allow for "determination of probabilities for GCR-related damage." Such specimens would be exposed to all types of space radiation and for months or years (whereas Earth accelerators specialized in one or two types, with "beam time" extremely expensive by the hour). The AMAC estimated that a full 30 percent of critical questions for SEI "would benefit from access to free flyers." It also sought knowledge about the long-term health of astronauts and their families. Those constituencies and US taxpayers would bear the financial burden of negative radiation outcomes such as cataracts, cancer, fertility problems, and birth defects.[58]

Thus, SEI got off to a reasonable start in terms of scientific support, but it had no political legs. Political scientist Lyn Ragsdale ascribed its failure in part to space overall becoming "ancillary policy" rather than "primary policy" by the Reagan years (1981–1989). Although space held primacy in the minds of NASA employees, aerospace contractors, and enthusiasts, for the president and for Congress, she said, it was simply not a top priority. Space policy no longer asked the question "What should we do?" but was instead part of the process of meeting continuing government commitments. As a result, it received incremental funding, always based on last year's budget, rather than the huge new cash influx the Moon Race had required. The only nonincremental money NASA got was to replace *Challenger*. Support overall for space was spotty within Congress (including staff) and among various executive branch personnel, although dissenters did not come right out and tell Bush, Quayle, or Truly what they thought immediately. Decentralization ruled in the "plural institutions" that make up two of the three federal branches, she explained, even more so within NASA itself. "Decision ambiguity results, yet [the decision-makers] *make decisions nevertheless.*"[59]

Thinking he was showing presidential leadership in submitting a large, grand, and expensive space agenda, Ragsdale added, George H. W. Bush had packaged SEI startup studies with the supersonic National Aerospace Plane (NASP), a space nuclear power study, and Mars Observer, and asked for $1.3 billion in his FY 1991 budget request, all as an SEI line item. Bush himself even pointed out that NASA's budget request contained the largest increase for any federal agency that year. NASA itself was beginning to think in terms of a $30 billion total FY budget in the foreseeable future, more than double what it had until then. Ragsdale called this an attempt "to stuff a primary policy into an ancillary box." Congress, on the whole, reacted as though the whole proposal were ludicrous, particularly in light of the unfinished (and expensive) space station and a shuttle that wasn't living up to its promised cost-saving, and therefore dismissed it out of hand.[60]

By 1993, with a new president directing national space policy, JSC had put in place a formal Spaceflight Radiation Health Program (SRHP) in alliance with contractor Krug Life Sciences, Inc. (acquired by Wyle Laboratories in 1998). The technical memorandum the program issued that year discussed hopes for station research, in particular better radiation monitors. The team also wanted to improve the 20-year-old AP8 and AE8 models of the trapped radiation environment, which did not account for dynamics within the radiation belts, and to incorporate the SAA drift into them. The team at JSC further expressed a desire to apply whatever was learned to development of countermeasures and a "biological dosimeter."[61]

The latter referred to the use of a living being to indicate the presence and magnitude of radiation. USAF and extramural tests on irradiated animals had done this already, using wound-healing ability as a reference.[62] Gemini studies of white blood cells irradiated in flight had proven the concept for microscopic biodosimeters.[63] In addition to studies important for the space station, in the report was a what-if to worry about in the more distant future: nuclear power plants on the Moon, and the accompanying radiation hazards.[64]

The SRHP also employed USRA and the contractors who comprised the Space Radiation Analysis Group (SRAG) in existence at JSC since 1962. The SRAG's responsibilities were to monitor space weather constantly, in order to project the expected radiation exposures during missions and also to interact real-time during missions with flight directors in the event of a solar particle event (SPE). The group also trained the crew in the operation of instruments related to radioactivity or detection, including isotopes brought on board for an experiment.[65]

Shuttle-Mir Radiation Studies

The Shuttle-Mir project, an antecedent to the ISS, began in early 1994 and ended four years later. Researchers within NASA, the Lawrence Livermore National Laboratory (LLNL), the Russian Space Agency (RSA), and ESA carried out radiation studies, consisting primarily of dosimetry inside and outside of the station. The objective was to better understand the radiation environment of Mir's 51.6-degree inclination, which was the same as that planned for the ISS. OLMSA created a Radiation Road Map that specified what NASA needed to know, how the Agency might obtain that information, and the incremental steps needed to advance the field.[66]

The first such experiment, Inflight Radiation Measurements, was a joint JSC-IMBP assessment during Mir 17, prior to the arrival of any NASA astronauts, and again during Mir 18 and 19 and the second shuttle docking mission, STS-74. JSC and the IMBP also collaborated on Cytogenetic Effects of Space Radiation in Human Lymphocytes during Mir 18, the first mission to include a NASA astronaut, Norman Thagard.[67] Other NASA experiments included the Real-Time Radiation Monitoring Device (RRMD) on ISS mission Increments 3 and 4, and Effective Dose Measurement during a spacewalk, the latter using a modified ESA device used on Euromir 95 during Increments 4 and 5.[68] NASA also brought along two DOD devices, CREAM and RME-III, on Increments 5, 6, and 7.[69] Ames sponsored an experiment by the University of San Francisco and the IMBP, Environmental Radiation Measurements on Mir Station, which used passive dosimeters, and the Tissue Equivalent Proportional Device (TEPD). The dosimeters were outside during Increment 4 and inside Mir during Increments 2–6.[70] The Russian and German space agencies collaborated on Active Dosimetry of Charged Particles during Increment 6. An examination of the variation of the radiation spectrum, it correlated Mir detector readings with NOAA Geostationary Operational Environmental Satellite (GOES) temporal data on radiation fields and events.[71] By the end of 1997, NASA had concluded that Mir exposure levels had been high enough (10.67–17.20 rem) that its cumulative effects could prove career-limiting to astronauts and affect their health. Dan Goldin asked the TransHab (inflatable crew module) development team at JSC to study and possibly modify ISS habitats to prevent this.[72]

Goals for such studies included verifying the results of one type of dosimeter against another, testing the predictive abilities of computer modeling of the space radiation environment, trying out new biodosimetry techniques, and gauging similarities between US and Russian dosimetry equipment, protocols, shielding materials, and placement. For example, the University of San Francisco-IMBP study used four Area Passive Dosimeters fastened to the internal walls of the Mir base block and two in the Kvant module. The External Dosimeter Array used one NASA and one RSA dosimeter, similar to a unit flown in 1991.[73] Other experiments used different types of dosimeters or placed them on the astronauts and cosmonauts themselves and consistency in placement proved to matter.

Radiation researchers at JSC used Mir 18 as an opportunity to compare passive dosimetry measurements against a promising new technique: chromosome painting. Fluorescent markers applied to chromosomes revealed breaks before and after radiation exposure, providing a bread-crumb trail of prior exposures and doses. Originally the technique worked only with very recent radiation exposures, on the order of hours or days, but by the early 1990s a refined protocol, Fluorescent In Situ Hybridization ("FISH-painting"), when combined with chromosome aberration simulation software, showed promise as a means to detect exposure much later than the actual event. Tests on the USAF-NASA monkey colony confirmed the validity of FISH-painting decades after exposure. This meant individuals could act as their own control, or biodosimeter.[74]

An important, but unexpected, lesson learned from Shuttle-Mir was just how complex an issue shielding was. NASA and DOE research had confirmed experiments dating to the 1930s that, contrary to popular imagination, lead was not the best insulator, because heavier atoms, when hit by a particle would emit a number of new and equally dangerous particles. The lighter the atomic weight the better, so hydrogen—H—would be the best insulator and ordinary water, H_2O—for drinking or as waste—would offer adequate protection, as would polyethylene, made of CH_2 units. ISS planners could use this knowledge in determining where to place the potable water supply and the urine containment fixture, and how to incorporate plastic into sleep stations to maximize "solar storm cellar" space.[75]

The Bioastronautics Roadmap and Radiation

In 1997 JSC's Space and Life Sciences Directorate initiated the Bioastronautics Roadmap, originally called the Critical Path Roadmap. Years of discussion had gone into the concept. JSC invited the NSBRI to participate in 1998 and in 2000 the hybrid Critical Path team established a baseline for the Roadmap. Planners began to reinvestigate risk in light of growing awareness of other health effects of radiation including cardiac, circulatory, and digestive disease. Over the next four years they would also examine risks in light of three different mission scenarios: ISS, 30-day Moon missions, and two-and-a-half year Mars trips, gathering input from internal and external advisors. As the Roadmap matured radiation began to move up in priority.[76]

In a July 1998 letter to NSBRI director Bobby Alford, Dan Goldin put radiation research at the top of his "high priority research effort" list and in a 1999 briefing he expressed "extreme satisfaction" with a new JSC radiation risk definition and mitigation plan crafted by the Center's radiation group. Alford had already made radiation one of eight NSBRI research areas, under the rubric of Radiation Effects, DNA Damage and Repair.[77] PIs at consortium universities evaluated chemoprevention, blood cell formation, DNA damage, and immune system function. The University of Houston worked on related technology, including newer in-flight monitors.[78] In response to continued enthusiasm, in mid-2000 the NSBRI changed its radiation team, dropping funding for one researcher's work on DNA structure and another's on cancer sensitivity, and picking up more operationally oriented PIs at different institutions, like a University of Pennsylvania scientist studying pharmaceutical countermeasures and another Brookhaven investigator.[79] At BNL's new Space Radiation Lab, NASA began to organize along cross-cutting lines to optimize ventures such as radiation research and technology development.[80] A year later, the NSBRI designated BNL neurobiologist Marcelo Vazquez sole team leader and liaison for radiation with projects such as a radiation detector under development at the APL in Maryland.[81]

Also in 1998, the Life Sciences Division of OLMSA distributed a new radiation strategic plan within the Agency and to a hundred extramural researchers at the Ninth Annual Space Radiation Health Investigators' Workshop in Loma Linda, California. The SRHP would follow the AMAC plan of 1992, but OLMSA labeled it "an organic evolution" of the Division's own 1991 plan. OLMSA, under AA Arnauld Nicogossian, intended it to define how radiation research would fit into the HEDS enterprise, especially in light of genetics advances over the past decade.[82]

JSC would be the lead Center, but the Life Sciences Division would continue to use a leveraging strategy, citing existing collaborations with the BNL, the AFRRI, the Deutsche Agentur für Raumfahrtangelegenheiten (DARA), the National Cancer Institute, Loma Linda University, the Japanese HIMAC (Heavy Ion Medical Accelerator in Chiba), a NASA-industry transportation initiative called the High Speed Research Division, and possibly the Gesellschaft für Schwerionenforschung (Society for Heavy Ion Research) in Darmstadt, Germany. There was still hope for small-scale orbiting research, too, in the form of Bioexplorer satellites with onboard centrifuges. The program would also seek to use radiation data from robotic precursor missions under the direction of JPL.[83]

The Division intended to conduct research in three phases, covering short, intermediate, and long-term progress. Phase 1 (1998–2002) would "exploit currently available science," including LaRC and JSC models, "to identify risks accurately," develop emergency measures, begin countermeasure research, and model the radiation environments of both LEO and "the near Mars environment." Phase 2 (2003–2009) prioritized refining models further with ISS data, especially given the expected station centrifuge as a control. NASA also would complete a database for HZE effects in a specific range, 50–600 million electron volts (MeV). Researchers would assess protective properties of Martian soil and atmosphere to design habitats and shielding that would last five years. They would engineer storm shelters for deep space missions and radiation protection for exploration spacecraft, and implement radiation forecasting, warning, and monitoring systems. Phase 3 (2010–2023) would "fully develop biomedical science and technology for radiation risk mitigation and to take advantage of . . . breakthroughs expected to lead to practical biomolecular radiation countermeasures." "Radiosensitivity diagnostics" would be ready for use in crew selection and after a decade of using a human centrifuge on orbit, scientists could expect to understand radiation's synergy with hypogravity. A forecasting, warning, and monitoring system for a permanent colony would exist.[84]

However, prognosticators pinned these hopes on the aforementioned "breakthrough strategy," expecting significant scientific paradigms to shift at a predictable rate. From a 1998 perspective, based on the "current rate of progress" the statistical *uncertainty* associated with radiation would thus decline to a factor of five (meaning five times greater than the actual risk) by 2023, down from a factor of ten. Or the breakthrough rate in genetic and other biological research, largely outside NASA control, might even double, thus enabling a factor-of-three radiation risk uncertainty. Or it might triple, allowing a factor-of-two uncertainty. The authors decided that it "was realistic to assume that at least one significant breakthrough, leading to a reduction in uncertainty by one-half, is likely to occur every 5 years. With this assumption, the goal of an overall uncertainty equal to ±50 percent can easily be reached within the life of the program at a 15-year rate of continuous improvement."[85] Essentially, it was an educated guess that after a quarter-century of further research and what it believed to be a statistically likely number of serendipitous discoveries, NASA might hope to improve its radiation risk-projecting skill from 1,000 percent to only 50 percent *inaccurate*.

In the near term, OLMSA opted to judge the program's success by the number of beam hours requested and used, number of peer-reviewed publications and patents, number of proposals received in response to a solicitation, and the number accepted. Program success also was to be judged by how accurate predictions of both radiation doses on orbit and the incidence of SPEs turned out to be, and improvements in shielding.[86] Phases 1 and 2 did achieve some success.

By 2000, the JSC Space Radiation Health Project was estimating radiation levels on the Martian surface based on altitude data gathered by the Mars Orbital Laser Altimeter aboard the orbiting Mars Global Surveyor, then fed to the HZETRN and QMSFRG (Quantum Multiple-Scattering theory of nuclear FRaGmentation) models, which calculated the protection a Martian CO_2 atmosphere at that altitude offered to 12 vital human organs. In 2001 the SRHP saw MARIE (MArs RadIation Environment Experiment) speed toward the Red Planet aboard Mars Odyssey. The original plan had been for its lander to send data back to JSC but funding curtailment left them to settle for the orbiter's.[87]

Launched March 8, 2001, Expedition 2 carried JSC's Phantom Torso, DLR's Dosimap, and the Japanese (JAXA) Bonner Ball Neutron Detector (BBND) packages to the ISS. Expedition 3 continued operation of the BBND. The crews of Expeditions 4, 5, and 6 took part in the CSA's Extravehicular Activity Radiation Monitoring (EVARM) study in 2002. Expedition 6 saw the initiation of a German (DARA) study, Chromosomal Aberrations in Blood Lymphocytes of Astronauts (Chromosome), which continued through Expedition 11 in 2005. ESA and the DLR collaborated on an experiment similar to Phantom Torso, Matroshka-1, which the Expedition 8 crew stowed outside the ISS and the Expedition 11 crew retrieved in 2005.[88]

In 2003 the NASA Space Radiation Laboratory (NSRL) at BNL opened. It was a dedicated radiobiology accelerator offering researchers three or four "runs" on the beam, each several weeks long. The accelerator was actually a "booster" for the existing Alternating Gradient Synchrotron where other NASA-sponsored researchers, including an SRHP-funded group, had gotten beam time since 1995. It more than doubled NASA's research output in heavy-ion and proton radiation. The primary goal was to evaluate and estimate "the risk from space par-

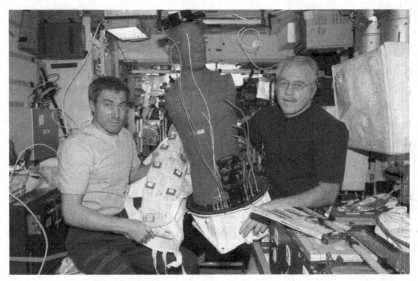

Matroshka-1 inside the ISS with its rescuers in 2005 (*L*, Russian cosmonaut Sergei Krikalyov and *R*, NASA astronaut John L. Phillips). This shot also shows its "poncho." Photo credit: NASA.

ticles and their fragmentation products to long-term space travelers and to mission-critical components. . . . Biological research will encompass studies of neurological effects, cellular and molecular alterations, including changes in DNA, in the cell matrix, and their short and long term consequences, as well as measurement of *in vivo* and *in vitro* oncogenesis," in other words, tumor formation.[89]

LSAH data gathered by JSC led to a preliminary conclusion in 2003 that astronauts exposed to radiation above a threshold of 8 millisieverts (mSV), the calculated lens equivalent dose, had a greater incidence of cataracts and at an earlier age.[90] Astronauts in that category included the Apollo, Skylab, Shuttle-Mir, and shuttle astronauts in 28.5° orbits 400 kilometers (about 250 miles) or more above the Earth.[91]

Overall, NASA's space radiation research effort was a quiet example of leveraging others' work and spending. It advanced a long way from sending photographic film up with a Mercury astronaut to verify the presence of radiation. Access to DOE beam time and data from defense projects on satellite component hardening provided useful information. Integration with NOAA satellites, mathematical modeling studies, and research aboard the shuttle, station, and on the ground and with NASA's international partners, exemplified the process of maximizing knowledge with minimal NASA money.

7

Design and Redesign

The Many Space Stations of NASA

The phrase "living and working in space" meant doing so aboard a station or vehicle that had to function safely and efficiently as workplace, laboratory, farm, transportation, and home. In the 1980s and 1990s, deploying a US space station was still the priority. As a research platform, it went through many evolutions and decision-makers considered life sciences research a low priority.

Space Station Designs 1963–1981

In the early 1960s, Langley Research Center and a few contractors sketched ideas for manned orbiting laboratories.[1] From April 1963 through February 1966 Langley funded a study for a Manned Orbiting Research Lab (MORL), which focused on "experimental and scientific data gathering objectives." Douglas Aircraft was selected for further development work. Its first version had artificial gravity created by the S-IVB stage of the Saturn 1B lifter serving as a counterweight to rotate the entire system. A later version was a zero-gravity habitat with a short-arm centrifuge.[2] In 1967, the McDonnell Douglas Astronautics Corporation tried to sell the federal government on "Big G," a gigantic version of its Gemini capsules, as a logistics module shuttling back and forth to a NASA station or to the USAF's Manned Orbiting Laboratory (MOL), or even to a small space station formed by two clustered Big Gs. Combining its Gemini and North American Rockwell's Apollo plans and hardware, MDAC was confident that by 1971 it could have a crew heading to a station for 28- and 56-day research missions. By 1975, Big G would be shuttling nine people or taking 12 on "special missions" aboard an even bigger "growth version." Big G could also function as an emergency return vehicle.[3]

In late 1971, the United States and USSR held talks on a vehicle to be made of anticipated spare Skylab, ASTP, and Salyut components. The cooperative facility would hold up to six and "support a meaningful space laboratory experiment program" of astronomical, biomedical, botanical, radiation, and fundamental biology research. It would also serve as a platform for Earth-resource surveys and a test bed for flight and navigation technology. The plan was to launch up to three 56- to 90-day NASA missions in 1976 (the US bicentennial), with 30-day "visits" by "no more than three" Soviet cosmonauts at a time. Though 50 percent of the hardware, launch capability, ground control, and rescue ability would have come from the Salyut program, by then in orbit for five years, the name was "International *Skylab* Space Station" (emphasis added).[4]

Another 1971 NASA-funded study was a Phase B plan for a Modular Space Station Program. In the MDAC version, it was to be assembled from segments brought up by shuttles, with add-ons and solar panels. The goal was still "an on-orbit research and development laboratory," hopefully "significantly" cheaper than free-flyers dedicated to individual disciplines. This ISS—Initial Space Station—would have been three modules occupied by a crew of six—two astronauts and four PIs with "a minimum of astronaut training."[5] It was an Earth-observation platform looking at fish populations, soil types, and the ionosphere. Life sciences research subjects were to be the astronauts, with some plant, microbial, and invertebrate specimens and possibly protein crystallization later. It could be enlarged to Growth Space Station (GSS) status with a crew of twelve and as many as nine attached research modules or free-flyers and three additional logistics modules, and its life expectancy would double from five to 10 years. Initial operations would have commenced in 1981.[6]

On a smaller scale, NASA awarded General Electric a study contract in 1975 for a human-tended flyer that was specifically life-sciences focused. By 1977 GE had produced a preliminary design for a shuttle-deployed lab called the Biomedical Experiments Scientific Satellite (BESS), to be launched in 1982 with two primates, twenty rats, and an onboard centrifuge for invertebrate research with in-flight controls. One design challenge was making BESS roomy enough to contain visiting astronauts in a shirtsleeve environment, but small enough to fit into the payload bay alongside Spacelab. Another issue was developing a robotic retrieval system to transfer the animals out of BESS and into the Spacelab for return to Earth.[7]

Marshall contracted in 1975 for study of a Spacelab-derived station, a four-module baseline platform requiring two shuttle flights to install in LEO. It was to be a permanent platform for missions longer than 30 days. Astronauts would visit every 90 days. Called the Manned Orbital Systems Concept (MOSC), Marshall envisioned using it for time-dependent phenomena. Four to 12 crew would occupy this stopgap station.[8]

Between 1976 and 1979, JSC and Marshall each sponsored two more studies on stations and manned platforms. JSC's Very Large Space Construction Base (1976–77) would have been essentially a factory with a materials-processing crew of seven. A 1978–79 JSC study examined an Orbital Service Module described in an MDAC document as "an evolutionary plan leading to a full space station operational capability." Its four components included a Power Extension Package (PEP) that a shuttle orbiter could use to power payloads and then stow in the cargo bay, a free-flyer platform for experiments, a separately orbiting power module for longer-duration or more sophisticated payloads, and a "permanently manned space operations center." MSFC let two study contracts for payload modules, one crewed, the other not, partly with an eye toward proving the commercial feasibility of electrophoresis-centered space manufacturing.[9]

By late 1981, the final cosmonauts had departed Salyut 6, the first space station with long-duration (six-month) crews, while NASA was finishing yet another station study. This was a million-dollar JSC Phase B study for what it called the Power System (renamed Space Platform), for space science research and materials processing. There was a $1 million study contract extension into 1982 and separate contracts of close to $1.5 million for "extensive unmanned and manned platform studies" with Marshall and Langley.[10]

The number of proposed payloads grew, requiring more trusses and modules, but the configuration, purpose—and downgrading of science research—reflected the Reagan presidency's Cold War focus.[11] McDonnell Douglas had already reported in a 1981 internal memo that "we are evaluating how and if a NASA-developed vehicle should be adapted for various military uses in different crisis levels (peacetime through post strike reconstruction)." These included a "weapon platform" for a "space laser," "surveillance and reconnaissance" with an "IR telescope," and a "space command post." The MDAC memo said "scientific experimentation continues to be stressed in large space station mission planning but more emphasis is now being placed on the operational potential," including national security, which "cannot be overlooked."[12] This came at the same time the National Security Council drafted a "Presidential Directive [on] Space Transportation Policy," classified as secret, that asked readers to "assess the implications of using the Shuttle as an integral part of a weapon system." Would America risk "losing all our space launch assets through attrition" and "accept a conscientious decision to expose the Shuttle flight crews to anti-satellite attac[k]"?[13]

Early Soviet Stations and the Reagan Response

Biomedical research had been an important goal with the Soviet space station program beginning with Salyut 1, launched in 1971, two years ahead of Skylab. Because cosmonauts were the first to experience Space Motion Sickness—Soyuz capsules were larger, allowing more sickness-inducing head movements—the Soviets had several years' start on SMS research. The USSR made Boris Yegorov the first physician in space aboard Voskhod 1 in 1964, nine years ahead of Joseph Kerwin on Skylab 2. Soyuz 9 in 1970, at 18 days nearly twice as long as any previous mission, saw the first case of deconditioning severe enough that cosmonauts had to be carried away from the landing site, spurring research into rehabilitation and prevention.[14] By the time the first NASA astronaut floated into a Russian space station (Mir) in 1995, the United States had accumulated 513 human-days in a space station (Skylab), while the Soviets/Russians and their Intercosmos partners had amassed over 10,000.[15] The United States and USSR had roughly the same number of physician-astronauts/cosmonauts by that time, though: some twelve or fifteen, and the CSA and NASDA each had two.

By the last crewed flight to Salyut 6 in 1981, the Soviets (with partners) had performed more than 1,500 experiments in space, including 900 in medicine and biology.[16] Two hundred materials-processing experiments on five Salyut missions included electrophoretically produced vaccines.[17] Half a dozen cosmonauts had approached or exceeded six months in space. Meanwhile, NASA's shuttle had achieved a single crewed test flight after a nearly six-year absence from LEO. NASA's only life science research in space had been carried out aboard three Soviet satellites, Cosmos 782, 936, and 1129, a total of forty-three biological experiments.[18]

In May 1982, NASA formed the Space Station Task Force and announced "science, applications, commercial, [US] national security and space operations missions" as the reasons for deploying a US station. The June 1982 Request for Proposal (RFP) for a $1 million, one-month study of "Space Station Needs, Attributes and Architectural Options" stipulated responses based on "the most efficient interaction with the user communities . . . [and] in close cooperation with the potential domestic and foreign users of the space station," including scientists. A requirement for "free flying platforms for experiments demanding both low gravity and low disturbance conditions" implied science research and pharmaceutical production runs. Physical sciences and commercial production were of higher stated priority than national security in the RFP, but 5 percent of the study funds were Department of Defense (DOD) money and contractors were asked to consider how a station would interact with the "total DOD space infrastructure envisioned to be in use in the later 1980s through the year 2000." This included how it would enhance the survival of military space systems and "be configured to reduce vulnerabilities and enhance military utility, responsiveness and endurance."[19]

It was to this vision that Reagan would ultimately commit the United States in January 1984. A few months after his State of the Union announcement, NASA organized the Space Station Program Office and asked contractors for detailed preliminary designs. Charles Walker of McDonnell Douglas had flown by then and the supposition was that more industry astronauts would be working on station, because the RFP stipulated that it be "'customer friendly' and dedicated to fulfilling customers' needs, to excess, if possible." Furthermore, "Once baselined, the mission capabilities . . . will not be reduced without full consideration of the customers' requirements and the understanding of the impact of the capability reduction." "Small-scale customers" were "most welcome."[20]

The station would "be a national facility and as such . . . available for use by all USG [US government] agencies, including those . . . responsible for insuring our national security." The NASA document added that "funding for the definition of such use would be the responsibility of the user," to underscore the separation of civilian and military customers. Possible nonmilitary government clients included the NIH, USDA, NOAA, US Geological Survey, and DOE.[21]

NASA wanted the station's technical specifications divided into four Work Packages (WPs) among four Centers, each to be bid separately, and totaling $70 million. JSC received the biggest contract, around $29 million, for WP 2, which included EVA and habitability structures and equipment. WP 1 was modules (i.e., labs) to be overseen by MSFC. This was the second largest contract, around $25 million. WP 3 was payloads, meaning the science equipment. NASA assigned it to Goddard, with a $10 million budget. Package 4 was primarily power-related, and the responsibility of Lewis Research Center for $6.25 million.[22]

The Space Station Mission Requirements Working Group (MRWG) analyzed requirements from the Science/Applications Technology Development and Commercial Advocacies group to determine the user communities and their needs. The MRWG identified three life sciences missions, seven science and applications materials processing missions, technology development, and commercial missions, to be performed in the station's commercial processing facilities, life sciences lab, and on habitable portions in general. Four other commercial missions were "place holders for industrial participation."[23]

The Statement of Work (SOW) made mention of using the "5th percentile oriental female" to "95th percentile American male" as the standard for strength, size, and reach parameters.[24] Interiors would use colors for visual orientation cues (up/down, escape, stowage) and aesthetic variety. Lighting was to be bright enough and of the right type to "facilitate human productivity." It would establish a night/day cycle and offer optimal contrast for CRT computer terminal use. Controls were to have visual, tactile, and auditory feedback, and those for emergency use must be readily identifiable in the dark and be protected from being inadvertently switched on or off. The designers were to designate fly-through traffic routes for safe and controlled translation so occupants could go through hatches and around corners without "unusual body contortions or major reorientations." There could be no sharp edges. Life support equipment was to control odor via "adequate body waste collection and disposal and through the treatment of biologically active trash . . . supported through . . . atmospheric revitalization and airflow." The station had to include toxicology detection and prevent off-gassing and microbial contamination. Radiation protection needed to be sufficient for one-year stays. The SOW limited noise levels in all areas to under 70 decibels (dB); those involving speech communication could be not more than 50 and sleep areas between 45 and 55 dB.[25] International Standards Organization (ISO) guidelines would set levels for whole-body vibration exposure. Gray water was to be recycled and the station would have a laundry facility to wash clothing and towels.[26]

Privacy also was a consideration. Each astronaut was to be given quarters with about 150 cubic feet (4.25 cubic meters) of volume, within the specified anthropometric range. Each was to have its own ventilation, temperature, and lighting controls plus a sleep restraint. Occupants needed to be able to dress and undress, carry out personal hygiene tasks, and stow personal belongings. The domicile had to have easy egress, be reconfigurable but acoustically isolated and alarmed.[27]

Sick-bay specifications implied its use for crew health maintenance and limited research. It should include a handwash station, drinking water, and room to don and doff a Lower Body Negative Pressure (LBNP) suit and to collect and isolate bodily fluids. Inhabitants would carry out routine medical tests there, including imaging with downlink, IVs, blood pressure, electrocardiograms, defibrillation, and pulmonary testing. Astronaut-physicians could perform minor surgery with local and regional anesthesia or use a two-person hyperbaric chamber. The design also had facilities for deceased crew remains and safe haven for radiation or meteorite emergencies.[28]

Crew would be limited to one 8-hour spacewalk or EVA each per day. EVA suits had to be resizable within 15 minutes and contain 1.12 liters (37.0 ounces) of drinking water and 750 kcal of food. A Manned Maneuvering Unit would be on hand for space rescue.[29]

In the spring of 1987, the White House, "with the support of the NASA Administrator," asked the National Research Council to study the whole program. The NRC supported science research in LEO but was not as enthusiastic about paying for Freedom specifically. It recommended "platforms, other than the Space Station . . . even after the Station is deployed," an "improved performance" shuttle, a heavy-lift launch vehicle, and expendables (both human-rated and for cargo). It also cautioned that more attention be paid to estimating and controlling costs, so that operating the station did not wind up consuming the nation's civilian space budget. It found fault with NASA's ability to estimate the cost of station development, coordination of the STS and station programs, and time spent managing the operational characteristics.[30]

The following year, the nonpartisan Congressional Budget Office (CBO) studied all NASA programs. It saw merit in NASA's argument that Congress did not give it enough funding to perform its mission regarding the station and that sometimes costs were outside anyone's control in a far-reaching, international, futuristic project. However, should Congress still underfund Freedom, the CBO saw two alternatives: (a) stretch out the station into the next century or (b) restructure, with "deemphasis of manned activities." While the latter would mean more money for free-flying science platforms, overall the CBO felt the growth of life sciences "would be unnecessary and could be reduced, since ambitious manned missions are not envisaged for this option."[31]

The space station debate became acrimonious in the late 1980s and remained so into the early 1990s. Reportedly "blood feuds" erupted within NASA's own science departments over the station's purpose and priorities, the need for its existence, and even the suitability of NASA as the agent to manage it and other programs. As antagonisms grew, so did fear about speaking out or with "opponents" in Congress. Senator Al Gore complained that it was "extremely difficult to get NASA witnesses to speak candidly in Congressional hearings . . . because it invariably leads to the long knives working the NASA budget." Several Ames employees asked an outside review team to meet with them separately, offsite. A NASA Lewis manager sent a three-page anonymous complaint to, among others, NASA's Inspector General about poor management at Lewis and a "gee whiz" attitude at Headquarters. Outside the Agency, "a large, influential and vociferous segment of the space community" opposed Freedom, believing it a zero-sum game, badly run, that would wind up achieving nothing.[32]

The 1990s: Station Redesigns

In spite of cost savings expected with ESA, Japan, and the CSA as paying partners and in the midst of Augustine Committee deliberations over the station and SEI, Congress ultimately demanded a major restructuring.[33] In October 1990 it cut Freedom's budget by almost 25 percent. Warning that "it is essential that the Agency recognize that the budget crisis is only beginning," the Committee of Conference alluded to the looming post–Cold War "peace dividend," meaning the transfer of defense-related spending to social programs.[34] The dividend became an across-the-board bloodletting for aerospace and defense. The US

Department of Labor estimated that the category of guided missiles, space vehicles, and parts manufacturing lost 54.6 percent of the highly skilled, well-paid workforce it had employed in 1989.[35]

The Committee of Conference "direct[ed] NASA to immediately implement a revised space station design and assembly sequence" that was "in useful phases" not dependent on each other. The first phase was to be devoted to microgravity, the next to life sciences, and only then should NASA proceed to permanent crewed capability. A smaller station that could be assembled more easily would also require fewer risky shuttle flights. Stand-alone small increments were necessary "if any station is to survive" and "should new funding threats evolve." The Committee still told the Agency it was "essential" to achieve "early manned capability."[36]

NASA's results, reported on March 21, 1991, included design changes that removed "almost $6 billion" in expenditures from the budgets of FYs 1991–1996.[37] Freedom looked to be a safer station in that it would require fewer shuttle missions and spacewalks but was still to be equipped with costly four-person assured-return vehicles. It would "accommodate" the materials and life sciences research called for in the Augustine Report with a microgravity laboratory downsized almost 40 percent in length. NASA life sciences would lose half their 24 racks, though the number of international partner racks would theoretically remain the same, 16.[38]

Documents disclose an unrealistic sense among some contractors and Agency personnel, accustomed to Cold War funding, that this was only a temporary setback. Proceedings of the Space Station Evolution Symposium, held August 6–8, 1991, reveal that some participants envisioned user demand for science and technology R&D in LEO being so high that eight or even sixteen crew in four habitats would be on station. Freedom would have 64 racks, with 10 for life sciences. The centrifuge could fit in any one of five places, including the planned "Italian mini-lab." One attendee pointed out that the international MOU already specified twice the crew (eight) and over 10 times the power (150 kW) allotted in the shrunken configuration, so *expansion* was crucial. Some justified optimism with expectations of the new SEI leaping congressional hurdles that Freedom by itself had been unable to clear—establishing a 30-year life span for the station and redesigning it for adaptation to a fuel depot, or at least to co-orbit with such facilities when the anticipated the Lunar Transfer Vehicles and manned Mars spacecraft were hangared there.[39]

In early 1993, after Richard Truly had left the Agency and Daniel Goldin became Administrator, more downsizing and shifting took place. Goldin ordered the Advisory Committee on Redesign of Space Station, consisting of 45 NASA employees and 10 delegates from the international partners (ESA, CSA, NASDA, and the ASI) to "significantly reduce development costs, while still achieving the goals for long-duration scientific research." Also known as the Vest Committee for leader Charles Vest of MIT, the group was aided by advisors from the CIA and defense contractor General Research Corporation. Representatives from the Russian space program also advised the group.[40]

The station would have to be complete in five years, one year after establishing initial on-orbit functionality in 1997. Materials and life sciences were "high priority goals" and "meaningful science and technology" were to be given "*primary consideration* in assessing each configuration option," but Goldin was willing to accept that they "meet minimum objectives" given "growth potential in out years at marginal cost." His position even included beginning the science early, aboard a Shuttle, Mir, or Spacelab, before the station was ready, but still counting it toward station objectives. The team was to cut operating costs in half, which might mean shortening the station's useful life to perhaps 10 years, extendable to fifteen.[41]

Canada, ESA, and Japan were incensed that the US president (Clinton) had singlehandedly ordered a major budget cut and a reduction of the station's life from 30 years to 10. NASA was making plans at meetings to which the partners had not been invited, sitting on cost data they needed in order to plan and sell the reconfiguration back home, discussing options without asking about their power and technical needs, failing to reference the previous year's agreement on an interim ESA ACRV, and unilaterally inviting the Russians to participate.[42]

International partners made their complaints known to multiple listeners. Margaret Finarelli, Associate Administrator for Policy Coordination and International Relations, briefed Goldin on April 19, 1993, that Japan's ambassador Takakazu Kuriyama would be expressing to him "dissatisfaction with their involvement in the Station redesign process" later that day. She referred Goldin to a March 31 article in the Tokyo *Asahi Shimbun* that reported "Japan and European nations seem to have been pushed around, with the United States failing to sound out their views on its decision in advance, and later calling them to attend meetings at short notice." Finarelli also included a recent *Los Angeles Times* article that reported Prime Minister Kiichi Miyazawa planned to "express displeasure at the United States' unilateral decision to reduce the SSF budget" at an April 16 meeting with President Clinton. Additionally, the State Department had alerted Finarelli that the Japanese Minister of State for Science and Technology, Mamoru Nakajima, planned to express "frustration with the redesign process" at a May 3 US–Japan Joint S&T High Level Committee Meeting.[43]

In a June 4, 1993, briefing, Goldin was told that ESA's ex-officio representative at the first closed-door meeting of the Advisory Committee on the Redesign of Space Station (ACRSS), Frederik Engstroem, noted "serious concerns about the final outcome of the process in terms of its ability to 'continue to accommodate and encourage international participation.'" However, "informally, they are even more bluntly pessimistic," the author of the brief, presumably Finarelli, wrote. "Because they claim to have lost confidence in the ability of the technical redesign to come up with a solution that will accommodate them, they [have] elevated the issue to the political level by calling for consultations under the Intergovernmental Agreement (IGA) among the United States and the thirteen Governments that signed the IGA."[44]

The Director General of Canada's Trade Competitiveness Bureau, Dennis Browne, had written to the State Department's Deputy Assistant Secretary for Science and Technology, Bureau of Oceans and International Environmental and Scientific Affairs, John Boright, on May 26. He claimed that "at the last minute, the USA made a deliberate decision that the non-USA partners were to be excluded from these important considerations." He was referring to a May 24 Vest Committee meeting that contained important costing data. "Canada and the other international partners face critical decisions concerning our continued participation in the space station program, which must be taken within a very tight time frame imposed by Washington. Our experts cannot provide a considered opinion to decision makers if they are denied access to fundamental information."[45]

The State Department had already responded with a "one-day intergovernmental consultative session on the space station redesign process" that NASA, OSTP, and OMB representatives attended. There partners including Japan, ESA, Belgium, France, Germany, Italy, the Netherlands, Norway, and the United Kingdom "expressed deep concern that the redesign options may not accommodate their existing contributions in a technically viable way . . . [and] . . . U.S. development cost savings should not result in cost increases for them." In particular, "Option C, the single launch large volume station, is not of the same technical maturity as the other options . . . and would make their modules inappropriate—or even unwelcome—due to its low power-to-volume ratio."[46] It did not help that the Japanese and ESA experiment modules were so heavy they would require the space shuttle be upgraded with Advanced Solid Rocket Motors (ASRM) and that project was being cancelled.[47]

NASA Director of Space Station Redesign Bryan O'Connor addressed Russian participation. Discussion was "limited to exploration of various ways to use Russian technology." Russians served as "consultants" between April 31 and May 5, and "no decisions were taken about how the Russians might participate." The unclassified report on the meeting also said that "OSTP rep Gerald Musarra emphasized that any decision on use of Russian assets would be made only after appropriate consultations with current partners."[48]

Overriding all objectives was the need to achieve maximum productivity and "assured early science," with a crew of just two—or less, as Goldin gave one option to be a station not permanently manned—and working with only 13 cubic feet (0.37 cubic meters) of payload space and 30 kW of power. The Agency suggested a reduced, 10-year life span for the station, extendable to 15. This would fulfill NASA's mandate to provide *a* station, just not the ideal one.[49]

The need to cut costs was actually dire. February 1993 meeting notes show that Goldin saw the station management structure he inherited as "terrible," that NASA had "an inefficient structure in place to do this job," with "no distinct accountability." He told managers bluntly, "If we don't fix it the program could fail." He wanted Centers under one NASA manager and contractors to "test the realism of their costing." Comptroller Mal Peterson told Goldin the estimating base was "flimsy" and that he had "'no confidence' in the station funding and schedule projections." Goldin also warned about internal "attacks on JSC," and that "fratricide must stop or we will lose the program." Saving the station was vital: "We can light up the sky with the inspirational work of Space Station Freedom or we can standby [sic] and watch the greatest technological bonfire of the century if it's cancelled."[50]

Goldin had found upon his arrival at the Agency that NASA had never priced Freedom in its entirety. There was a timeline but no costing beyond six flights. The Critical Design Review (CDR) included no flight

Appendix F

SPACE STATION OPTION RATING BY RESEARCH DISCIPLINE

Discipline	SSF at PHC	Option A HTC	Option A PHC	Option B HTC	Option B PHC	Option C PHC Solar Inertial	Option C PHC Local Vertical
Scientific & Commercial Microgravity							
Biotechnology							
Protein Crystal Growth (c)	3	2	3	2	3	3	3
Cell Tissue (c)	3	2	3	2	3	3	3
Materials Science							
Electronic & Photonic Materials (d)	3	3 / 1 (g)	3	3 / 1 (g)	3	1	3 (h)
Metals & Alloys (d)	3	3 / 1 (g)	3	3 / 1 (g)	3	2	3 (h)
Glasses & Ceramics (d)	3	3 / 1 (g)	3	3 / 1 (g)	3	2	3 (h)
Fluids	3	2	3	3	3	2	3 (h)
Combustion	3	2	3	3	3	2	3 (h)
Life Sciences							
Grav. Bio. (short-term) (a)	3	3	3	3	3	3	3
Grav. Bio. (long-term) (a)	3 (b)	0	3 (b)	0	3 (b)	3 (b)	3 (b)
Human Physiology (short-term) (a)	3	3	3	3	3	3	3
Human Physiology (long-term) (a)	3	0	3	0	3	3	3
Radiation Biology (short-term) (a)	3	3	3	3	3	3	3
Radiation Biology (long-term) (a)	3	0	3	0	3	3	3
Controlled Ecological Life Support (a)	3	0	3	0	3	3	3
Environmental Health (short-term)	3	3	3	3	3	3	3
Environmental Health (long-term)	3	0	3	0	3	3	3
Operational Medicine (short-term)	3	3	3	3	3	3	3
Operational Medicine (long-term)	3	1	3	1	3	3	3
Human Factors (short-term) (a)	3	3	3	3	3	3	3
Human Factors (long-term) (a)	3	0	3	0	3	3	3
Exobiology (short-term)	3	3	3	3	3	3	3
Exobiology (long-term)	3	0	3	0	3	3	3
Engineering Research							
Robotics	3	3	3	3	3	3	3
Envir. Effects (atomic)	3	1	1	2	3	1	2
Envir. Effects (orbital debris)	3	1	1	2	3	1	2
Structures	3	2	3	2	3	1 (e)	1 (e)
Communications & Information Systems	3	3	3	3	3	3	3
Propulsion	3	3	3	3	3	3	3
Fluid Management	3	3	3	3	3	3	3
Human Support	3	3	3	3	3	3	3
Space Science							
Sensor Development	1	1	1	1	5	2	5
Atmospheric Science	3	1	1	3	3	1	1 / 3 (f)
Earth Observing Science	2	0	0	2	2	0	2

Scale: 0 to 5

		Notes:
0	no capability (a)	Requires normoxic and 0.3% CO_2
1	significantly degraded capability	(b) Requires Centrifuge/Habitat Holding Facilities
2	degraded capability	(c) Requires approximately 30-day increments with stable microgravity environment
3	meets minimum guidelines capability	(d) HTC ground-tended ops assumed
4	enhanced guidelines capability	(e) Large truss structure unavailable
5	significantly enhanced guidelines capability	(f) LVL-x/LVL-y attitudes
		(g) Utilization flight/ground tended

Station redesign and capabilities. Appendix F, Space Station Option Rating by Research Discipline, in *Space Station Redesign Team Final Report to the Advisory Committee on the Redesign of the Space Station,* TM-109241 (Washington, DC: NASA, 1993), 295.

drawings or hardware. Goldin had thought the science budget was included in the $30 billion station budget but discovered that it was not. To learn all that, Goldin said, he had had to dig through the archives, because the information was nonexistent or not forthcoming.[51]

In the end the redesign team presented three alternatives: Options A (modular), B (Freedom-derived), and C (single-launch core station), with A and B further subdivided into human-tended and crewed stages. Options A and B would have provided "limited on-orbit research capability" initially, particularly because electrical power availability had been downsized. They would have met the $9 billion (1994–1998) budget goal but be unable to accommodate ESA, the Japanese, or the Italians. For life sciences, these three options were no better than, and in some cases much worse than, Freedom. As demonstrated in the chart in Figure 7.1, some studies could still be carried out in the short term but would no longer be possible over the long duration.[52]

In line with recommendations of the Advisory Committee on the Redesign of the Space Station and concurrent with those of his own redesign team, Goldin restructured station management. JSC became "host center" for the program. One prime contractor, Boeing Defense and Space Group, would manage the others as subcontractors. That way, Goldin stated, NASA could "[get] out of the way and [let] industry do the job" but "provide the proper checks and balances to ensure that the contractors deliver what they promise."[53]

It was primarily money plus a major geopolitical event that killed Freedom yet revived it as the ISS: the December 1991 dissolution of the Soviet Union. In September 1993, Vice President Gore and Russian Prime Minister Chernomyrdin agreed on a collaboration that would put Russian cosmonauts on the shuttle, NASA astronauts inside Mir, and bring Moscow in on a US-sponsored station. NASA's newest orbiting habitat would not be Freedom or even Mir 2, but a new ISS.[54]

An Opportunity Lost: Artificial Gravity

On the eve of manned spaceflight, aerospace medicine specialists had identified areas of concern when it came to weightlessness, or what they called "subgravity," and had been studying its effects on animals and humans for two decades. With humans the exposure was limited to seconds, rather than hours or days, because it was achieved by means of parabolic flights. Enough had been learned to feel certain that short-term exposure would present no real hazards to health or obstacles to performance. What was less well understood were the combined effects of weightlessness with radiation, acceleration, deceleration, and other environmental factors. Speculation about extended stays in space (trips to Mars, space stations, etc.) led to consideration of artificial gravity.[55]

In planning for a USAF MOL expected late in the decade, Wright-Patterson AFB's Aerospace Medical Laboratory conducted parabolic flights in 1960 using a C-131 transport aircraft, trying to establish the minimum threshold at which enough gravity existed for test subjects to stand and walk unaided. They found that only 0.2 g's were needed. The study was done to develop a "design envelope" for a space station, not for medical reasons per se.[56] Doctors knew nothing for certain in 1960 about the loss of calcium and bone mass after even short stays in space, so the question of whether 0.2 g would allow an astronaut to get enough weight-bearing exercise to maintain bone health did not arise.

Within NASA, Langley Research Center was the focus of interest in artificial gravity. Researchers there examined rotating toroids (doughnut shapes), spoked wheels, and spoked hexagons in 1960 and 1961, as the best means of producing a workable orbital habitat with artificial gravity. President Kennedy announced his Moon landing goal in May 1961 and NASA considered options for getting personnel, vehicles, and equipment to the Moon. One concept, Earth Orbit Rendezvous (EOR), would have put hardware into Earth orbit, where astronauts would bolt together a lunar vehicle and dispatch it to the Moon. That appealed to the space station enthusiasts but in July 1962, NASA management decided on another option, Lunar Orbit Rendezvous (LOR), meaning straight to the Moon, no station.[57] With its original raison d'être thus gone, Langley designers regrouped and came up with the Manned Orbiting Research Laboratory, designed to mesh more closely with Apollo objectives. One of those objectives was studying the effects of weightlessness on astronauts over days, weeks, and months. A very small station, MORL could be designed without artificial gravity or perhaps with a separate centrifuge module for comparison studies or astronaut reconditioning. In 1963 Langley dropped its rotating space station work.[58] Artificial gravity by means of spinning an entire vessel would languish at NASA for decades.

Thinking ahead to post-Apollo missions in 1964–1965 the idea of some kind of station reemerged and this time the Agency looked at rotating only the astronaut, producing an individualized artificial-gravity-on-demand. Under contract to NASA's Crew Systems Division, Douglas Aircraft Company tested a 4.59-foot (1.4-meter) centrifuge as a means to produce short periods of gravity aboard a space station so that returning pilots could maintain fitness for reentry maneuvers. They found that a compact short-arm centrifuge might be better than a long-arm ground-based centrifuge at stimulating the baroreceptors, the body's blood pressure sensors.[59]

The NRC's Space Studies Board recommended a human centrifuge on station six times in the 1970s. The LSAC asked for one as early as 1978, the AIBS in 1983, and the International Conference on Environmental Systems in 1984. The Developmental Biology Workshop that Ames organized in 1984 and the Paine Report in 1986 were two NASA-sponsored "voices" calling for a human-rated centrifuge on Freedom. A 1986 panel of scientists met at the University of California–Davis at the invitation of Ames specifically to consider the question "Does NASA need a centrifuge aboard Space Station Freedom? If so, why?"[60] Engineers knew that a centrifuge would be a good thing for them as well, because controlled-g experiments with small mammals in sufficient numbers and over a long enough period to generate reliable data meant someone could finally tell station, planetary habitat, and extended-mission vehicle designers how much artificial gravity, if any, was needed to keep astronauts in working order.[61]

In 1987 the National Research Council published a report reminding policymakers that ambitious plans were afoot for NASA, but that key technologies in life support, propulsion, space medicine, and other vital areas had been "severely restricted" during the past decade and a half "and mainly focused on relatively modest advances in state-of-the-art support of near-term NASA missions." NASA spent less than 3 percent of its budget on space technology research and "of that, virtually none has been spent on technology development for missions more than five years in the future." In other words, for nearly two decades the Agency had put most of its money into massive operational programs like the shuttle, with so little given over to R&T, particularly in the OAST, that NASA was "no longer a strong technical organization." When the NRC set out to quantify just what amount of investment was needed, "the result was depressing." No new launchers had been on the drawing board for 17 years. There was no orbital transfer vehicle. Powering the station was still problematic, and issues of vibration dampening, necessary to make any station a viable research lab, were not even well understood, let alone remedied. Progress in closed-loop environmental systems, which could hold down weight and costs, was "desultory," and "the same can be said for spacesuits." Astronauts were still wearing stiff suits and spending precious space hours in prebreathing before spacewalks. In the shortsightedness of the late 1960s and 1970s, NASA had terminated projects because of "budgetary pressure from the operational programs or because no programmed mission had been defined." Overall, the Council concluded, advanced space R&T at NASA "continues to be seriously underfunded—by at least a factor of three."[62]

One of those undernourished technologies was artificial gravity, especially the ability to study gradations of g-force from zero to one-sixth to one-third to one. Thought not to be critical for shuttle missions or even months-long station visits (i.e., the near term), NASA had discontinued artificial gravity research in the late 1960s "as not being necessary for planned missions."[63] It had taken nearly 15 years for the medical experts to recognize a definitive connection between increased calcium in the urine postflight and bone loss, and then see that link's relation to the weightless condition. The occasional medical book or journal article had made the connection between immobility (often in relation to polio epidemics of the 1940s-1950s and to wounded military veterans) and increased calcium excretion.[64] In late 1963, the Soviets had mentioned an increase in urinary calcium that showed up the previous year after the Vostok III (Nikolayev) and IV (Popovich) missions. While NASA flight surgeons had conducted pre- and postflight urinalysis and blood tests that included calcium levels beginning with the first suborbital flight, it was not until Gemini IV, the first multi-day mission in early June 1965, that they felt a crew had been in space for sufficient time for such data to be meaningful. Gemini flights IV (McDivitt and White) and V (Cooper and Conrad) showed nothing unusual in the urine, though a radio-densitometry examination of the hand and foot showed decreased bone density. That was downplayed when a study in the early 1970s showed that the decrease had been overstated. Higher levels of calcium excretion in the urine did turn up in December 1965 with the Gemini VII crew (Borman and Lovell). JSC flight surgeon Charles Berry observed it again in Apollo missions VII (Eisele, Schirra, and Cunningham) and VIII (Borman, Lovell, and Anders) in 1968, noticing also that the decrease seemed unaffected by exercise, but attributing it later to "electrolyte imbalances or hormonal changes."[65]

Throughout the 1970s medical articles dismissed these changes as being "of no clinical significance," and researchers misunderstood evidence from the first two Skylab missions. First, they had examined just one weight-bearing bone, the os calcis (heel), and two nonweight-bearing bones, the radius and ulna (forearm). Second, although bedrest studies begun in the 1970s had (correctly) predicted spaceflight bone loss, ground-test subjects had lost bone throughout the body while astronauts had not. Scientists had not recognized that weight-bearing bones in the lower body, such as the heel, would show demineralization, but nonweight-bearing bones in the upper body, such as the forearm and the skull, could actually show an *increase*, confounding the blood and urine calcium level tests.[66] Perhaps the duplicate test results of Frank Borman, Jim Lovell, and Pete Conrad (Skylab 2 commander), the only repeat flyers, also made the phenomenon look like a personal physiological quirk.[67] The final results from all three manned Skylab missions showed losses ranging from 0.9 percent to 7.9 percent, averaging 5.1 percent, and those who gained bone mass averaged 1.5 percent. Two JSC life scientists, Carolyn Leach Huntoon and Paul Rambaut, did make the link between heightened levels of excreted calcium and bone loss, correctly arguing it was due to weightlessness.[68] A joint study by JSC, the USAFSAM, and the US Public Health Service Hospital in 1979 got closest to the truth in a test that involved *all* the Skylab astronauts plus all their backups (as controls), six years after flight. All but one of the nine flown plus two of the seven controls showed less bone density in the os calcis. The nine who had spent one to three months in orbit had a notably higher loss. The author couched his conclusion cautiously, stating that "the data are consistent with a conclusion that space flight leads to a statistically significant long-term loss of bone mineral."[69] Ultimately, Skylab gets credit for fixing NASA's attention on bone loss and the role of microgravity. The number of flight and ground studies increased dramatically at both JSC and ARC, and the JSC mathematical modeling program included bone density, building on a decade of Skylab and bedrest data.[70]

Still, no one had proven what the full implications of living in weightlessness were for long-term health and productivity, nor whether exercise countermeasures then in favor at NASA and in the USSR even worked. The 1987 NRC report stated that "if none of the simpler countermeasures prove adequate . . . to permit work on the surface of Mars and to permit healthy readaptation to Earth's gravity, then serious consideration must be given to the provision of artificial gravity by a rotating spacecraft or pair of tethered vehicles." This would be a complex engineering task, the council admitted, but if the United States required a Mars mission, that would "force reconsideration" of the idea. Work had been stalled for almost 20 years, so no one at the Agency knew the minimum level of gravity required for good conditioning or had fully worked out the length of centrifuge arm required to generate that many g's for adult humans.[71]

While the NRC's committee was doing the background research for their report, Ames had long since taken the lead within the Agency to secure a human-rated centrifuge for Freedom. In the late 1970s, it had solicited input from science advisors as to useful equipment for the shuttle and for the station later. It then had TRW and Lockheed submit equipment proposals for evaluating vestibular reactions to acceleration changes, including balance problems and space sickness, at the time considered the most daunting issues in operational medicine and for long-duration missions. The two contractors produced concepts in 1980 for animal centrifuges generating 0.1 to 20 g's of artificial gravity and human-rated modular "chairs" accelerating occupants at 0.001–0.2 g's. One depicted a sophisticated Vestibular-Variable-Gravity Research Facility aboard a Spacelab with a containerized animal centrifuge to spin monkeys, cats, rats, quail, frogs, and even a tank of fish for Ames researchers to observe by television. This did not come to pass, but in 1986 Ames did open a Vestibular Research Facility on its own grounds. It would come to include two linear sleds and a multi-axis centrifuge.[72] Ames also in 1988 formed the Centrifuge Facility Science Working Group. Members included animal and plant physiologists and others in the biosciences. The SWG was to determine specific science needs so the two companies given study contracts, Lockheed Missiles and Space and McDonnell Douglas Space Systems, could come up with appropriate conceptual designs. This they did, in February 1991.[73]

The assessment of the 1986 UC Davis conference attendees, most of whom had centrifuge experience and many who also had conducted research in space, was that a centrifuge on station was "an absolute requirement," both as a control for plant and animal research and as a means of assessing artificial gravity as a countermeasure for humans. Ground tests on centrifuged animals hinted that in addition to bone and vestibular effects, such physiological functions as red blood cell production and fat synthesis could be gravity-dependent. ESA experiments on orbit had shown that white blood cell production was reduced. Only via a centrifuge in space could life scientists test for *all* human physiological reactions to micro-g, with exact control of variables and the ability to separate out results *not* due to weightlessness, and test for lunar and Martian gravity levels, and conduct multiple iterations of the same experiment. The group was so enthusiastic that it recommended developing a centrifuge of the "largest possible diameter," which could spin fast enough to generate 3 g's (to simulate launch and reentry forces).[74]

By mid-1990 engineer Larry Lemke of the Space Projects Division had concluded from the Lockheed data that "a large centrifuge attached to SSF is technically feasible, but I'm not sure I recommend it. The interface problems with the space station program might make a free-flying facility much more attractive." The chief problems were vibration and a place to attach a centrifuge module. A forthcoming report also showed that "the internal rotor approach . . . is incompatible with a standard Shuttle launch and was groundruled out." In other words, without a heavy lift vehicle (HLV) to launch it, a 32.81-foot (10-meter) rotor (the part of the centrifuge that spins) could not be placed in orbit.[75] When Congress mandated a station redesign in 1990, the new concept NASA proposed in March 1991 contained no definite commitment for a human centrifuge at all.[76]

Tethers and Artificial Gravity

The National Commission on Space, concerned about anyone committing to artificial gravity for a long-distance flight to Mars, recommended in 1986 that NASA orbit a test facility shortly after the station came online, then expected to be 1995.[77] NASA authorized "preliminary feasibility studies" of the "Artificial Gravity Research Facility," or AGRF, led by Marshall and Ames.[78] Marshall led another study in which JSC, KSC, Langley, Lewis, Ames, and Headquarters participated: a side-by-side analysis of sending humans to Mars in a vehicle with artificial gravity versus in one without.[79]

The Mars test was based on using two slowly rotating components, a habitation module and a counterweight, linked by a tether. Marshall engineers believed they could generate a 1 g environment by rotating a 734.91-foot (224-meter) habitat at 2 rpm. Ames designers working on the AGRF looked at a larger envelope of g-ranges, but rotating at just 1 rpm would require their vehicle to have a radius of 0.72 miles (1.16 kilometers). Disorienting Coriolis forces, creating a feeling of walking at an angle or being heavier when moving in one direction than another—generating nausea at the turn of a head—were a concern, as were details of how narrow to make the spokes for the wheel configuration while still allowing crew passage.[80] Generating different levels of g at various points on a ship meant that translating frequently from one area to another might present a problem. Perhaps naively, the MSFC team wrote, "At the present time, there appears to be a fairly high degree of optimism among many life science people that 'countermeasures' can be found to offset or prevent the deleterious physiological effects of zero-g," countermeasures meaning something other than artificial gravity. Actual experiments in space did not happen but the author pled a case for continuing tether and artificial gravity studies by noting that "the high commonality potential of systems between zero-g and artificial-g vehicles would allow development and production to begin early on the common systems and a decision to be made later on the gravity question."[81]

As NASA narrowed its focus to the shuttle and the station, it put artificial gravity and tethers for life science purposes on the shelf. Relegated to gee-whiz scenarios, K-12 curricula, and historical studies, they remained mostly in stasis. Ironically, the topic would come up near the end of the first decade of the twenty-first century as space proponents debated plans for other missions after the demise of the ISS.

8

The Cold War
and Its Aftermath

Scientific Exchange, Social Change

The end of the 1980s and the 1990s were pivotal for international relations as the Cold War and the Soviet Union came to an end. Political warming affected the conduct of space life sciences research and relationships being carefully negotiated among US, Russian, Canadian, Japanese, and European partners. NASA managers found themselves trying to keep programs alive and bargaining with international partners building their own launch systems, station components, and research facilities. It was a time marked by stumbling and progress alike.

The Cosmos/Bion Flights

In the 1970s NASA and the Soviet space agency had cooperated on the ASTP, which included some life sciences experiments, and the United States flew animals, insects, eggs, and plant cells on three Cosmos/Bion biosatellite missions.[1] John M. Logsdon, Director of the Space Policy Institute at George Washington University, noted that "cooperation fell prey to a deterioration in the overall state of US–USSR relations during the presidency of Jimmy Carter and the first White House term of Ronald Reagan. The Carter White House by 1978 was questioning whether it was in the [country's] interest to be seen as a highly visible cooperative partner with a Soviet Union that it was accusing of human rights violations, and the Shuttle-Salyut project was set aside. As part of the U.S. reaction to Soviet involvement in the declaration of martial law in Poland in 1981, the [1972] U.S.–U.S.S.R. space cooperation agreement was allowed to lapse when it came up for renewal in 1982. . . . Any cooperation in space between the United States and the Soviet Union had to be on a scientist-to-scientist basis, with no formal government involvement or funding."[2]

Despite its public posturing and private policy musing, the Reagan White House did not forbid scientific exchange or want to see it end. "The U.S. should not further dismantle the framework of exchanges," reads National Security Decision Directive (NSDD)-75, dated January 1983. "Indeed those exchanges which could advance the U.S. objective of promoting positive evolutionary change within the Soviet system should be expanded." Another motivation for continuing to allow interaction was the worry that otherwise "the Soviets will make separate arrangements with private U.S. sponsors, while denying reciprocal access to the Soviet Union." NSDD-75 also encouraged scientific cooperation with China "to reduce the possibility of a Sino-Soviet rapprochement."[3]

NASA life scientists lobbied internally to keep the exchange going. Kenneth Souza, Cosmos Project Manager at Ames, wrote to Robert Dunning, Manager of the US/USSR Biological Satellite Program at Headquarters, in early 1980 to present the case. "Due to the present political unrest between the U.S. and USSR and the consequent uncertainty of the future of our Joint U.S./USSR program in space biology and medicine, I thought it would be prudent to provide you with reasons for continuing this program," he began. Souza listed 11 justifications for the Cosmos program under way in 1980. Among them: US flights with primates were five to nine years away, but the Soviets were already orbiting rhesus monkeys; the expected lower cost of biosatellites versus the shuttle; the opportunity to screen potential shuttle or Spacelab experiments; improved US understanding of Soviet spacecraft systems, mission operations, and hardware; and the exchange of ideas and data with biologists and radiation specialists from both the USSR and Soviet bloc nations.[4]

The Soviets launched Cosmos 1514/Bion 6 containing US life sciences experiments on December 14, 1983 and retrieved it five days later. In a letter to Harold P. Klein, Director of Life Sciences at Ames, Arnauld Nicogossian, Life Sciences Director for NASA, called the project "a long and difficult task . . . particularly with the expiration of the Space

Agreement and tense [US] and Soviet relations." The team was to be "truly commended for succeeding in spite of the difficult constraints on travel, communications, and general support."[5]

Cosmos/Bion missions were not optimized for US goals, but were scheduled around the Soviets' needs and capabilities. Conditions inside the satellites were not ideal for life sciences research, either. Cosmos 1514/Bion 6 in 1983 pulled an estimated 20–60 g's on reentry, enough to cause significant injury or death to rats and monkeys. One primate died of "congenital causes" that included an impacted intestine, a hernia, and damage to the esophagus. Although the report mentioning this did not point to excessive g-force as a factor, something during flight might have caused those injuries.[6]

NASA-sponsored US scientists learned they needed to be painstakingly specific in preflight agreements with the Soviets. The University of Louisville's J. Richard Keefe's Rat Neuro-Ontogeny Experiment involved animal breeding aboard Bion 6, so he expected postflight data on all the rats returned, including the newborns. Working with rat ontogenesis in space was not new to the Soviets. Souza had seen a paper at the XIIth Joint USA/USSR Working Group Meeting in November 1981 that described a similar experiment done on five pregnant rats during the 1979 Cosmos 1129/Bion 5 mission. Still, postflight Keefe received the heads of the infant rats but nothing indicating which mother went with which offspring.[7] Jeffrey R. Alberts, an Indiana University researcher, found that the USSR team had used a different breeding procedure than the one specified at a meeting the previous year. It produced baby rodents, but he needed detailed information on what the procedures had been for his publications to stand up to peer scrutiny.[8] In an Ames researcher's primate study, the Russians left a transducer in one control rhesus on the ground unconnected. In the corresponding flight animal it was functional, but placed in the path of the waste management fan and so exposed to constantly changing air pressures. The Ames team also believed data from one of the flight monkeys to be "compromised due to deterioration (loss of signal amplitude and waveform integrity) of the implant preparation." In addition, the PIs reported, "no [postflight] studies were conducted, precluding any endpoint calibrations and/or evaluations. No vivarium [animal enclosure] control data was provided, which limits the physiological data base."[9]

NASA moved ahead with Cosmos 1667/Bion 7 and Cosmos 1887/Bion 8 in 1985 and 1987, the same year the US-USSR Civil Space Agreement created the Space Biology and Medicine Joint Working Group (JWG). In 1989 NASA was part of Cosmos 2044/Bion 9 and began preparing for the liftoff of Cosmos 2229/Bion 10 in December 1992. After meeting five times the JWG had dissolved in the early 1990s (along with the Soviet Union) but reconstituted itself as the Space Biological and Life Sciences Studies Joint Working Group in late 1992, comprised of special topic teams cochaired by US and Russian leaders: Carolyn Huntoon and Eugene Berejnoy for Space Medicine, Kenneth Souza and Eugene Ilyin for Space Biology, John D. Rummel and Eugene Zaitsev for Life Support Implementation, Donald Stewart and Alexander Kiselev for Telemedicine Implementation, and Donald Robbins and Vladislav Petrov for Radiation.[10] The last four Cosmos/Bion missions went essentially as planned, with dozens of experiments returning data on physiological responses to space flight.

Shuttle-Mir: Phase 1 of the International Space Station

That same summer the United States and Russia embarked on a joint project that foreshadowed the great difficulties and opportunities to be faced with the ISS. The Shuttle-Mir program was an idea proposed by George H. W. Bush's White House, authorizing one astronaut visit to the Soviet Mir Station in return for one cosmonaut aboard a shuttle, but later the same year Bush allowed for expanding the initial agreement to encompass more crew exchanges and a vehicle docking.[11] By the time it ended in 1998, seven NASA astronauts would undertake a Mir mission and nine cosmonauts a shuttle.[12]

NASA life scientists were not only willing to resume the relationships in the new Russian Federation, by 1992 some were actively planning to do so or had quietly found a way. Planners at Headquarters looked at collaboration as an opportunity to "help Russia make it as a democratic nation; Establish [it] as an international space partner; Provide capitalism experience; Help Russian economy (infuse $); Help establish Russian Space Agency (RSA) credibility/ authority in Russia."[13] In April an ad hoc group of life scientists from JSC and the NASA-funded Space Biomedical Research Institute consortium issued a proposal to use Mir or the expected Mir-2 replacement core module as a laboratory while they waited for the US station by "acquir[ing] Mir consultants" to "develop Mir utilization plans . . . modify Mir protocols . . . and implement [them]" with US hardware "to conduct long-term Life Sciences experiments which would accelerate the [US] space station program." The bioradiation team at JSC had secured funding in January 1991 to put an instrument on Mir and on a Russian Mars satellite in 1994 and 1996, through the Solar System Exploration Division. The Biomedical Operations and Research Branch had loaned the Russians an ionchromatograph, chemicals, and assay kits in 1990 for endocrinal studies of Mir cosmonauts in exchange for data. JSC's Cardiovascular Lab had provided Holter monitors, data recording and analysis equipment, training, and medical data on US astronauts in exchange for Mir cosmonaut results. There were plans to exchange information on behavioral and performance issues to plan future collaboration, lend the Russians "a complete LIDO Isokinetic Dynamometer," and supply "the Neurocom system for measuring posture equilibrium and supporting hardware and software."[14]

The Shuttle-Mir agreement ended the need for backroom planning. The two sides set to work in a series of meetings in Russia during the summer and fall of 1992 to develop a quick list of experiments that could be ready to fly in just two years, less than half the time shuttle PIs normally had. Because of the compressed schedule and lack of firm knowledge about Mir's onboard hardware, power supply, and other vital technical matters, planners decided to fly continuations of earlier life sciences experiments, primarily medical, since the only certainty was a human subject. Also, the 26 experiments would be under the supervision of NASA PIs instead of extramurals to save time.[15]

Republican George H. W. Bush was not reelected in November 1992, and Democrat William J. Clinton assumed the presidency in January 1993, retaining Goldin as Administrator. Clinton directed NASA to invite Russian participation in the US space station, at that point called Freedom. NASA, ever since the Eisenhower presidency "an arm of American diplomacy," did not have autonomy in its Soviet and later Russian dealings. This can be seen in many of the talking points prepared for Administrator Daniel Goldin when he interacted with foreign officials.[16] He spent much of his nine-and-a-half-year tenure both hand-holding and pressing the Russians while assuring US executive branch and congressional personnel that everything would turn out well in terms of safety and finances.[17]

Americans appreciated that the Russians were the world's experts in long-duration space flight and that ex-Soviet expertise would be available (at low wages) to help build the US station "faster, better, cheaper,"

per Goldin's favorite saying. There was some enthusiasm among the broader US science community for cooperation with and monetary support for their counterparts in the new Russia.[18] Some sensed value in keeping whole scientific institutions, such as the IMBP, in place to retain collective institutional memory.[19] One option was to merge US efforts with the follow-on Mir-2. A Lockheed study issued in September 1993 evaluated configurations that would incorporate Mir-2, saying that it would "enhance and [provide] earlier life sciences where Russia excels: long-duration human research; regenerative life support."[20] A July 1992 status report had already predicted that Mir-2 would not be ready for launch until 2005, however, and even then was dependent on the not-yet-operational Russian Buran shuttle. (Buran had not flown since 1988 and would soon be formally cancelled.)[21]

If Mir-2 was that far away, the alternative of an international space station might be appealing to the Russian Space Agency. Plans for Shuttle-Mir expanded to 10 missions and in December 1993 NASA and the RSA made it a formal, stand-alone, specifically preparatory component of the ISS called Phase 1. Legally, NASA would lease space on Mir.[22] The Implementing Agreement stated: "Special emphasis will be placed on science, particularly life science, as well as engineering & operations objectives." Phase 1 was to provide practice in organizing, managing, and operating a safe, productive, long-term international space program.[23]

Per Russian request, Phase 1 was also to be a lesson in Capitalism, specifically cost estimating and accounting.[24] Money continued to be in every respect a critical issue for Russia, with extreme inflation and a sharply devalued ruble. NASA's life sciences personnel offered to stock the IMBP with the latest equipment, calling it a loan, but knowing it would not be returned.[25] The JWG had reported in 1992 that the United States installed a Hologic 1000/W bone mineral analyzer at the Russians' Bone Lab, did maintenance on the bone densitometer at Star City, and supplied hardware for the IMBP's MicroVAX computer and an Ingress database management system plus software. NASA had set up a Local Area Network at the IMBP using Ethernet and the JWG wrote that the Agency would "consider lending their Russian colleagues portable personal computers with built-in modems and several stand-alone modems to test potential communications capabilities" among IMBP facilities, laboratories working with joint data bases, and remote operations sites. Computers and software were critical to working over long distances, so the NASA team rationalized the loan/gift as being something that would advance the project or would cost too much to ship back.[26]

In return, the Russians had provided transponder time for a telemedicine project and paid for five US specialists to fly between Moscow and the launch site it still used at Baikonur, in the newly independent Kazakhstan, and to stay at accommodations there. The Americans' Telemedicine Implementation Team had agreed to consider the possibility of equipping at least one site in Moscow with a studio and Russia would purchase and deliver one of its Satellite Data Relay Network (SDRN) satellites, similar to the United States' TDRSS, to NASA Lewis Research Center by January 14, 1993. The SDRN might "become the property of NASA at the completion of initial TDP [Telemetry Data Processing.]" The United States would do any maintenance and repair; Russia would provide spare and replacement parts. Afterward, Russia could keep the video equipment.[27]

NASA astronauts on Mir would be test subjects for studies of the physiological effects of four- to six-month sojourns in microgravity and would generate medical data that met NASA standards for scientific rigor. The flight opportunity was scientifically appealing enough that in 1994 German and Canadian space agencies expressed interest in *their* astronauts taking part and NASA planners talked about making 14 flights to Mir.[28] PIs inside and outside of the Agency expected also to carry out an array of animal and plant experiments that would take advantage of the (near) absence of gravity for such a long time. Someone even proposed a human-tended free-flyer during 1993 negotiations, based on a Russian design. The US and Russian space agencies formed a Joint Working Group to plan the in-flight science in 1995, and the life sciences group grew from a single team headed by astronaut Peggy Whitson at JSC to several teams, including microgravity in Huntsville and fundamental biology at Ames. Counting nonlife sciences, the list of experiments for six astronauts to fly after Norman Thagard's initial Mir stay expanded to 75.[29]

Within the $400 million Shuttle-Mir contract NASA had identified money specifically for the RSA to support its domestic science community, in particular the IMBP.[30] The space agency (and others) had predicted to Congress and the White House that working for the peaceful exploration of space would keep unemployed Russian rocket scientists from going over to the Dark Side of weapons design for rogue nations.[31]

Billionaire financier and philanthropist George Soros had privately paid $40 million on a January 1993 pledge of $100 million to support Russian science, Goldin's advisors told him, and Soros would be very angry if the US and Russian governments did not find the money to match that sum.[32] The RSA delayed opening a Houston office and asked NASA to pay for one in Washington, DC so that in addition to placating Congress, the White House, and one wealthy philanthropist, in 1994 Goldin had to explain that NASA could not underwrite a Russian office because all the other partners paid for their own Washington facilities. Even transferring money from the US Treasury to the RSA bank account had taken months. (First the Russians had to open an account.) One of Goldin's advisors wrote before an April meeting that RSA General Director Koptev was "very worried about his $400M, *not* about Congress." Goldin was to "hold [Koptev's] feet to the fire. Russia must show it's serious. RSA must take control of its contribution. NO talk about $1 billion."[33]

Operationally, Shuttle-Mir's implementation and operation were by all accounts fraught with frustrations.[34] A chief problem was that Mir, launched in 1986, was by 1995 aging and required constant maintenance in orbit. Also, the Soyuz-TM crew transport and Progress-M supply vehicle docked at the Kvant module, which was not accessible to the shuttle. Mir's manipulator arm, Lyappa, could only operate a few more times, but NASA negotiated to use it to reposition the Krystal module to make room until a new shuttle docking module arrived that would attach to Krystal.[35] Medical planners warned that Mir's atmosphere was contaminated with benzene beyond US Occupational Safety and Health Administration (OSHA) safety limits.[36] Over the course of the program, in-flight emergencies, including the crash of a Progress supply vessel into Mir and an onboard fire, both in 1997, changed the environmental parameters significantly and sometimes made research impossible.[37]

The first NASA Phase 1 astronaut, Dr. Norman Thagard, had little to do aboard Mir until the Spektr module with science experiments arrived, just one month before his July 1995 return. Part of that delay was the slowness of Russian customs in processing US contents at Kaliningrad; some items had to be replaced because they had spent a month in a subfreezing warehouse. Some only made it in the carryon baggage of an employee who didn't know any better. At times Headquarters did not even know where Spektr and its gear *was*.[38] Another factor was Russian reluctance to reveal the inner workings of what was formerly a classified

Shuttle-Mir program cosmonaut Gennadiy Strekalov demonstrates the technique for capturing air samples. Photo credit: NASA-JSC.

Cold War module to former adversaries.[39] Some problems, however, were simply a logical outcome of pulling together a science program on short notice. For example, the logistics of shipping eliminated the GN-2 Dewar protein crystal experiment because no company would ship liquid nitrogen from the United States to Baikonur. Moreover, the Agency did not know at first if the old experiments it was reflying would even fit into Spektr or meet Russian flight-test requirements. Translators put volume after volume of technical specifications, test procedures, and other documents into Russian, then edited and rewrote them as Phase 1 progressed.[40] NASA did not finalize the manifest until Spektr was at Baikonur. As the time and space crush became worse, NASA pulled some experiments, including a rotating chair, for launch later.[41] The 2,000 kg (4,409 lbs.) of uplift NASA thought it was getting with Spektr would prove to be just 705 kg (1,554 lbs.).[42]

Ultimately, 7 astronauts and 17 cosmonauts conducted 100 experiments proposed by about 150 investigators from the United States, Russia, Canada, the United Kingdom, Japan, Germany, France, and Hungary, according to the Phase 1 Program Joint Report. There were some significant successes, such as growing three-dimensional tissue cultures that lived for months rather than days, cultivating the first generation of plants derived from the seeds of a space-grown crop, and producing the most biomass (plant matter) ever in orbit. Authors reported there were 18 peer-reviewed papers and 11 symposia presenters in the life sciences (six based on passive radiation measurements) by 1999, five years after the first of the seven mission increments.[43] There were disappointments, too, such as Super-Dwarf wheat growing extra leaves but no seeds in the wheat heads.[44] Discouraging or unexpected results are as valid as those hoped for, but require further investigation to explain the anomalies. CELSS researcher Frank Salisbury, in fact, noted this as a strength of Shuttle-Mir. Having earlier experiments on Mir and continuing with similar studies during the joint program, he wrote that the ups and downs of multiple equipment malfunctions over several missions revealed "the importance of being able to do follow-up experiments based on results from previous experiments."[45]

Mir from the shuttle, 1997. Photo credit: NASA.

NASA's official final report on the program rated it a success but contained some qualifiers. For example, Russia canceled some experiments on NASA's initial list, declaring they "did not fit with their national science agenda."[46] As the guest, NASA had to acquiesce on combining some experiments so they could be performed with less effort or training time.[47] Being agreeable, however, meant that US scientists got the opportunity to study some of the equipment used aboard Mir, including a Japanese quail incubator, a plant chamber, and a device for an experiment on "biogrowth." However, there were few science simulators at the Gagarin Cosmonaut Training Center in Zvezdny Gorodok (Star City), and transportation difficulties resulted in training for procedures that were not what would be done in orbit.[48] A very encouraging outcome was that Russia by then published almost 90 percent of its medical data in the open literature, versus 5 percent in years past. At the same time, they clung to countermeasures data as intellectual property not open for US research aboard their station.[49]

The same qualified success could be claimed for personnel matters. Sometimes the Russian group did not schedule its cosmonauts for training on an experiment they were scheduled to carry out. NASA instructed cosmonauts in Houston on procedures for drawing blood but once back in Moscow, they could find few volunteers on whom they could practice.[50] Russian cosmonauts were paid bonuses per task, which included taking part in experiments, and that led to some misunderstandings.[51] Russian medical officials performed tests that the United States did not consider prudent or valid and exposed their cosmonauts to more radiation than NASA felt was safe.[52] They did not compile, keep, or process statistics the way US researchers did, nor did they believe in repeating experiments to verify the results.[53] Russian ideas of what constituted interference with a test subject were different, too. One NASA researcher reported that the results of his experiments on orthostatic intolerance went out the window when a Russian colleague began massaging the carotids of a cosmonaut attempting to stand postflight, because the man was dizzy.[54] Among NASA astronauts the biggest problem was the difficulty of working in Russian, and over the program's lifetime, language training was doubled while science training was cut in half.[55]

In spite of increased Russian language training, communication with the ground was of poor quality at times, of short duration, and overall was spotty. Astronauts reported that this added to their feelings of isolation and frustration and lessened their willingness to put up with the imposed eating and exercise regimes, invasive medical procedures, and disparities in culture, including the command and organizational structure aboard Mir and on the ground. Shuttle-Mir also revealed how unreliable previous Russian data and reports had been, as astronauts saw that their counterparts did not stick to the exercise regimes Russian flight surgeons had stipulated, did not adhere to a diet, did not do all the medical tests, and were overworked—sometimes to the point of exhaustion and breakdown.[56]

What became most controversial about Shuttle-Mir were insinuations that the Russian station had never been safe and that politics had put astronauts' lives at risk and wasted US taxpayer dollars on a boondoggle with little or no payoff. NASA had signed over its astronauts to be guests on a vessel under another nation's control, one argument ran, then washed its hands of responsibility for their safety by refusing to demand full partner status. Both agencies had operated Phase 1 ostensibly to "try out" East-West cooperation and technology integration for Phase 2 (the construction of the ISS), but Russian systems that had behaved catastrophically on occasion, specifically the oxygen canisters and automated Progress docking system, were later still in use on station and issues like noise and ventilation were still safety concerns, giving rise to the accusation that neither side had acted on its new knowledge. Continuing the partnership into Phase 2 was a White House foreign policy and national security decision. Some thought it not a safe or sensible one, and that once top Agency management had bought into it, the mindset and atmosphere was antagonistic toward any NASA employee who might question a decision, reminiscent of the situation that contributed to the *Challenger* disaster. At this rate, neither the RSA nor NASA could be fully trusted, or so all these arguments concluded. The supermarket tabloid media even resurrected one old claim with a new spin: NASA was actually collaborating with Moscow so that it could carry out cruel, needless animal experiments in space, beyond the reach of US law.[57]

Representative James Sensenbrenner (R-WI), Chair of the House Committee on Science, requested that NASA's Inspector General (IG) examine Mir safety, US research productivity, and the cost effectiveness of the Phase 1 partnership. His committee held hearings in September 1997, after a Mir fire, Progress-Spektr collision, and a computer failure that temporarily cost Mir its attitude control. Both IG Roberta Gross and Congressional Research Service analyst Marcia Smith noted that

NASA had "sold" Phase 1 as a way for NASA to get a leg up on in-flight science research, particularly in the life sciences, but the Spektr loss alone had "severely curtailed" science on Mir.[58]

Drawing on NASA's own internal documents and reports to Congress, correspondence from NASA astronauts who had served aboard Mir, confidences from scientists and engineers involved, and contemporary press reports, space journalist and former NASA engineer James Oberg concluded that the Agency itself had realized as early as 1991 that Mir presented a challenge for science. It was noisy, which constituted a safety hazard and might affect the results of animal tests. It produced too much vibration for successful science (especially bio- and materials processing). It was underpowered, and its crew had to "devote most of their time to housekeeping, maintenance, and physical countermeasures." Briefings by German mission specialist Ulf Merbold, who had visited Mir in 1994, revealed his opinion that "Mir was no place for science. . . . Many parts of the day there is no telemetry and no communications at all. . . . The Principal Investigators don't know anything [about what is happening], they record data onto a diskette, and don't see any results until post-flight. Often there was a problem, and nobody detected it, and the experiment failed." A NASA flight controller told Oberg privately that European counterparts "had told him that their whole project was 'a joke.' . . . It was all for show." Some items said to be on Mir were never found; Russian equipment had broken and not been repaired. A Huntsville scientist told Oberg that quality of the microgravity was "not too good" because Mir's reaction-wheel attitude control system created the space equivalent of a bumpy ride. It had no air cooling for equipment or real-time data and ground command. Worse, for biologists and doctors, "not only does [atmospheric composition regulation] not exist . . . there isn't even any data available on what the atmospheric conditions are, which tends to invalidate results of life science experiments."[59]

Oberg also reported a statement by NASA contract historian David Portree, who reviewed the records after the mission and told him, "Don't believe NASA propaganda to the contrary. . . . Science was used during Shuttle-Mir as a justification for policy, so NASA made a big deal about it to cover that fact." A "veteran doctor-astronaut" told Oberg in private that "when it became obvious that the scientific goals of Norm Thagard's flight [in 1995] could not by any stretch be achieved . . . we were told that the requirements would be changed such that whatever he achieved would become our goals. . . . We'd declare anything as 100 percent success." Oberg's conclusion was that the Agency had tried so hard to keep the floundering Russian space program afloat that it damaged its own reputation for scientific and engineering excellence and willingness to be honest in spite of unpleasant internal or external consequences. Earlier Freedom designs had taught NASA what it needed to know to proceed with the ISS, he concluded. The Agency "learned nothing useful" from the Shuttle-Mir collaboration.[60]

Some discrepancies and shortfalls in terms of Shuttle-Mir's scientific results were a matter of how to compile, measure, and report outcomes. JSC physiologist John Charles reported "a lot of interesting discussions with our management about how best to quantify science," especially in the face of unexpected results or outright failures. The science community and management debated questions like: If 80 percent of a plant crop survives, does that mean 80 percent of the science had been done? What if the 80 percent turn out to be sterile so that 0 percent could be used to produce a second crop in space? In human experiments, like a sleep study, the questions might be: Does one lost measurement render the rest of the data completely meaningless? What if the participants had done 99 percent of the work? Could that make the experiment 99 percent successful anyway? If a PI said, "Well, they put an awful lot of work into *not* getting the right data," how would that be quantified? Likewise, how does one compare the success of an experiment that only required a switch to be turned on at the beginning of the mission, then off at the end, against one requiring blood samples to be drawn several times day, then centrifuged, then frozen, and the freezer had to be maintained when the power went out time and again?[61]

Management "was always trying to find the silver lining," Charles added, which was helpful when testifying before Congress but not so helpful to the science teams. After the Progress collided with Spektr, which had nearly all of astronaut Mike Foale's experiment packages inside, Charles said, "[Management] kept saying, 'Well, if you gave up that requirement, then that means it was never really on the books anyhow. So we can say he got . . . 100 percent of his stuff done.' And we'd say, 'Look he didn't do half the experiments. [They] are in the module that's blocked off. You can't say he did 100 percent of his payload.'" Management's rejoinder would be an attempt at negotiating the quantification. "Well, can we say he did 80 percent of his payload?"[62]

Some of the medical experiments were only delayed when Spektr was closed off.[63] The Mir Operations Working Group and Joint Science Working Group scrambled to regroup and reschedule experiments, and two cosmonauts from the next mission and Foale conducted a suited Internal Vehicular Activity (IVA) and spacewalk to assess and repair the damage as much as possible, then boost power by manually realigning some solar panels. In his statement during the September 1997 congressional hearing on Mir safety, Phase 1 Program Manager Frank Culbertson commented that after all this, "Although the mix of science has changed from what was originally planned, the science value of the 6th long duration increment is expected to be approximately 80% of the original. . . . The final plan now includes 35 experiments for the increment [mission]—more than any other to date!"[64]

Overwhelmingly, the NASA scientists, flight surgeons, astronauts, managers, flight controllers, and engineers who spoke on the record for the JSC Oral History Project in 1998 felt the experience had been extremely positive and in fact vital for Phase 2, the ISS. They cited learning to integrate negotiating, engineering, management, and training styles between the two cultures as the most difficult task they had to do but also the most rewarding. The experience gave them an invaluable boost in preparing for the ISS, not only because they would have this previous experience with the Russians, but because they had an opportunity to interact with scientists from other space agencies as well.

Phase 1 Working Groups made recommendations for future scientific collaboration. Some had to do with upfront screening of proposed experiments, which had been done in great haste because Shuttle-Mir had been initiated not by scientists, but by politics. The two years between the call for proposals and launch were not enough to narrow the focus of experiments, which generally proved too broad, or to uncover software and hardware incompatibilities, or to have crew practice enough that PIs could differentiate between the science they desired and the science that could actually be done in space and so reprioritize or rewrite tasks accordingly.[65] A 1994 statement that much of Shuttle-Mir's purpose was to verify that the station environment, equipment, procedures, and management could come together to produce good science proved true and in this respect, Phase 1 was a successful dry run for the ISS.[66]

Phase 2: Beginning Construction of the International Space Station

In 1998, as Shuttle-Mir wrapped up and the Space Station Intergovernmental Agreement had been signed, Russia defaulted on its national debt and devalued the ruble. Someone there had the idea of charging NASA an $80 million Value Added Tax (VAT), essentially a $1.2 million sales tax, and renegotiating the currency exchange rate to get another $479 million.[67] This did not happen, but on the promise of an on-time first element launch by Russia, Dan Goldin petitioned Congress for an additional $60 million, granted on October 1, 1998 as a contract modification, paying the RSA for "provision of Russian crew time and Service Module stowage during the assembly phase of the International Space Station Program."[68] The timing of the launch, at Baikonur on November 20, 1998, depended on the Russians having deorbited Mir so as to focus attention and resources on Zarya, the name given to the first element. It was to be built by the Russians but owned by the United States. Administrator Goldin also worried about the Russians being able to fabricate enough Progress tugs for both ISS assembly and Mir deorbit, and if talk of going ahead alone with their earlier plans to build a Mir-2 was just bluff or were they truly behind the ISS.[69]

Money turned out not to buy exclusivity. Moscow had taken time away from dealing with the aftermaths of the Mir fire and collision to earn some European Currency Units (ECUs), predecessor to the Euro, by training French astronaut Léopold Eyharts for his February 1998 Pégase mission. The RSA had already squeezed in two paying German astronauts (Thomas Reiter and Reinhold Ewald) and another CNES "spationaut" (Claudie André-Deshays) between US flights. NASA learned in January 1999 that RSC Energiya had asked CNES to double the planned stay of Jean-Pierre Haignere aboard Mir to 180 days. Adding guest cosmonauts alarmed Goldin, fearing Moscow might balk on the agreement to deorbit Mir because it was producing a stream of hard currency revenue. Russia's agreement with the Americans was to concentrate on the ISS.[70]

Goldin warned RSA General Director Yuri Koptev in a teleconference that continued ambiguity over Mir's deorbit, combined with program delays and the Russian government's lack of financial support had "grave implications" because it had created "growing concerns within the U.S. Government that NASA-provided money for ISS has in some manner assisted in Mir's extension." When he went before Congress on

February 24 for hearings on NASA's FY 2000 budget, Goldin would not ask for funds for additional purchases from the Russians if the situation was the same.[71]

Documents show that by 1998, the year before the first element lifted off from Baikonur, NASA's Administrator and international relations executives had "just about had it" with the Russians.[72] Koptev was asking Goldin to help him get his own government to release US money promised to him, also to phrase NASA statements and contracts so that the Russian government would not back out of their duly negotiated agreements.[73] (A congressional committee had visited the Kremlin in early 1996 to convey such concerns and Vice President Gore had negotiated personally with Russian Prime Minister Viktor Chernomyrdin, both at NASA's request.) In 1999, Goldin went so far as to request that a team be formed to "determine whether NASA funds paid to Russia for joint human space flight activities were reaching their intended destination." Individuals from the Financial Management Division at JSC, the HEDS Audit Program within the Inspector General's office, the ISS Resources Office, the accounting firm Arthur Andersen, and an Administrative Contracting Officer at JSC tracked NASA money to verify that it went to Russian subcontractors. In the interim, Congress did provide $60 million and in November 2000 Expedition 1, made up of US commander William Shepherd and two Russian crewmembers, Yuri Gidzenko and Sergei Krikalyov arrived for a 136-day stay.[74]

Bions 11 and 12: The End of US-Russian Biosatellite Cooperation

Amidst the hubbub over Shuttle-Mir safety, it would be easy not to notice that US-Russian biosat efforts had come to an abrupt end. Months before the Mir fire and collision, another space voyager had died after a joint RSA-NASA mission. It was a Russian monkey who returned from a Russian flight in a Russian satellite and died on a Russian operating table, surrounded by Russian doctors—but US taxpayer dollars had helped fund his misadventure.[75]

In February 1994 NASA had cancelled SLS-3 and so owed a team of French researchers some flight time.[76] NASA signed an agreement with the RSA in December to fly the US-French experiment package on a Russian satellite, Bion 11, with a Bion 12 flight to follow soon after.[77] The French had planned to study the physiology of rhesus monkeys in weightlessness. Coincidentally a group from the IMBP had an equivalent number and type of experiments in mind and a dedicated colony of monkeys raised, maintained, and trained to function in space.[78] NASA Ames would manage the project and contribute angular motion hardware and circuitry to measure yaw and pitch motions of the monkeys' heads, acquiring data related to Space Motion Sickness.[79]

The AIBS had reviewed the original US-French experimental proposals for scientific merit in 1988 and approved them. A "second independent review panel" had done the same in March 1992 and concluded the study would address problems of human adaptation to space, "Earth-bound disease processes" and "areas of general scientific interest" and thus should go ahead. A third "independent Science Critical Design Review" had given the thumbs-up for inclusion on SLS-3, as had various internal Animal Care and Use Committees (ACUC) and legal personnel.[80] NASA had objected to an experiment that would have invasively measured intracranial pressure and to any chronic cardiovascular instrumentation at all, saying it would not pass muster with their ACUC.[81]

US PIs inspected the IMBP facilities in August 1995 and assured NASA that the Russians understood US standards for animal care and would adhere to them.[82] The NASA Advisory Council in 1996 sent Joan Vernikos, Chief Veterinary Officer Joseph Bielitzki, US Program Scientist for the primate experiments Victor Schneider, and Yale University surgeon Ronald C. Merrell, chair of the US Bion Task Force, to evaluate Russian primate facilities and the monkeys themselves.[83] The Russian Academy of Sciences reviewed the proposal for compliance with the World Health Organization that December via their institutional ACUC, a section of the Russian National Committee on Bioethics, and included NASA-Ames members in the process.[84]

NASA had planned to ask the NIH to select a panel of outside primate experts to do a cost-benefit analysis to determine whether the quantity and quality of science attainable on this one flight sufficiently advanced knowledge. The panel was to convene with its findings in December 1995, but sometime in 1994 or early 1995 PETA allegedly targeted the project, possibly in response to the death of an Ames rat at the hands of an inexperienced postdoctoral researcher. The incident resulted in Goldin's office suspending all animal research at Ames immediately, "except for those few cases in which suspending research would increase the pain and suffering to animals (or require additional animals to be used to repeat tests in progress)." The Agency appointed

CSU veterinary pathology professor Martin Fettman, a payload specialist on STS-58 (SLS-2), to head a review panel and look into matters. His
panel found that the position of Center veterinarian had been held by
a contract worker with "inadequate authority and excessively complex
and ill-defined lines of reporting." That could be linked to "events" that
"revealed serious problems with the procedures for dealing with allega-
tions of noncompliance." The contract veterinarian had resigned and
Ames management moved to hire an interim veterinarian who would
report directly to the Center director while Headquarters searched for a
"senior veterinarian" to oversee animal research once OLMSA Director
of Life Sciences Vernikos allowed its resumption. Vernikos warned Ames
of permanent research curtailment if it did not comply and pledged to
conduct monthly site audits with Fettman.[85]

Vernikos and Fettman began the audit process in July but also had
to address concerns raised in two June and July letters from PETA to
Goldin and the OLMSA Associate Administrator, asking for a halt to
the Bion program. In a 2006 interview, one of the original PIs stated
that PETA had specifically planted an individual in the Ames animal
care facility after hearing about the plans for primate research, and
this person had stolen records and generally "made a whole lot of very
muddy water." Headquarters cancelled some of the Bion experiments
and changed project management but the danger of the plans for immediate postlanding surgery, to include muscle and bone biopsies, went
unrecognized.[86]

Fettman's panel by late July 1995 found that Ames was "satisfacto-
rily implementing all of the recommendations" of its April 13 report.
Personnel there were cooperative and open, providing documents "expeditiously." Panel members "remain convinced that the highest quality
scientific research with animals can continue to be conducted at NASA
Ames . . . and that the procedures necessary to insure rigorous attention
to animal welfare will continue to undergo revision and improvement
as required by changes in legal requirements and public expectations."
PETA's concerns were "isolated incidents" that had occurred "over the
past six years" and had been "addressed as they occurred." Ames management was improving its procedures specifically in pre-, intra-, and
postoperative protocols, record-keeping, and in animal holding (cag-
ing), care, and euthanasia. Still, PETA sent another protest to OLMSA
in September.[87]

Word of the experiment had also gotten to the media in the United

States and abroad. Representatives Tim Roemer (D-IN) and Greg Ganske (R-IA), had tried to amend NASA's FY 1997 appropriations bill to
cancel Bion funding and Senator Christopher S. Bond (D-MO) introduced a bill to do the same.[88] The Headquarters FOIA officer responded
to a PETA request for Bion funding information and Arnauld Nicogossian, Acting AA of OLMSA, replied to a PETA complaint sent to Goldin,
confirming that there was no launch escape system for the Bion occupants. The RSA and Russian Academy of Science wrote to NASA about
the complaints *they* had had from PETA and the replies they had drafted
to a protest PETA wrote to President Boris Yeltsin.[89]

The launch went ahead as planned. The Bion 11 capsule bearing two
juvenile rhesus monkeys, Lapik and Multik, lifted off December 24,
1996 from Plesetsk, in Russia. Landing was at 8:05 a.m. Moscow time
on January 7, 1997. At 5:30 p.m. on January 8, or R + 33 hours, surgeons anesthetized "animal #357" (Multik) for a bone density scan and
a biopsy. Within four hours he was dead.[90]

It is unclear from the records who the actors actually were in Multik's
final drama: Russian, US, or French. Six months before Vernikos had
written the French PIs, very worried because the Toulouse Space Center
had not provided a name or credentials of the surgeon performing the
biopsies, "a clear surgical description of both biopsy procedures . . . and
the associated postoperative care plan," or even confirmation they were
still willing and able to do the procedures. She offered to provide a NASA
surgeon but "at this late date will not accept responsibility should problems arise obtaining approval . . . or with the quality of the biopsy samples." If they did not reply promptly she would recommend "deselecting"
the experiment.[91] The trip report for a November 1995 NASA visit to the
IMBP Planernaya animal facility mentioned a private meeting that Bion
Program Manager Lawrence Chambers had with director Anatoly Grigoriev about the "apparent lack of technical support and management
leadership" in the animal handling, data collection, and data recording,
specifically. "No one seemed to be in charge," and "there seemed to be a
lack of concern by some of the key people involved in monitoring and
maintaining the animals," the report claimed. "To[o] many mistakes
were being made by responsible people both with the animals and the
equipment."[92]

Still, the mission had gone ahead. Muscle biopsies were to be done at
R+ 12–24 hours and the more invasive iliac crest (hip) bone biopsy, a 30-
to 40-minute procedure under general anesthesia, at R+ 24–36. This was

slightly sooner than the Cosmos 2044/Bion 9 biopsies with which they were to be compared and was to include all the ground control monkeys as well.[93]

The postevent analysis that Ronald Merrell did at the request of Nicogossian concurred with the immediate report of Joseph Bielitzki, who arrived on the scene as revival efforts commenced. Multik had died after postoperative extubation, choking on vomit and not responding to resuscitation attempts. Neither report stated who, specifically, was in the operating rooms: Russian, French, or US surgeons or veterinary technicians. Merrell's report added the detail that Lapik, subject #484, "also experienced difficulty." He and his committee "believe[d] that the death of #357 [fell] outside the expectations of the research protocols approved by the Ames ACUC" and two other review committees. The duration of anesthetization noted in Bielitzki's report and his remarks about moving to a different building and reintubating during this time raised red flags at Headquarters. Why more than three hours to essentially take an x-ray, perform a "30–40 minute" bone biopsy, and a theoretically shorter muscle biopsy? The cause later was revealed to be the lack of redundant anesthesia equipment and the poor layout of the IMBP, which meant the Russians had moved the monkeys (a "logistical issue . . . never apparent in any prior review") and had to remove and reinsert different tubes. The report also noted a "lack of preanesthetic laboratory testing and intraoperative monitoring." On the return flight to Moscow Multik had had two episodes of shaking, a possible sign of dehydration, and an (apparently) postmortem blood test suggested the same. Aspirating his stomach would also have been "prudent since reward juice had been offered in the hour prior to anesthesia." Multik was "not immediately attended by an [sic] veterinarian at the time of extubation," which was "in retrospect . . . worrisome as was the lack of physiological monitoring." He was not "maintained in a position that allows vomitus to drain out of the mouth." Two days after the report, Vernikos notified Ames life sciences personnel that she was canceling all nonhuman primate (i.e., monkey) activities associated with the scheduled Bion 12.[94]

Original plans had been to consider Multik's death a "lesson learned" about anesthesia use so soon after return from flight, and to proceed with Bion 12 in 1998. The Russians had also been prepared to offer NASA, ESA, CNES, and NASDA a set of new biosatellite flight opportunities for 1998–2005.[95] After Vernikos cancelled primate flights, however, both sides tried to work out a plan for a Bion 12 that used mice or rats, incorporated Neurolab work or LifeSat payloads, or helped with international partner obligations. Given that the satellite and all the animal test equipment were geared for a payload of two rhesus monkeys, this proved to be too expensive and time-consuming for any group to benefit. As Russia had already done 80 percent of the work on Bion 12, including training, care, and selection of more monkeys, the United States had to pay $1.168 million dollars in addition to the $4.45 million already spent to fulfill its legal obligations and exit the contract.[96]

After NASA cancelled participation in Bion 12, a *Space News* editorial scorned the action as "pure defensive politics" that "sets a precedent that will come back to haunt researchers" in government. "NASA's main accomplishment has been to send a message to groups like . . . PETA that their particular brand of street theater protests will achieve the desired results." The *New York Times* called PETA management "thrilled with [the] news."[97]

Few people may have known that primate research had not really had full support from senior management for some time before Multik died. Goldin met with presidential science advisor Jack Gibbons about primate testing in early 1996, and it is clear, from Gibbons's memo shortly thereafter, that Goldin (and other federal research organizations) had wanted nothing more to do with monkeys. "I sympathize with your concern that the era of need for primates in NASA's research is now behind us, and that it may be time to retire those animals," Gibbons wrote. "I should point out that the Air Force is also interested in options concerning their primates, and that the National Institute of Medicine [*sic*] is planning to do a related study under NIH sponsorship."[98] As those groups were also targets for protestors, it is possible that Goldin's aim was to stop generating negative publicity that siphoned off energy, resources, time, and public support. PETA's unrelenting criticism, directed at an organization used to being celebrated for bona fide heroics, finally wore down the willingness of those who were the public face of NASA to run the shame gauntlet yet again.

9

More People, Less Science, Less NASA?

International Participants, Centrifuge, and Nongovernmental Organizations

As the third millennium of the Common Era approached, it became clear there would be a bare-bones space station. Consequences for NASA life sciences included intensified collaboration with foreign space agencies and shuttle orbiters no longer functioning as research facilities because missions were redirected to building the ISS. The flame of artificial gravity studies flickered out and work on the station centrifuge came to a halt. There was even talk of offloading some of NASA's responsibilities onto the shoulders of nongovernmental organizations (NGOs) in an attempt to return the space agency to its NACA research lab roots.

Other Foreign Collaboration

In addition to its affiliations with the space agencies of Canada, Russia, Europe, and Japan, NASA worked independently and via International Space Life Sciences Working Groups (ISLSWGs) in the 1990s to increase participation in space life sciences research by other nations. The Agency signed agreements to fly radiation monitoring equipment aboard shuttles, launch science labs, conduct telemedicine operations, create aerospace medicine training centers, share K-12 curriculum, and demonstrate science to schoolchildren worldwide. Administrators met with industry and government representatives of long-time partners such as France, Italy, and Germany and with smaller nations or

Appendix E: International Priorities in the Space Life Sciences Disciplines incl. Definition of Thematic Areas

RELATIVE EMPHASIS IN COMMON LIFE SCIENCES RESEARCH AREAS

(0 = No Activity, 1 = Minor, 2 = Nominal, 3 = Major Activity; N/A = not applicable

+ = increasing, - = decreasing)

RESEARCH AREA	CNES	CSA	DLR	ESA	NASA	JAXA	NSAU
Biological Materials Science	1	0	1	0	0	2	2
Molecular and Cellular Biology	2	2	2+	3	2	2+	3
Developmental and Reproductive Biology	2	1	1+	2	0	1+	2+
Plant Biology	1	0+	3	2-	2+	1	3
Cardiopulmonary Physiology	3	2	2	2	2	2	1
Musculoskeletal Physiology	2	3	2+	3	3	2	2
Neurosciences	3	3	2	2	2	2	2+
Regulatory Physiology	3	2	1	2	2+	1	1+
Behavior, Performance, and Human Factors	2	1+	1	1+	3	2	1
Medical Support Systems	3	0	1	1	3	1+	1
Life Support Systems	1	0	1	2+	2+	1+	2
Environmental Health	1	0	0	1+	2	1	1
Radiation Health	2	3	2	1	3	2+	1
Exobiology	1	0	1	3	2+	1	1
Biospherics Research	0	0	0	1	0	1	0

International Space Life Sciences Working Groups (ISLSWG) chart of strategic plan for 2004, international life sciences research priorities by space agency/nation. ISLSWG, International Strategic Plan for Space Life Sciences, 1995, Rev. 1, Oct. 2004, Appendix E.

countries that had hitherto not worked with NASA. With long-time partner France, NASA tried something new to both, having astronaut Charles Precourt, fluent in both French and Russian, take part in training for the CNES Cassiopée mission to Mir. Both NASA and sponsored institutions such as the NSBRI signed cooperative agreements with foreign partners, related to medicine, bioresearch, human spaceflight, and life sciences technology.[1]

Israel

One newcomer was the Israel Space Agency (ISA), which in 1986 had secured a spot on STS-47 in 1992 for a colony of oriental hornets (*Vespa orientalis*) in a Tel Aviv University-Israel Aircraft Industries, Ltd. neurovestibular experiment and flight hardware demonstration. More than half of the Israeli Space Agency Investigation About Hornets (ISAIAH) insects died in flight. The likely causes had been overcrowding and too much humidity from the hornets' water-supply system.[2]

Israel established a series of three knowledge centers, beginning in 1996, one of which was the Israel Space Weather and Cosmic Ray Center, with its Emilio Segrè Israel-Italy Observatory. In collaboration with TAU and Technion University in Haifa, ISA set up the facility, donated by the Italian Space Agency on Mt. Hermon in the Golan Heights. The center compiled information on solar radiation levels and events, such as large unexpected flares, to send to NOAA's World Data Center in Boulder, Colorado. NASA mission planners accessed that data to ensure astronaut safety on station, in particular during spacewalks.[3]

Judging by the Israeli press, the nation was not eager to put one of its citizens in space and ISA said that it had no funds for payload specialist training, but with the 50th anniversary of the founding of modern Israel approaching, politics won out over any lack of national interest in human space flight.[4] In late 1997, the Israeli Air Force selected Col. Ilan Ramon to be his country's first astronaut, and he went to JSC for payload specialist training the following year.[5] Ramon's primary assignment was to operate Tel Aviv University's Mediterranean and Israeli Dust Experiment (MEIDEX) that looked at atmospheric particles, or aerosols, over the Mediterranean and the Atlantic coast of the Sahara, as well as water and land surface reflectivity, and atmospheric phenomena. He was assigned to the *Columbia* mission STS-107, which broke up on reentry February 1, 2003. Col. Ramon took part in life sciences experiments on orbit, including ESA's Advanced Respiratory Monitoring System (ARMS), the two astroculture plant growth experiments, and tasks for the Microbial Physiology Flight Experiments and Physiology and Biochemistry teams. Ramon also helped with Freestar's STARS (Space Technology and Research Students) projects inside Spacehab, six experiments designed by children in Israel, Australia, China, Japan, Liechtenstein, and the United States.[6] The crew filmed ants, bees, silkworms, fish eggs, spiders, and growing crystals for daily downlink.[7]

Brazil

In March 1996 NASA and Brazil signed a general agreement on cooperation and the peaceful uses of space, the South American nation expressing interest in flying microgravity experiments.[8] The National Institute for Space Research (Instituto Nacional de Pesquisas Espaciais, INPE), a unit of the Brazilian Ministry of Science and Technology, and Brazsat Commercial Space Services Ltd., the country's first private-industry off-Earth venture, arranged for Brazil to participate in a Latin American working group called ChagaSpace.[9] The working group's concern was investigating treatments for Chagas disease, an insect-borne parasitic disorder infecting millions of people in South and Central America and Mexico, killing tens of thousands annually. Founded at Earth University in Costa Rica in 1995, ChagaSpace was originally a consortium of biomedical researchers from Chile and Costa Rica. The consortium had formed at the suggestion of mission specialist Franklin Chang-Diaz, a native of Costa Rica, and former payload specialist Larry De Lucas, then with the Center for Macromolecular Crystallography at the University of Alabama at Birmingham.[10] Two ChagaSpace experiments with Brazilian participation flew in 1997, on board STS-83 in April and STS-94 in July. By then the consortium had grown to include experimenters from Uruguay, Argentina, and Mexico.[11] Then in October 1997, Brazil signed an agreement to participate in the ISS.[12]

Some legal and policy analysts saw the invitation as a reward given by the Clinton administration in exchange for Brazil's new compliance with US and international demands concerning intellectual property rights and arms control. It might also have been a tool to curry Brazilian favor for the Free Trade Area of the Americas proposal the United States had set forth in late 1994.[13] Moreover, Brazil had been seeking a cessation of sanctions that since the early 1990s prevented it from marketing commercial space services, including satellite construction and launch services at Alcântara on its northern Atlantic coast, to the United States.[14]

NASA was to transfer responsibility for construction of six ISS items to Brazil, designated a "participant" rather than a "partner." These were the Express Pallet (a small payload-US truss segment interface),

Technology Experiment Facility (TEF) for long-term space exposures, Window Observational Research Facility Block 2 (WORF 2) for remote sensing, Unpressurized Logistics Carrier (ULC), and Cargo Handling Interface Assembly (CHIA) and the Z1-ULC attached system (Z1-ULC-AS) for mounting external passive payloads and experiments. Under the Implementing Agreement, Brazil would retain ownership of TEF and WORF 2. Its right to use the ISS pressurized volume and external payload area, to manifest its own experiment locker, to TDRS services, and to shuttle launch and return of 120 pounds (54.43 kilograms) of payload, primarily as two 50-kilogram (110.23-pound) TEF trays, would begin to accrue a year after NASA deployed Brazil's contributions, and extend for 10 years. Brazil also could choose to market its allotment and access to another partner.[15]

Things started out well for Brazil. In June 1998, STS-91 carried three protein crystal experiments, the research of Glaucius Oliva of the University of São Paolo. He hoped to crystallize several versions of glyceraldehyde 3-phosphate dehydrogenase, an extract of a Brazilian plant that might be a treatment for Chagas. De Lucas's Commercial Protein Crystal Growth (CPCG) apparatus held Oliva's medical experiments and nine others from Japan, Germany, Australia, and the United States.[16] The Brazilian government budgeted $120 million over three years for the ISS project and aircraft manufacturer Embraer agreed to serve as general contractor for the Brazilian ISS Project, subbing the work out to 15 local firms. INPE received $8 million to spend hiring Boeing to provide technical definition.[17] Also in 1998, the Brazilian Air Force selected the nation's first astronaut candidate, Lt. Col. Marco Pontes, who went to JSC to train and qualified in 2000 as a mission specialist.[18]

In 2001, however, things began to unravel. Early in 2001, Embraer told INPE that for $120 million it could only build one of the six station components. By June 2002, Administrator Sean O'Keefe gave Brazil one month to deliver the Express Pallet, which NASA needed to maintain its schedule and responsibilities to other partners and participants. Receiving no formal response, NASA issued an RFP in August, with proposals due in October, the contract to be awarded by March, and the first Express Pallet due at JSC two years later.[19]

Marco Pontes's flight status was in limbo, underscored by the loss of *Columbia* in 2003. Late in 2004 Brazil indicated that it could spend just $10 million on ISS work, and over four years.[20] However, in October 2005, Brazil signed a $20 million contract with the Russian Space

Agency for Pontes to train at the Gagarin Cosmonaut Training Center for a flight to the ISS aboard a Soyuz the following year.[21] So in April 2006 there was indeed a Brazilian on the ISS, taking part in the Chlorophyll Chromatography (CCM) experiments on Brazilian plants, carrying out research in the US Destiny module, and taking part in NASA physiological experiments—but his roundtrip flight was courtesy of Russia, not the United States.[22]

Ukraine

A former member of the USSR, Ukraine was in the unique position among new post–Cold War countries of producing its own launch vehicles and satellites, and many Ukraine-born cosmonauts had been in space. Three years after once again becoming a sovereign state, it established the National Space Agency of Ukraine (NSAU), with the goals of participating in manned space flight and using LEO as a place to conduct life sciences research. Because agriculture made up a significant part of the Ukrainian economy, the government planned to include plant studies as well as human physiology and medical research.

On November 22, 1994, at a White House summit, President Clinton and the new republic's second president, Leonid Kuchma, signed the Agreement on Cooperation in the Exploration and Use of Outer Space for Peaceful Purposes. It promised to explore bilateral efforts, including "exchanges . . . in the fields of science, technology and education." One of these was a plan to build, launch, and service orbital and ballistic biomed satellites with payloads such as a centrifuge, virus specimens, fish, insects, newts, tissue samples, and ESA's Biopan. Another was to be a shuttle flight teaming Ukrainian and US scientists for life science experiments, accompanied by the first cosmonaut of the new republic.[23]

Leonid Kadenyuk, a former Soviet Air Force pilot who qualified as a test cosmonaut for Soyuz, Soyuz-TM, Salyut, Mir, and Buran went to JSC with civilian physicist Yaroslavl Pustovyi as candidates to become Ukraine's first official space flyer. In November 1996, Kadenyuk was chosen as prime candidate, and began training for STS-87. This may have been a foregone conclusion, as his official NASA biography indicates that the military flyer had earlier been "transferred to the Institute of Botany, National Academy of Sciences of Ukraine, Kyiv, as a scientific investigator developing the collaborative Ukrainian-American experiment [CUE] in space biology."[24]

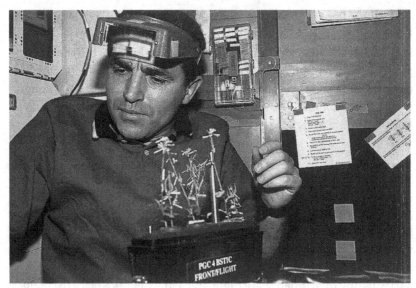

Leonid Kadenyuk of Ukraine experimented with *Brassica rapa* plants on STS-87. NASA-MSFC image. Photo credit: NASA-MSFC.

The CUE consisted of 10 experiments in plant reproduction and biology, carried out in the Plant Growth Unit (PGU), a Biological Research in Canisters (BRIC) container, and BRIC-LED (canister with light-emitting diodes) on *Columbia*'s middeck. Five of the experiments, two with *Brassica rapa* plants, two with soybean seedlings, and one with *Ceratodon purpureus,* a moss, were to fulfill the primary objective of finding which parts of plant cycles failed to function normally in the absence of gravity. Another five were to compare pollinization and fertilization processes in space with ground controls, a secondary objective.[25] The PIs were a half-dozen US scientists from three universities and more than a dozen Ukrainian scientists. During the flight, Kadenyuk would manually pollinate plants using a "bee stick." Normal pollination cannot take place in space, earlier researchers had observed, because of the lack of air convection to move the pollen from one plant to another.[26] After Kevin Kregel successfully demonstrated the technique on STS-78, Mike Foale had been the first to grow plants from seed to seed on Mir, using a dead bee on the end of a toothpick.[27] In keeping with the 1994 summit agreement that there be educational exchange as well, over half a million school children from the United States and Ukraine would be taking part in a *Brassica rapa* experiment along with Kadenyuk. Thirty US students and 32 in the Ukraine held a teleconference with Kadenyuk, aboard STS-87, and compared notes and observed how his plants were doing.[28]

After the flight of STS-87, NSAU's plans for further collaboration with NASA in life sciences research remained in limbo. Ukraine maintained a "tenuous but existing" interface with NASA. Beginning in 2004, however, political turmoil in Ukraine took over the headlines, with contested elections for the country's leadership, violent protests, and the Russian takeover of Crimea.[29]

Delegating NASA Tasks to Others?

In the 1990s USRA almost had an opportunity for the biggest contract of all: managing the science on the space station. This would not have been the first such arrangement. The Space Telescope Science Institute, which AURA operated since 1990, for example, scheduled the observing time on Hubble, including unaffiliated or amateur astronomer use.[30] Over the years, interested parties had also studied the concept of ISS commercialization in some form. In the spring of 1994, the joint R&D efforts of satellite manufacturer Loral Corporation and Deutsche Aerospace AG (DASA) led to a plan to "develop, build and market space-borne microgravity laboratories and equipment for government and commercial customers . . . including biosciences and protein crystal growth."[31] Two years later, the Potomac Institute, a Washington think tank, hired former Administrator James Beggs, James Rose (ex-MDAC pharmaceutical electrophoresis project executive), and several others who wrote a 98-page study of commercialization options under a NASA grant. They interviewed a number of companies, including some in the life sciences, about their interest in working in space.[32]

Industry's outlook in that report was essentially that such an opportunity would be nice, but would never happen because NASA was uncooperative, suspicious of privatization, and funding for the offices tasked with it had largely vanished.[33] The report pointed out that civilian contributions to a US space program had been specified in the NASA Act of 1958, something only altered in 1984 when Congress added the proviso that the "NASA Administration . . . seek and encourage to the maximum extent possible the fullest commercial use of space activities." Rose's report identified three ways of making this happen: via NASA, another government agency, or privatization. Most profit-seeking collaborations were startups coupled with academia, such as the joint venture between Vanderbilt University in Nashville and a private company to make raw materials used in treating diabetes, and the University of Alabama's hopes, backed by private investors, of producing marketable space-grown protein crystals for medical applications, and for a design for a portable bone-scan device.[34]

Ideas for ways to privatize and commercialize station (and perhaps the shuttle) began to include an NGO in 1998. Arnauld Nicogossian, then AA of OLMSA, and Joe Rothenberg, AA of Space Flight, signed off on an enthusiastic Commercial Development Plan for the ISS, which "in concert with the 1998 Commercial Space Act, represents an unprecedented initiative to stimulate business growth in the space sector." It called for employing the private sector "to break down market barriers in the near term and open the path for economic expansion." This would "begin the transition to private investment and offset a share of the public cost for operating the space shuttle fleet and space station" and "establish the foundation for a marketplace and stimulate a national economy for space products and services in low-Earth orbit, where both demand and supply are dominated by the private sector." Several longtime contractors, NASA's Commercial Space Centers (primarily educational institutions) and "their existing 135 industrial affiliates," "a nationally prominent business school with recognized high technology acumen," and accounting firm KPMG Peat Marwick's "Space and High Technology Practice," would be "tasked with" evaluating the idea. A new Senior Assistant for Access to Space, the Offices of Policy and Plans, General Counsel, Public Affairs, "an experienced professional economist," "a recognized firm in the practice of name brand management," the OMB, and a new NGO would also play roles.[35]

Two months later Nicogossian requested the NRC, which had often urged the Agency to plan for the needs of scientists who would use the ISS, to examine the idea of an NGO managing the facility.[36] Four of fourteen individuals who comprised the Task Group to Review Alternative Institutional Arrangements for Space Station Research were from the life sciences. The Task Group met in May, June, and September 1999 and finished its report by the end of the year. It sought less to turn the ISS and shuttle combo into a profit center than to give scientists "early, continuing, and substantive involvement in all phases of planning, designing, implementing, and evaluating the research use of the ISS." The result would be shorter time between selection and flight, lower costs, and simpler procedures and interfaces. An NGO would be a single point of contact, a facilitator for investigators working with NASA for the first time, and an advocate for their interests throughout the lifetime of a space experiment. Said NGO would select and train ISS crewmembers, recruit potential commercial users, broker funds, establish procedures for the protection of proprietary information, coordinate international and joint payloads, conduct payload planning, testing, and integration, and "take the lead in identifying new technologies and approaches" for enhancing ISS research use, including equipment and support upgrades. Headquarters would conduct peer review, set policy, define strategies, and defend annual budgets, but the NGO "would play a key role in assisting headquarters in these activities," just as it would "play an active role on behalf of the user community in areas where other organizations may have the lead," like station operations, maintenance, payload safety, education, and outreach. Because NASA was already planning and manifesting station research, the task force wrote, "it will be very important to move expeditiously in FY 2001 to begin the transition and implementation process."[37]

NASA publicly announced a "master plan for the economic development of the ISS" in early February 2000 at an industry forum in Albuquerque. Mark Uhran, NASA Director of Space Utilization and Product Development, was quoted in news reports as stating that the new plan "tackled a set of thorny and persistent obstacles" to profitability in space. It offered a "firm" pricing policy, a "rational and stable" process for engaging with the private sector, and protection of intellectual property. This enthusiasm and self-confidence was contrasted with the conclusion of a December 1999 report by KPMG: "Future commercial markets for the ISS are still too premature, and any market study would be wholly speculative. In the larger sense, markets for the ISS must be nurtured, rather than studied." The same article quoted former MDAC payload specialist Charles Walker as being "skeptical, if not a little . . . cynical." Walker, by then senior manager for civil space business development at Boeing, had seen NASA's last attempt at firm pricing reduce his employer's commercialization plans to ashes with the explosion of *Challenger* in 1986.[38] There were also concerns that "severe resistance" from JSC could short circuit the formation of any new organization that required Congress's approval, as the Houston Center had historically strong ties with the Texas congressional delegation.[39]

Several members of Life and Microgravity Sciences Applications Advisory Committee (LMSAAC) questioned the NGO concept and its implementation. At an October 2000 meeting, Kathie Olsen of the Office of Biological and Physical Research pointed out that "the role of the federal government is to set the science and research policy; part of that role is determining priorities on science, selection of science, and oversight of the science." Walter Hill of Tuskegee University offered the opinion that Headquarters must maintain "vigilant oversight" and "research selection should be maintained at NASA HQ via [the] peer review process." Sam Coriell of the National Institute of Standards and Technology cautioned against using the STScI as a model for an ISS NGO because the Hubble user community was "more focused." Elsa Porter of Meridian International Institute in Oregon recommended that commercial activities be organized as their own division because "the management of commercial activities requires competencies quite difference from [those] needed for management [of] scientific research." The LMSAAC further recommended that any NGO "have independent advisory boards and undergo periodic review by both NASA and the user community," and that "its Director should have a scientific background sufficient to facilitate optimal involvement of the entire research community in ISS activities."[40]

By February 2001, the OBPR had conducted an internal study but put the NGO idea on hold until details on cost and organization could be examined more thoroughly. If there were to be an RFP, it would come no earlier than fall 2001.[41]

NASA missed Congress's September 30, 2001, deadline for submitting an NGO implementation plan to the Appropriations Committee but regrouped to try for November armed with two scenarios: "conservative," with "few functions transferred" to the NGO and "aggressive," with "many functions transferred." If both NASA's administration and Congress were supportive, the Agency would proceed with the procurement phase.[42] By November 2001, however, the ISS Management and Cost Evaluation (IMCE) Task Force Daniel Goldin had established that July was reporting to the House Science Committee that Core Complete was "not credible for the $8.3 billion budget." However, "given the science interest, international partner interest, and the credibility issue that OMB and the Congress have with NASA concerning ISS, *neither* [Core Complete nor full up] is appropriate, meaning doable" (emphasis added). The Task Force recommended NASA do the best science it could at Core Complete, then reassess in 2003. The quality of science achieved—and they assumed a station centrifuge—would then determine at what "end state" Congress ought to fund the ISS. There was no mention of an NGO.[43]

By February 2002, Sean O'Keefe, like Goldin said to be "very 'pro-NGO,'" was the new NASA Administrator.[44] In his first appearance before Congress since leaving OMB, he testified that NASA had been working on the NGO option for three years, "as previously recommended by the National Research Council and directed by both the White House and Congress." The Agency was "enthusiastic" about "the potential it holds for achieving the fullest possible engagement of our nation's intellectual resources across the Government, academic, and industrial sectors."[45]

NASA did establish the "ISS Utilization Management Concept Development Team" the following month, under Mary Kicza of the OBPR, complete with "red" and "blue" teams. The latter classified the management tasks associated with the ISS as "inherently governmental," "appropriately governmental," or "other." Red Teams I and II looked at technology and personnel matters. One conclusion was that moving to an NGO would take three to four years and involve the transfer of 50–250 employees and full-time equivalents (FTEs) to carry out the duties of the estimated 2,000 Center personnel who performed some sort of ISS utilization task. To secure a contract, the winning bid would have to show efficiency gains to offset negative impacts.[46] After further study the development team concluded that a nonprofit, such as USRA, was the most desirable organization. Headquarters would still do the peer review.[47]

In June 2003, Congress wrote language into its 2003 omnibus spending bill authorizing NASA to establish an ISS NGO, and on September 9 the Agency issued a draft Statement of Work for what had been renamed the ISS Research Institute (ISSRI). Based on outsider comments received from the SOW, advice from the Space Station Utilization Reinvention team created in January, the Space Station Utilization Advisory Subcommittee of the Biological and Physical Research Advisory Committee (BPRAC), and others, Kicza's OBPR team began work that summer on a draft RFP to be released in late 2003 or early 2004.[48] The BPRAC had warned in February 2003 that "NASA still needs to articulate more clearly the division of research management between OBPR and the proposed ISS NGO as well as with other research institutes involved with OBPR. The manner in which NASA Headquarters intends to handle the NRA [NASA Research Announcement] and grant selection processes for ground-based and flight research is a particular concern."[49]

By late spring someone had moved to assuage such concerns but there were still issues the BPRAC felt worthy of referral to the NAC. One was that NASA would select commercial research projects based on prioritization standards the ISSRI and independent outside reviewers had set. The BPRAC was also concerned about funding sources, international conflicts—based on current ISLSWG guidelines—and ownership of the data archives. Sharing equipment no longer in use by PIs might be cost-efficient but by widening the user base to guest investigators, their work would have to be prioritized in with that of OBPR. The ISSRI would create a new class of Guest Investigator, working for it rather than for NASA, and "ISSRI, not NASA, will be responsible from start to finish. The Institute would select the investigators and the projects . . . manage specific flight hardware[,] the research conducted with it, [the] archived data and samples, and [the] commercial hardware." ISSRI, "as the process owner, [would] develop and approve . . . research announcements, select panels, monitor peer panels, partner with OBPR to develop recommendations, develop and present final recommendations to [the] selecting official, and make final selections."[50] "The proposed NGO process is alarming in having a Guest Investigator (GI) do what a Principal Investigator (PI) should," the Committee concluded. "An NGO should not manage the GI, should not be involved in peer review process, and should not run independent research in their own organization."[51]

USRA had indicated an interest in bidding, teamed with Battelle Labs, which had decades of experience managing DOE facilities. Associated Universities, Inc., which operated the National Radio Astronomy Observatory on behalf of the NSF, had also sent a representative to a NASA ISSRI workshop in the fall of 2002. Reportedly, 120 attendees had come to the event in Cocoa Beach, Florida, indicating a high level of interest in the ISS NGO concept. Attendees expected the money to be good, on the order of $90 million by 2007, "potentially worth several times more," per *Space News Business Report*, once the ISS was operating as a full commercial lab in space. The only dark cloud seemed to be the contract's four-year life span (in contrast, for example, with the NSBRI's five-year contract with extensions to 20 years). This would make it very difficult to attract tenure-track university scientists.[52]

The announcement of the Vision for Space Exploration hit the brakes on the ISSRI effort even before the draft RFP could be released. In a January 22, 2004, press release Mary Kicza, NASA's Associate Administrator for OBPR, stated: "NASA has reassessed its plan for an ISS Research Institute, and determined that the scope of and need for an ISSRI needs to be reevaluated." An ISSRI would be postponed for a year, and the Agency also would maintain the option to cancel it outright, pending further consideration of the station's "more focused research agenda."[53] Eventually, in 2010, Congress would tell NASA to select an organization to manage the ISS National Lab and the Agency would commission an outside consulting firm, ProOrbis of Malvern, Pennsylvania, to "formulate a reference model for an enterprise to manage those uses." In 2011 NASA would give a contract to the Center for the Advancement of Science in Space (CASIS), created by the state of Florida's Space Florida and ProOrbis, to manage non-NASA research on the ISS.[54]

Opportunity Lost: Space Station Centrifuge

An onboard rotor did survive the station's change from Freedom to Alpha to ISS, but only for plant and animal use. Studies and plans for the device progressed inside and outside the Agency. In the early 1980s University of California–Davis neurobiologist Charles Fuller had examined the adaptability of squirrel monkeys to chronic centrifugation in long-arm (25-foot) centrifuges and a decade later he studied the same thing in rats, using 1.8-, 2.5-, and 6-meter (5.91-, 8.20-, and 19.69-foot) centrifuges to achieve 1.5 g's, stopping each daily for one hour or once a week for seven hours, then testing the animals' righting reflexes. He found the frequency of stopping essentially did not matter, but the size of the centrifuge arm (and corresponding Coriolis force) did.[55] Ames investigators studied the physiological impact of centrifuge-related hardware noise in 1994, focusing on animal life support equipment including air handlers and temperature control mechanisms. Neither the prototype nor the flight test version of the Animal Enclosure Model, which had four large fans, met the standard for middeck noise allowances.[56]

At the end of August that year, a team of NASA personnel and outside experts reviewed progress. They judged the science requirements—which they pointed out had been worked on for 10 years—to be "reasonable and attainable." NASA had "reduced" the performance

United States Patent [19]

Mulenburg et al.

[11] Patent Number: 5,616,104

[45] Date of Patent: Apr. 1, 1997

[54] HUMAN POWERED CENTRIFUGE

[75] Inventors: Gerald M. Mulenburg, Mountain View, Calif.; Joan Vernikos, Alexandria, Va.

[73] Assignee: The United States of America as Represented by the Administrator of the National Aeronautics and Space Administration, Washington, D.C.

[21] Appl. No.: 513,263

[22] Filed: Aug. 10, 1995

[51] Int. Cl.⁶ A63B 69/00
[52] U.S. Cl. 482/57; 472/21; 472/35
[58] Field of Search 482/51, 57, 148, 482/110; 472/1, 14, 16, 28, 17, 18, 21, 35, 36

[56]   References Cited

U.S. PATENT DOCUMENTS

629,746	7/1899	Grosset	.... 482/57
1,174,544	3/1916	Bursteen	.... 482/57
1,409,071	3/1922	Chakiris	.... 472/28
1,887,410	11/1932	Holt .	
2,497,372	2/1950	Pricer .	

3,083,037	3/1963	Gordon et al. .
3,209,468	10/1965	Frisch .
3,216,423	11/1965	Blonsky et al. .
3,467,373	9/1969	Justice .
3,602,501	8/1971	Garner .
3,663,016	5/1972	Morris .
3,675,259	7/1972	Gilchrist .
3,677,541	7/1972	Race .
3,826,488	7/1974	Hall, Jr. .
3,936,047	2/1976	Brandt et al. .
4,147,343	4/1979	Hyde et al. .
4,428,576	1/1984	Fisher, Jr. .
4,620,700	11/1986	Snarr .
5,050,865	9/1991	Augspurger et al. .
5,378,214	1/1995	Kreitenberg .
5,395,290	3/1995	Knijpstra .

Primary Examiner—Stephen R. Crow
Attorney, Agent, or Firm—Kenneth L. Warsh; Harry Lupuloff; John G. Mannix

[57]   ABSTRACT

A human powered centrifuge has independently established turntable angular velocity and human power input. A control system allows excess input power to be stored as electric energy in a battery or dissipated as heat through a resistor. In a mechanical embodiment, the excess power is dissipated in a friction brake.

15 Claims, 3 Drawing Sheets

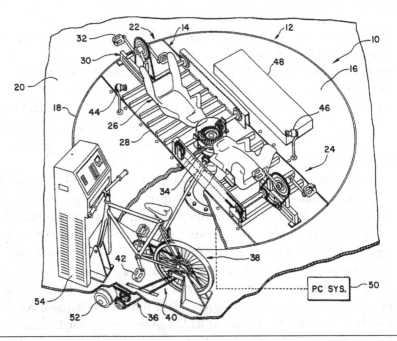

Ames Research Center's human-powered centrifuge was a low-cost proposal for lessening or eliminating "General Reentry Syndrome." Mulenberg et al., Pat. No. 5616104, assigned to NASA, filed Aug. 10, 1995, granted Apr. 1, 1997, US PTO, https://www.uspto.gov/ (accessed Sep. 22, 2020). United States Patent and Trademark Office.

requirements in the previous six months, simplifying the device, but designers were still not sure where to put it. Plans then called for a 2.5-meter (8.20-foot) diameter device, possibly smaller if NASA placed it in a Spacehab vessel. There had been "numerous perturbations" affecting costs, including a request to develop a biotelemetry system and a "service rack," and to conduct "three verification flights" on station to verify that the centrifuge would behave as expected and that "viable science" could be accomplished with it.[57]

The final report stated that this would place "a severe strain on the early-year budgets (FY-96 through FY-98) in that both Center reserves and Headquarters APA [Allowances for Program Adjustment] were proposed to solve the shortfall." Recently NASA had transferred responsibility for the Life Sciences Glovebox and lab support equipment from the station to OLMSA "without any budget transfers," the report added. An estimated $4 million would be needed but had not been budgeted by the ISSA for "payload to rack analytical integration," a cost that would transfer to OLMSA. Funding for "two elements within the transferred lab support equipment line have yet to have cost estimates," and items critical to the centrifuge facility must be funded. One hopeful note was that the schedule for both centrifuge and Glovebox design seemed "longer than necessary" and slippage could allow for work to be done in series, rather than in parallel, a less costly option. If OLMSA would provide "guidance on how often and for what length of time the Centrifuge Facility can be expected to operate," a reduction in "some of the additional pieces of hardware elements presently in the proposed baseline" might be possible.[58]

The review team questioned the need for a costly waterproof EVA mockup since the centrifuge would be assembled and disassembled *inside* the station—that is, crewmembers would not be wearing their cumbersome EVA suits. Other recommendations included "more realistic operational scenarios" to avoid overplanning in terms of the number of racks, refrigerated space, and shuttle transport. This would also make for better understanding of the impact on the thermal and electrical load. Based on experience with "significant problems during the early Spacelab years" reviewers also advised common connectors and fasteners, which might also allow for a more efficient storage of spare parts.[59]

Laurence Young of MIT's Man-Vehicle Laboratory confirmed in a separate 1994 report that the assessment of readiness was accurate. He reiterated, "The scientific importance of the research that will be conducted with this equipment is paramount—indeed it represents the future for gravitational biology. All the major reviews of space life sciences over the past two decades have stressed the need for long duration exposure of humans, animals and plants, and for the provision of a variable gravity on-board centrifuge to study levels of gravity below 1-g and to isolate microgravity from other aspects of space flight." Young also referenced compromises: the willingness to periodically stop the centrifuge and to begin experimentation with rodents only, not primates. Ground tests, he said, had justified the centrifuge size and acceptability of periodic stopping. He also said that delays had been beneficial in that workable LED lighting for plants had advanced, tests on cabin air had proven a need for additional CO_2 controls, and NASA had gotten better data on vibration tolerance. A contractor had by then "adequately demonstrat[ed]" that micro-g requirements on station could still be met given "animal movements." Like the NASA review team, he faulted space station program management for "a continued absence of commitment as to location of the centrifuge and associated racks." Without stating whether these would go in a "module, half-module, node, or Spacehab," downstream planners could not tell whether the centrifuge would launch "substantially assembled" or require assembly on orbit. No one had designed in "sufficient health monitoring" to detect if the animals were even alive without stopping it to look. He also pointed out that electricity was scarce enough that scientists might only have full use of the centrifuge during "rare 90 day increments dedicated solely to CFP [Centrifuge Facility Project] requirements."[60]

Japan and the Space Station Centrifuge

By early 1995, the Office of Biological and Physical Research (OBPR) Research Facilities Assessment team was becoming increasingly frustrated at the lack of space allocated within the ISS for life sciences components. Functional performance requirements for the centrifuge, Habitat Holding Rack, Glovebox, Advanced Animal Habitat, and Plant Research Unit were complete and Requests for Information (RFI) released. The Mouse Development Insert (nursery), Egg Incubator, Cell Culture Unit, and Aquatic Habitat were nearly done. However, no one had assigned any space to the critical freezers and racks, nor even addressed the logistics of getting all that mass up and down. Neither was the team sure there would be sufficient resources, primarily power, to run the gear once it was installed.[61]

Money was an issue partly addressed by combining the Gravitational Biology Facility (GBF) and Centrifuge Facility (CF) into a single program by purchasing a few items such as a freezer, microscope, and dosimetry equipment from partners, and by modifying a few devices in use elsewhere at NASA.[62] The Ames team also asked Headquarters to consider recalculating reserves to reflect equipment's recurring cost versus development cost, to move money around to different codes, to stretch out deployment of the animal habitats, and to slip the Glovebox and rotor timeline. These latter two actions would "lower FY96 and 97 requirements yet not alienate the science community unless they ultimately don't get the full complement."[63] At the same time, the Italians were lobbying Headquarters to exercise the option in its Multi Purpose Logistics Module (MPLM) MOU to build "a mini-laboratory to house the centrifuge in exchange for additional utilization and crew opportunities on the Station." However, Italy also asked that NASA guarantee funding first.[64] None of these efforts were enough, and on March 10, 1996, Headquarters had the Procurement Office put all related Space Station Payload Facility activities on a 90-day hold while it reconsidered.

Reminiscent of the denial reaction a few years earlier to "peace dividend" budget cuts, centrifuge program managers at first reacted as though the 1996 freeze did not really affect them, sending each other e-mails asking that particular items be exempted from the hold on the basis that they were close to contract award. Dozens of prototype and off-the-shelf items had such status and there were task orders with multiple contractors, so surely Headquarters *must* mean new procurements only. "Isn't this a reasonable view?" one manager queried.[65] Apparently not. Associate Administrator for Space Flight, Wilbur C. Trafton, extended the hold another 90 days in June.[66]

At a June 13, 1996, teleconference Ames Research Center, JSC, and Headquarters project leaders argued that further delay could cost NASA those contracts and even the availability of the contractors. The Agency had already recompeted once and scaled down the scope "by a factor of two." In addition, potential contractors had stated that their contracts were only for the baseline period. Some contractors harbored ill will as it was, after the earlier procurement had been cancelled. Furthermore, layoffs and reassignments among civil servants and contractors would lose the availability of the skills needed to design and build the equipment. Some were also perturbed about the science community's concern, industry's unhappiness, and a possible congressional inquiry if they cancelled or delayed another 90 days—or conversely, top management's wrath if they proceeded on schedule.[67]

In a memo containing talking points for the telecon dated June 10 and marked "Negotiation Sensitive," was a statement that NASDA and ESA had both "expressed an interest" in providing the Glovebox and Centrifuge Accommodation Module (CAM) as a cost offset for shuttle launches of their own labs.[68] NASA had floated the idea with the Russian Space Agency but in February had been turned down. In March 1996 NASDA's Science and Technology Agency (STA) had approached the US space agency. The subject of Japan came up during the teleconference, and participants debated that option. Who would review a NASDA proposal, and how? Would Japan propose all or just some centrifuge project construction? Working with NASDA could save money, but "experience shows interfaces between habitats . . . and Glovebox will [be] difficult to work. Trying to do this with [an] international will be a major challenge." Furthermore, it was late in the game to be bringing in foreign contractors little known to NASA. There would be no time for "Phase B type studies."[69]

However, a Headquarters team decided to invite the Japanese later that year to build as much centrifuge equipment as possible.[70] They had actually made overtures nearly a year before, in December 1995, and Randy Brinkley, Manager of the Space Station Program Office, had extended a more formal inquiry in April to a Mr. M. Saito, and in late May two NASA Office of Space Utilization System officials had met with counterparts at Headquarters. The goal had been to "coordinate technical matters" and "programmatic issues," including the schedule, "NASDA's utilization right of the Centrifuge," and "definition of offset completion" (i.e., money).[71]

Days later Space Station Program Manager Brinkley set essentially the same offer before ESA at a meeting in Houston. The Europeans faxed a decision in early June that they did "see potential merit" and that "an interest exist[s]."[72] Soon after that, the Russian Space Agency impacted the deal when expected tardiness in readying its own station module pushed the Space Station Project Office to shift plans for equipment that was equivalent to five life sciences racks into the CAM. There would not be enough room even to connect rotor to glovebox if that happened.

To further complicate matters, the Japanese had issued press releases domestically announcing that they were developing the centrifuge facility on the ISS. They had also taken preliminary remarks made in December out for firm bids. At the same time, they were still at the beginning stages in terms of figuring cost, asking perhaps naively for a "fixed price, no risk contract."[73]

NASA life sciences personnel, particularly at Ames, reportedly had many misgivings and a lot of anger about turning their project over to Japan.[74] At an August 1996 meeting, NASA life sciences personnel expressed their concerns about what they foresaw as a loss of leadership for the United States in gravitational biology. The US science community would feel NASA had betrayed them, especially if their status suddenly became that of "guest investigators" on their own station. Furthermore, the meeting participants felt Tokyo would feel no need to leap at the command of the US Congress and might be insensitive to issues of animal care and use. There was skepticism about a foreign entity coordinating with other NASA functions and departments, especially program element integration, ground and flight operations coordination, and budgeting.[75]

The Space Station Program as a whole had been proceeding regardless, but in September 1996 Headquarters asked that the program desist until after a formal meeting between the Japanese Science and Technology Association (STA) and the ISS Program Coordination Committee. Centrifuge was still on the agenda as only a "potential offset" for the cost of launching Japanese Experiment Module (JEM), which NASA recently had been forced to slip due to "insufficient power between Flight 12A and 13A & launch of SPP [Science Power Platform] on Space Shuttle." The two sides agreed to a 90-day study of "the feasibility of such cooperation." Two weeks later NASDA sent Brinkley an offer to enter into formal negotiations. The meetings, held at JSC at the end of October, produced a NASDA commitment "to seek a[n] 'agreement in principle' by April 1997," with intermediate programmatic negotiations to be held in December or January. With this, NASA handed over the centrifuge project to Japan.[76]

The Japanese agency agreed to develop nearly the entire station centrifuge package: the Life Sciences Glovebox (LSG), Centrifuge Rotor (CR), and Centrifuge Accommodation Module (CAM), the pressurized station segment into which the rotor, specimen habitats, and eventually the LSG were to fit. The CAM was to also have a freezer, habitat holding racks, general stowage, and enough empty volume for two or even three floating crewmembers to work inside.[77] NASA would still produce the animal and plant habitats and Canada would continue with its contract to build the insect habitat.[78]

The Life Sciences Glovebox was larger than the versions used on the shuttle, at 500 liters (132.09 gallons), and could hold two habitats for small animals and allow two crewmembers at a time to handle the specimens inside. The plan was to launch it separately in 2004, use it on the ISS, then relocate it to the CAM when that was placed in orbit two years later. The Centrifuge Rotor was smaller than original plans of the 1980s at 2.5 meters (8.20 feet) in diameter, large enough to spin plants and small animals, to serve as controls for weightless counterparts on board. NASA required it to generate variable gravity, in increments from 0.01 g to 2.0 g (simulating the g's of reentry), adjustable in 0.01 g increments. Life scientists could compare Earth and ISS gravity (1.0 and essentially 0 g), to the Moon's 0.17 g or Mars's 0.33 g.[79]

The centrifuge had already proven a design challenge for NASA and US contractors. For reasons of access and vibration- and gravity-control, it could not be put just anywhere on the station.[80] Other technical challenges the Japanese faced had to do with making the rotor spin efficiently, consistently, and for essentially the entire 10 years of its expected life in orbit. "Slip ring technology" involved the segment bridging the part that spun and the part that did not, while allowing passage of air and liquid for the specimens inside. Electricity and means to send video signals also had to make that link between static and moving components. A technique called "active balancing technology" was JAXA's hope for maintaining rotation at an even level in spite of the movements of the animals inside their cages. "Precision vibration isolation technology" was to be the means for dampening vibration generated by the centrifuge so it did not compromise the microgravity environment of the ISS as a whole, ruining other very delicate experiments going on at the same time.[81]

There were preliminary design reviews on the CAM and Glovebox in 1999 and on the rotor in 2000, with another scheduled for October 2001, but by then integration issues were putting the projected launch date in jeopardy.[82] At the opening negotiations in late 1996 NASA had asked for completion and deployment of the Glovebox in 2001 and the centrifuge in 2002. The Japanese offered late 2003. In 2002, the date slipped to April 2007.[83]

The centrifuge project would continue for several years beyond those first preliminary design reviews, but in all that period it never worked up sufficient momentum in any aspect (save that of scientific need) to propel itself across a finish line. The centrifuge concept was well understood and supported by the scientific community but had never been reframed to resonate with the space policy community, to make it credible, even desirable as an organic part of the ISS.[84] It would ultimately slow to a standstill, with the Vision for Space Exploration the final roadblock. There were various reasons for this: political, economic, technical, and some just plain bad luck.

It was never clear that Japan (or NASA, for that matter) knew just how complex it would be to design, build, and deploy the centrifuge and its support gear in a dedicated module.[85] The mid-1990s was the beginning of a bad patch overall for Japan's space endeavors, a period that would last at least until 2005, according to a RAND report. Beginning in 1994, Japan's three space agencies at the time, overseen by the prime minister and five other ministers, experienced a string of engineering failures with launchers, satellites, planetary probes, and aeronautical vehicles. That suggested there was a lack of cohesive oversight and "a pervasive lack of rigorous testing, quality control and quality assurance." Those would not bode well for an orbiting centrifuge. Even though Japan merged space and aeronautics functions into one new agency, JAXA, it continued to underfund it, at the rate of one-third ESA's budget and one-tenth that of NASA. Culturally, the old organizations did not merge into a whole, so Japan struggled with internal dissension, just as subsets of NASA sometimes acted against each other's interests. Japan lacked a single, overarching national goal in space, and though one of its objectives was building and deploying JEM, *also* building the centrifuge, its module, and equipment was an arduous and expensive way of getting that done.[86]

In 1993, when the Russians were trying to market their high-tech know-how and become a capitalist society, Dan Goldin had suggested taking advantage of their engineering expertise by having the RSA build the station centrifuge, or at least put a US centrifuge in a Russian module.[87] At that time, doing either of those might well have been the best idea for achieving a centrifuge in orbit. Some problems might have been resolved by making the CAM a free-flyer.

As for NASA, the centrifuge's fortunes depended on those of the ISS as a whole. The budgetary pie-slice allotted for centrifuge shrank as the station's appetite consumed more and more of the whole dessert. A big blow came with the FY 2002 budget in early 2001. It prioritized core completion, with the Gravitational Biology Facility, including centrifuge, to be added "late in the assembly sequence."[88] Life scientists who had eagerly anticipated the centrifuge for two decades, inside and outside of the Agency, were seething. "World-class research" would "not be possible" without the centrifuge and animal habitats, and if the ISS could not be fully outfitted, some believed, it shouldn't be doing any fundamental biology at all. Former astronaut Story Musgrave told the NAC Biological and Physical Research Advisory Committee, "There was great despair in the [Life Sciences Advisory] Committee that we are making a major step backward. If full funding is not restored, the LSAC recommended termination of all flight hardware. Bion and biosatellites should be discouraged. Cost savings should be transferred to a robust program on the shuttle that complements an enhanced ground-based research program. Future ISS activities should only be considered when funding is restored." The BPRAC concurred, "condemning" the loss, and called for all ISS bioscience money to be reassessed, with "savings" transferred to fund shuttle missions. However, it disagreed with the LSAS as to stopgap measures, calling for more Bions and even picosatellites.[89] A few GBF components, such as the Cell Culture Unit, were still funded in FY 2002, however, and that gave some at Headquarters hope money could be quietly found for plant and rodent units, even if it came from industry or a partner.[90]

By February 2002 Headquarters life scientists had found enough money for the Space Station Biological Research Project (SSBRP). Under an incremental approach, money would be available for two Habitat Holding Racks, one incubator, the Cell Culture Unit, the vital Life Sciences Glovebox, and software to go with the centrifuge they assumed NASDA would provide. The Canadians were still committed to the insect habitat. There was hope the FY 2004 budget, to be proposed in the coming year, might have enough give to "buy back" the rodent and plant units. The real rate-determining step was the fact that funding cuts meant only three crewmembers would be onboard long term. That left only minutes, or at best a few hours, each week for experimentation. In spite of allies like the OMB, which had directed OBPR "to create a $275 million science 'wedge' for FY 2003–FY 2007 from the research and technology budget," it would do no good if there were too few hands to do the labor.[91] Shuttle animal experiments had demonstrated how

much work beasts could be. They invariably required human interven-
tion, particularly the neonates, and experiments could not be turned on
and off to fit a schedule.[92]

That August, the OBPR's Research Maximization and Prioritization
task force, known as ReMAP, reported to the NASA Advisory Council on
its vision of prioritizing and goal-setting the Office's ISS research across
disciplines "for the first time." It pointed out the need to "resolve the
upmass and crew research time issues" and urged the OBPR to include
in its "high-priority research portfolio outstanding basic scientific research programs that address important questions in the physical and
biological sciences, and which require long-term experiments on the
ISS, based on their intrinsic scientific value." The centrifuge and habitats were "high priority facilities." The ReMAP team drew a line in the
sand for NASA: "If enhancements to ISS beyond US Core Complete are
not anticipated, NASA should cease to characterize the ISS as a science
driven program."[93]

The science and space medicine community still saw the centrifuge as
the top priority for the station. In 2003 the Bioastronautics Roadmap
named bone loss as risk number one for long-duration space missions
and pointed out that artificial gravity, which might include centrifugation, had countermeasure potential in that and other areas, including
"susceptibility to muscle damage," when combined with exercise.[94]

ReMAP and Bioastronautics Roadmap testimony may have shaken
loose some money (and prioritization) because the fundamental space
biology program got its plant and animal habitat budget a little over a
year later.[95] However, the price of prioritizing the habitats may have
ironically been the centrifuge, because its launch was slipped to September 2007 six months later.[96] As the station design shrank, without
a heavy lifter and heavier budget (and as it became more obvious that
the ISS could not accommodate all scientific activities anyway), projects judged extraneous began to disappear: CELSS, primate flights, a
non-Soyuz CERV, artificial gravity, a human centrifuge, and a crew of
more than three. Protein crystallization, which took up little room, a
small plant-animal centrifuge, a Canadian insect habitat, and a workstation were left hanging on for dear life.

10

The Vision for
Space Exploration

In January 2004, President George W. Bush came to NASA Headquarters to announce formally his goal for the nation's space program. In his "Vision for Space Exploration," Bush acknowledged the necessity of both human and robotic exploration and then defined the benefits in terms of gains in knowledge, international cooperation, national security, economics, and motivation. "Like the explorers of the past and the pioneers of flight in the last century," Bush stated, "we cannot today identify all that we will gain from space exploration; we are confident, nonetheless, that the eventual return will be great. Like their efforts, the success of future U.S. space exploration will unfold over generations." He committed the nation to a completed ISS, a return to the Moon by 2020, and robotic exploration preparatory to human missions to Mars and other planetary bodies over the next two decades.[1]

Origins of the Vision

Rumors of a presidential mandate regarding Mars had been circulating among the community of space proponents and journalists. Bush senior's 1989 SEI and studies from the 1980s and 1990s, including the National Commission on Space's 1986 *Pioneering the Space Frontier,* the 1987 Ride Report (headed by astronaut Sally Ride), and the Synthesis Group report, *America at the Threshold* in 1991, had all urged NASA to send humans to the Red Planet.[2] Former NASA Exploration Systems Mission Directorate Development Programs Division manager Craig

Cornelius wrote in the journal *Space Policy* that the "infrastructure investments" for manned missions to the Moon and Mars built upon "a pre-existing strategy for discovery-driven robotic investigation of Mars' geophysical parameters and potential habitats." These had been outlined in 2002 by the Pathways Science Steering Group Report of the Mars Exploration Payload Assessment Group within the Space Science Enterprise. The underlying philosophy, which was "influential in the formulation of the Vision," Cornelius said, was to structure manned Mars exploration as a process, discovery-driven but with long technology development lead times and high costs. The commitment would require judicious use of resources and careful planning.[3]

Presidential declarations on matters of what political scientist Lyn Ragsdale called primary policy versus ancillary policy (i.e., paradigm-shifting versus incrementally changing policy) usually mirror the Chief Executive's personal mindset. Some presidents had made reluctant decisions that resulted in direction changes for NASA (Eisenhower and *Sputnik,* Nixon and STS), but four presidents actively pointed the Agency in a new direction of their own volition, based on their own economic, political, or diplomatic philosophy: Kennedy (Moon), Reagan (SDI), George H. W. Bush (SEI), and George W. Bush (VSE). The latter was the sole president with an MBA (Harvard 1975) and the only one besides his father to have significant entrepreneurial experience (energy exploration and major league sports).[4] Thus it should not really have come as a surprise that the second Bush's particular vision had economics as an underpinning principle.

"The fundamental goal of this vision is to advance U.S. scientific, security, and economic interests through a robust space exploration program," Bush wrote. This meant an "affordable" program, human and robotic, that would "extend human presence" beyond Earth, but *after* coming up with the "innovative technologies, knowledge, and infrastructures both to explore and *to support decisions about the destinations*" (emphasis added). Finally, the Vision was to "promote international and commercial participation in exploration to further U.S. scientific, security, and economic interests."[5] There was a touch of the wonder, but the overall tenor was down-to-Earth, literally. The United States wasn't going anywhere that it did not make scientific, military, and economic sense to go, and only when it was the most productive *time* to go and the most effective *means* existed by which to go.[6] A student of history as well (Yale 1968), Bush also touched on a point important to America's self-image: inclusivity.[7] The United States would take anyone who desired to come along on its voyage of exploration: the business person, the soldier, the bureaucrat, the foreigner, the academic, and the serious nerd. He or she just might be the linchpin who could make the whole thing succeed.

For many, the Vision for Space Exploration was inspirational. For employees and contractors in the life sciences, however, it gave pause for concern. The Vision did not use the word "biology" (or any derivatives) anywhere, and the only references to "life" were in the astrobiology and life-support technology sense, not fundamental research. Bush used variations of the word "science" six times: expressing a desire to "explore scientifically valuable destinations," "to seek answers to profound scientific and philosophical questions," "to advance U.S. scientific, security, and economic questions," and "to further" same, then twice in reference to specific destinations, none of which had life.[8] Many life scientists believed themselves to have no place in this new plan, especially when it became clear that there was scant new money for it. In the accompanying FY 2005 budget Bush stated that $11 million of the planned $12 million spending increase over the next five years for "exploration" would come from reallocating funds, primarily retiring the shuttle fleet.[9] NASA would have to fund its future by shutting down part of its present.

Directed by the president to "plan and implement an integrated, long-term robotic and human exploration program structured with measurable milestones and executed on the basis of available resources, accumulated experience, and technology readiness," NASA Administrator Sean O'Keefe pledged the Agency to "explore in a sustainable, affordable, and flexible manner." A former business professor, DOD Chief Financial Officer and Comptroller, Senate Appropriations Committee staffer, Secretary of the Navy, and Deputy Director of the OMB, O'Keefe saluted the importance of the economics component to the VSE. He had the Agency's new plan published "simultaneously with NASA's FY2005 Budget Justification," calling the former "fiscally responsible, consistent with the Administration's goal of cutting the budget deficit in half within the next five years."[10] As part of this cost-trimming, the president had told NASA to return the shuttle to flight "as soon as practical," refocusing its purpose on station completion, then retire it quickly thereafter. NASA was to meet its obligations to ISS partners, but concentrate research efforts on supporting Moon and Mars exploration with medical countermeasures. These would be needed for "extended human expedition to the lunar surface" by 2020 and for trips to Mars sometime after that. Harkening back to Reagan pro-privatization days, Bush also directed NASA to "pursue commercial opportunities for providing transportation and other services" related to ISS and the lunar and Mars missions.[11] The transportation element of the Vision was named Project Constellation.[12]

Headquarters declared the FY 2005 budget ready for release just 16 days after Bush's speech and O'Keefe presented it to the media February 3, 2004. Its authors declared that the budget already "align[ed] with the goals" of the VSE, but functions and funds remained organized largely as before. Biological Sciences Research still existed as an entity requesting $491.5 million but had "pending Exploration Replanning" written in capital letters. The station centrifuge, Life Sciences Glovebox, and CAM remained in the FY 2005 budget but "as part of a barter offset with the Japanese Aerospace Exploration Agency (JAXA)."[13]

Sean O'Keefe resigned as NASA Administrator ten months later, on December 13, 2004, for a position in academia.[14] He had four days left in office when the Agency released the proposed FY 2006 budget on February 7, 2005.[15] On April 14 Dr. Michael D. Griffin inherited that budget struggle as NASA Administrator. An aerospace engineer with graduate degrees in physics, both electrical and civil engineering, and business administration, he had previously worked on Space Defense Initiative (SDI) projects at the APL in Baltimore and later as head of the space department there. He had been Chief Engineer and AA for Exploration at NASA Headquarters at different points in his career, held several management positions at Orbital Sciences Corporation, and was president of a CIA-established venture capital firm focusing on the very-high-tech sector. He taught university courses and coauthored a textbook of the same name on space vehicle design.[16] This eclectic mix of science, engineering, defense, industry, and education experience the Bush administration found ideal for the technically oriented Administrator it wanted to lead the way for the Vision.

By the end of 2005, Griffin would make made major changes to the way the Vision would be carried out and how the Agency would be organized, managed, and run. Some of those who had longed for a commitment to privatization of orbiting laboratories and crewed vehicles or to permanent human settlement in space would be buoyed by his moves. However, he had significant clashes within the Agency, with

Artist's concept of the Orion Crew Exploration Vehicle and lunar vehicle for the Vision for Space Exploration. Photo credit: NASA-JSC.

Congressional expectations, with that portion of the pro-space community that wanted more public input, and with the larger science community for a perceived slighting of "pure" research in both the life and physical sciences.

The FY 2006 budget ultimately turned out to be more than eight months from approval when Griffin took over. NASA had requested $16.456 billion dollars, 1.6 percent more than would be actually granted in late December, three months into the new fiscal year, even with $126 million supplemental funds for hurricane damage in the summer of 2004. It was a 2.4 percent increase over the prior fiscal year's appropriation, but the Agency had been counting on a 4.7 percent boost to continue progress with the VSE. There was intense pressure within the Agency while Congress debated this budget throughout 2005, because of restraints implied in the VSE and others emplaced by its new Administrator. Griffin almost immediately accelerated development of the Crew Exploration Vehicle (CEV) element of Constellation to narrow the time gap without shuttle service to the ISS. Research on station not directly advancing the Vision would have to go, as would roughly 2,500 civil servants. OBPR, which had planned to request $1.004 billion, merged with Exploration Systems, and from there Griffin redirected hundreds of millions of those science R&D dollars toward CEV.[17]

In numerous statements and speeches Mike Griffin expressed empathy with the desire of both intramural and extramural NASA scientists to remain fully funded and thus able to carry on research they believed crucial. However, he also said flatly that fundamental biology ranked so low that he believed it had to go. Testifying before the House Committee on Science in late June 2005, Griffin said about the ISS centrifuge, "that sort of research at the cellular level is not directly applicable, and would not be for many years, to problems of flying humans on voyages back to the moon or Mars." Griffin added, "I cannot responsibly prioritize microbiology and fundamental life science research higher than the need for the United States to have its own strategic access to space." Thus, the fate of scientists employed in those areas was "fairly obvious to any of us who have ever been grad students in our lives. . . . If we are not able to fund all of the work in fundamental life science, the researchers who were doing it will go elsewhere to other occupations, other research endeavors that are being funded, and we will have to put the program back together later."[18]

Clearly, the Bush-Griffin priority was not fundamental biological or medical research on station or aboard the shuttle, where manifests were being rewritten to focus solely on ISS completion. Congress, however, with its institutional memory of having approved various research programs over the past decade, became concerned that in NASA's attempt to meet Bush's challenge, it was "becoming a 'single mission agency' devoted to Vision while sacrificing its other responsibilities."[19] The larger life sciences community rallied to the cause and lobbied hard in 2004 and 2005 to stay alive, finding some support both in NASA and on Capitol Hill.

Response from the Life Sciences Community

The Life Sciences Technical Committee of the American Institute of Aeronautics and Astronautics (AIAA) went on record against Griffin's plan to speed up the CEV at the expense of science. "The NASA administrator has stated that life sciences will be deferred and 'put back together later' in order to accelerate CEV development to reduce the gap in access to space when the shuttle is retired in 2010 (or sooner)," the committee wrote. "The NASA administrator is not in an easy position, but it is doubtful that this research and its community will suddenly reappear later and escalate the research required for traveling beyond the Moon."[20]

The AIAA also formed the Exploration Life and Medical Sciences (ELMS) Coalition in March 2005 with the ASGSB and cross-pollinated it with entrepreneurial firms, enthusiast organizations, and other professional associations for life scientists "in order to retain our national space biology capability."[21] ELMS lobbied NASA and a fairly sympathetic Congress for a restoration of basic life sciences research aboard the ISS, which had been "repurposed" and in the process became an orbiting white elephant in some eyes.[22] After meetings with congressional members and staffers, the House (H.R. 3070) and Senate authorization bills (S.1281) for FY 2006 contained "revised language" calling for the ISS not to be solely utilized as an exploration-systems test bed, and to fulfill international partner obligations, provide diversity of research, and restore animal and cellular-level research on station. Congress asked Griffin to consider reinstituting the centrifuge and set aside at least $100M in FY 2006 for this in-flight research.[23]

ELMS also found strong support among overseas counterparts. In September and October 2005, ELMS and the Japanese Society for Biological Sciences in Space, the Japan Society of Aerospace and Environmental Medicine, the European Low-Gravity Research Association, European Space Science Committee Office, and ESA Life Sciences Working Group collaborated on letters to Griffin, concerned that some within NASA thought a full commitment to VSE meant gutting basic plant and animal research on the ISS.[24]

A contingent of space supporters, including active and former NASA managers and astronauts, responded to the Vision by likewise rejecting lunar exploration, and quietly making plans to substitute asteroid missions and visits to the moons Phobos and Deimos as preparation for Mars exploration. Their strategy was to wait out the lame-duck Bush administration, then begin pressing their case at the start of the next presidential campaign, in 2007.[25]

Griffin made major changes to how NASA would be organized, managed, and run in order to carry out the Vision. For the most part, those who had longed for a commitment to privatization or to permanent human settlement in space were buoyed by Griffin's moves. However, he had significant clashes within the Agency, with congressional expectations, with the portion of the pro-space community desiring more public input, and with the global life sciences community for canceling an estimated 75 percent of NASA's research capability in that field. Overall, it would be reasonable to say that Griffin carried the flag very well for the Bush economic agenda in the way he directed the VSE during that time. Nor was he alone as standard-bearer. In September of the following year Griffin would give a speech at Goddard, referring with some bitterness to the negative reaction his actions had generated among the science community at large. He also praised and echoed a speech made in March 2006 by John Marburger, Director of the White House OSTP. At the 44th Robert H. Goddard Memorial Symposium, Marburger, a physicist, had given a keynote address to an audience of engineers and scientists that starkly demonstrated how central economics was to the Bush space plan.[26]

"As I see it," Marburger remarked, "questions about the Vision boil down to whether we want to incorporate the Solar System in our economic sphere, or not." He quoted Bush's policy: "The fundamental goal of this vision is to advance U.S. scientific, security, and economic interests through a robust space exploration program." This policy "subordinates space exploration to the primary goals of scientific, security, and economic interests." Such a goal "identifies the benefits against which the costs of exploration can be weighed." Reducing space exploration to an economic formula, Marburger said, would allow other federal agencies to likewise weigh the costs of space program participation against "competing opportunities" to achieve like benefits.[27]

"The ultimate goal is not to impress others, or merely to explore our planetary system," Marburger added, "but to *use* accessible space for the benefit of humankind. It is a goal that is not confined to a decade or a century. Nor is it confined to a single nearby destination, or to a fleeting dash to plant a flag. The idea is to begin preparing now for a future in which the material trapped in the Sun's vicinity is available for incorporation into our way of life." In short, the Moon landing, which produced a plethora of tangible and intangible benefits for the United States (and the world) would prove a trifle in comparison to making the entire solar system an economic engine.[28]

The OSTP Director painted a picture of space science under the Bush Vision as part of "a balanced portfolio of public investment." And to be affordable, the fraction of the domestic discretionary budget used for space exploration "must be *small* enough to be stable against competition from other parts of the budget, and in particular those that are perceived to serve a wider variety of societal needs."[29]

Writing for a 2007 NASA collection of essays titled *Societal Impact of Spaceflight*, longtime planetary habitation scientist Wendell Mendell called the idea of an off-Earth economic sphere "a little-understood element" of the George W. Bush Vision. He described people who would come to be termed Ultra-High-Net-Worth Individuals (UHNWIs) but who were from outside the traditional community of NASA employees and aerospace contractors as being akin to religious "'believers,' who happen to have the money to pursue their dreams." For some, that dream was vacations in space or new homes on other planets. "If the business case is so weak," Mendell challenged, "why are these people in the game?"[30]

Since then, UHNWIs have actually built and deployed space vehicles. They have transported NASA crews to the space station. They launched one of the Mercury 13 (Mary Wallace "Wally" Funk) and Captain Kirk (William Shatner) into space. Some have expressed interest in Moon and Mars colonies. It remains to be seen what they—or NASA—are willing to spend on the life sciences research needed to make a permanent human presence in space and on other planets a reality.

Francis Haddy of the Mayo Clinic, which had ties to aerospace medicine dating back to the 1930s, asked rhetorically in the FASEB *Journal* of March 2007 if the United States would be ready to go back to the Moon by 2015 or 2020, and then on to Mars, given the cuts to NASA's life sciences research.[31] No, Haddy concluded, and pointed out that this was the same opinion, voiced in the same journal, about Bush Senior's SEI, and for the same reason: insufficient funding and time for preparatory research needed for a survivable trip. The earlier author, Robert W. Krauss, past president of the AIBS and retired executive director of FASEB, had written in 1991: "No scientific apex has ever been achieved without a pyramid beneath it, nor will humans long exist in space without a science to support them."[32] By the time Haddy had reached this conclusion two shuttle accidents had created such a backlog in station assembly flights that ISS research had been cut to nearly nothing; the LifeSat program had been scrubbed along with its space radiation studies; the ISS centrifuge had been delayed and Advanced Animal Habitat deleted, and the station could not even support more than two maintenance astronauts at a time. FY 2006 funding "promises to eliminate the majority of biological flight research . . . and reduces ground-based research to levels that essentially represent a phasing out of the program." Given this track record, NASA would not—could not—be ready to embark on a Mars trip "for some time."[33]

Parting Thoughts

In the bleak interval of the early 1970s, when the Apollo program was scheduled to end soon, leaving only ASTP manifested, and the Agency was pinning all its hopes on President Richard Nixon approving a new space shuttle program, Acting Administrator George M. Low contemplated the idea of turning the US space agency into a component of a broader "Aeronautics, Space, and Applied Technology Administration." Spaceflight and aviation would have still been 90 percent of the new agency, but the "surplus of talent" among aerospace contractors would have gone to work under "ASATA" supervision studying energy, transportation, healthcare, and other "difficult technological problems." What had been NASA would either redirect 10 percent of its personnel for in-house efforts on these non–aerospace projects, or accept additional money from other government agencies to do their research.[1] Not surprisingly, since Low had begun his career at the National Advisory Committee for Aeronautics, this sounded something like the premise for NACA's establishment by Congress back in 1915: "to supervise and direct the scientific study of the problems of flight, with a view to their practical solution, and to determine the problems which should be experimentally attacked, and to discuss their solution and their application to practical questions."[2] Low's reorganization idea went nowhere, however, after the space shuttle program was authorized on January 5, 1972.

Something like it came back later in the form of a suggestion by the Aldridge Commission in the summer of 2004 that the Agency consider restructuring its Centers as Federally Funded Research and Development Centers, or FFRDCs.[3] Post–World War II creations, FFRDCs were essentially government labs with a national security mission, run by private organizations.[4] The many national laboratories operated on behalf of the Department of Energy (Brookhaven, Los Alamos, Lawrence Livermore, Oak Ridge, etc.) and some of the military labs managed for the Department of Defense would be comparable to what could be done with NASA Centers. FFRDCs were not strangers to NASA, which had inherited one in the form of the Jet Propulsion Laboratory (JPL), managed by the California Institute of Technology.

An alternative Commission suggestion was the University Affiliated Research Center model, an academic organization that did research for the DOD, such as the Applied Physics Laboratory at Johns Hopkins University, which had designed numerous science satellites and experiments. Another was the Institute model, such as the Space Telescope Science Institute, also at Hopkins, or the NASA Astrobiology Institute (1998–2019) at Ames.[5] This linked different science organizations and universities, in a virtual sense, to one small central group within the space agency.

The FFRDC idea in particular was not met with an outpouring of support anywhere. To the extent that websites such as NASAwatch.com are typical, the idea was rejected as being open to "revolving door" abuse in terms of NASA employees going to industry and vice-versa, or being likely to increase the number of no-bid contracts and thus Agency costs. Furthermore, what would be the point in contracting out essentially all of a Center's jobs when NASA already spent 95 percent of its money in the private sector? Similar sentiments could be found in a 2007 CRS report that cited conflict-of-interest issues such as interlocking directorships, special relationships with the managing organizations, and unfair advantage over private corporations in their lack of taxation, insider information, and risk-proof status.[6]

At the end of 2005, it remained to be seen what, if anything, NASA would do with such suggestions. If Michael Griffin still stood at the helm past the end of the second Bush administration in 2009, and if he or any other NASA administrator had ejected from the NASA lifeboat all of those programs with the weakest link to the Vision, something like a "re-NACAization" might come about in the process. Functions such as life sciences research or medicine might conceivably be given over to other entities, perhaps analogous to amputation saving a life.

By 2020, NASA was still standing and intact, even after the one-two punch of the Great Recession and the end of shuttle access to LEO.

Several presidential administrations and Congresses had made their marks in terms of policy and the ISS was celebrating its twentieth anniversary of human occupancy. In some respects, however, the ISS had evolved into an entity that harkened back to the 1980s, a platform through which the United States might commercialize space research in low-Earth orbit. This had come about in part because wealthy entrepreneurs who wanted to experience space themselves had, over the course of the 2010s, created new, private means to access the station. Additionally, generous government subsidies had allowed many business startups and universities to try out LEO as a laboratory. Some of those entities were addressing issues that had long interested NASA, such as bone and muscle loss, but with the goal of creating a product that could profitably solve problems for the Earthbound as well.[7]

Looking back through budget proposals and annual reports, one can conclude that each administration (Bush, Obama, Trump, and Biden) concurred that NASA ought to return Americans to the Moon and then head toward Mars. It should continue to explore the solar system, albeit in search of economic resources as well as scientific knowledge. It should seek profit in LEO. The question of what kind of governmental agency could accomplish all of these objectives (including the increased military presence in space that President Donald Trump envisioned) became a topic of policy discussion. The National Space Council in 2020 saw government as a partner of industry, perhaps only as an advisor, a patron of R&D, or an "anchor tenant" of US facilities in space. Certainly, the government should also regulate safety issues and fulfill international agreements.[8] NASA's Aerospace Safety Advisory Panel noted in its 2020 annual report that the Agency had "evolved"—it was buying more and building or executing less. The growth of commercial space had pushed NASA to "tactically adapt," but it needed to "strategically and thoughtfully position itself and the nation for success and safe human exploration of space." Although it recommended a separate government agency for "civil space traffic management," meaning orbital debris, satellite traffic, and micrometeoroids, the Advisory Panel's chief solution was to refine the Agency on every front.[9]

NASA itself seemed to be pondering a retreat, announcing in early 2021 a huge increase in prices charged for work done on orbit and for training and hosting expected industry astronauts.[10] If that was a signal that it believed there to be too much on its plate, another option may yet emerge, the more narrowly defined National Aeronautics and

Space Administration that George Low considered. NASA—forever and always a leading-edge engineering organization—would identify health, habitation, and safety issues, but let industry find and implement solutions. It would deploy complex transportation and exploration systems, including habitats and spacesuits, that it had not created. It would send a fleet of robots to scrutinize the sterile environment of space. It would maintain a national astronaut corps that would take the United States to the Moon and Mars, and would continue to inspire, but with a narrow focus that advanced the interests of the United States and its allies.[11] If that came to pass, where would the components of what had always been a loosely conglomerated space life sciences program go besides into the history books?

Notes

NASA creates and has been the subject of many publications not listed in the bibliography. Examples include contractors' reports, press kits, internal newsletters, brochures, educational material, oral histories, meeting minutes, PowerPoint files, and email. Some can be located or accessed online. For example, the JSC History Portal https://historycollection.jsc.nasa.gov/JSCHistoryPortal/history/index.html links to oral history transcripts, NASA image galleries, and a searchable index for the JSC History Collection at the University of Houston–Clear Lake. The NASA History Program Office's Document Management System at https://historydms.hq.nasa.gov/ allows access to press resources and mission transcripts. Much original material, including papers of NASA Administrators, are in government or university archives. Items collected by an individual employee might be in corporate, local, or specialty archives. The NASA HQ Archives houses some collections compiled by historians doing contract research for the History Division, including material used in this book. Files are accessible by appointment. Websites here were accessed Nov. 2, 2021.

Introduction

1. Doc. II-16, "President's Science Advisory Committee 'Introduction to Outer Space,'" Mar. 6, 1958, pp. 1–2, 6, 13–15" and Doc. II-17, "'National Aeronautics and Space Act of 1958,' Public Law 85–568, 72 Stat., 426," July 29, 1958, in *Exploring the Unknown: Select Documents in the History of the U.S. Civil Space Program*, Vol. I: *Organizing for Space*, SP-4218, edited by John M. Logsdon et al. (Washington, DC: NASA, 1995), 332–334 and 334–335.

2. See for example Steven J. Dick and James E. Strick, *The Living Universe: NASA and the Development of Astrobiology* (New Brunswick, NJ: Rutgers Univ. Press, 2004).

3. See Joseph Corn, *The Winged Gospel: America's Romance with Aviation, 1900–1950* (New York: Oxford University Press, 1983) for public "airmindedness," progressivism, and federal policy on commercial aviation in the early years of aviation and NACA.

4. Joan Vernikos, "Life Sciences in Space," in *Exploring the Unknown*, Vol. VI: *Space and Earth Science*, edited by John M. Logsdon et al. (Washington, DC: GPO, 2004), 274.

5. John A. Pitts, *The Human Factor: Biomedicine in the Manned Space Program to 1980* (Washington, DC: NASA, 1985).

6. Framing first the shuttle as new and progressive but reliable and practical enough for the business community, and secondly the space station as offering productivity and presence for business and science is discussed in Valerie Neal, "Framing the Meanings of Spaceflight in the Shuttle Era," *Societal Impact of Spaceflight*, NASA SP-2007-4801, edited by Steven J. Dick and Roger D. Launius (Washington, DC: NASA, 2007), 67–88.

7. Louis Ostrach, interview with the author, June 13, 2006.

8. Judy A. Rumerman, comp., *NASA Historical Data Book, SP-4012*, Vol. VI, *NASA Space Applications, Aeronautics and Space Research and Technology, Tracking and Data Acquisition/Support Operations, Commercial Programs, and Resources, 1979–1988* (Washington, DC: NASA, 1999), 370, "Cooperative Agreements Between NASA and the Private Sector" (Table 5–15); Astronaut Bio: Charles D. Walker, MDC Payload Specialist, Feb. 1999, NASA, https://www.nasa.gov/sites/default/files/atoms/files/c-walker.pdf (accessed May 10, 2020).

9. "Untrained" here means individuals who were not professional astronauts. This included the first teacher in space (Christa McAuliffe), the proposed journalist in space, and the senator (Jake Garn, R-UT) and congressman (Bill Nelson, D-FL) aboard shuttle flights. Chapter 1 discusses in detail the experience of the first "industry astronaut," Payload Specialist Charles Walker.

10. This 1990s management ideology began with the goal of making and deploying satellites and planetary rovers in timely fashion and with no "bloat," gathering significant new scientific data as inexpensively as possible. Howard McCurdy outlined the implementation and outcome of this approach in *Faster, Better, Cheaper: Low-Cost Innovation in the U.S. Space Program* (Baltimore: Johns Hopkins University Press, 2003).

11. *Bioastronautics Roadmap: A Risk Reduction Strategy for Human Space Exploration*, NASA/SP-2004-6113 (Houston: NASA-JSC, 2005), "previously published under JSC 62577." A note stated: "This is a living document that undergoes periodic revision" and gave a URL that referred visitors (as of Sep. 19, 2011) to http://humanresearchroadmap.nasa.gov for the renamed *Human Research Roadmap*.

12. Vernikos, "Life Sciences in Space," 296.

Chapter 1. Everyone's a Scientist: Students, Industry, and Partners in Space

1. Folder "NASA Daily Activities Reports—1981," box 1 (hereafter b.), Ctr. Files, NASA Mgmt. Docs., NASA Dly. Actv. Repts. 1980–81, JSC History Coll., UHCL.

2. Emily Morey-Holton et al., *NASA Newsletters for the Weber Student Shuttle Involvement Project*, NASA-TM-101001 (Moffett Field, CA: NASA-ARC, 1988), 1.

3. Morey-Holton et al., *NASA Newsletters*, 6, 9, 11. An adjuvant enhances natural immune response to an antigen. Freund's serum provoked polyarthritis, a multijoint autoimmune disorder including rheumatoid and psoriatic arthritis and lupus erythematosus.

4. Morey-Holton et al., *NASA Newsletters*, 2, 11, 309; Emily Morey-Holton, interview with the author, Sep. 16, 2005.

5. Morey-Holton et al., *NASA Newsletters*, 1, 11, 51–52, 106, 128, 135.

6. Morey-Holton et al., *NASA Newsletters*, 9–10; Morey-Holton et al., "The Hindlimb Unloading Rat Model: Literature Overview, Technique Update and Comparison with Space Flight Data," in *Experimentation with Animal Models in Space*, edited by G. Sonnenfeld (Amsterdam: Elsevier, 2005), 7–40.

7. Morey-Holton et al., *NASA Newsletters*, 9, 235, 237, 241–242, 247, 255, 277.

8. Ibid., 9–10.

9. Technically it was listed as an SSIP *and* a DSO. As DSO 421, results were compiled and posted on JSC's online Life Sciences Data Archive (LSDA), showing results of blood tests and changes in body weight. See https://lsda.jsc.nasa.gov/Experiment/exper/554 (accessed May 26, 2020).

10. Morey-Holton et al., *NASA Newsletters*, 71, 84–86, 88, 90–91, 95, 104, 129, 130, 154, 188; Morey-Holton interview.

11. Morey-Holton et al., *NASA Newsletters*, 52–53, 75–76, 80, 104–106, 123, 130, 137, 244.

12. P. D. Sebesta to John Bryant, Dec. 22, 1983 in Morey-Holton et al., *NASA Newsletters*, 136–137, 150.

13. Morey-Holton et al., *NASA Newsletters*, 75, 139.

14. Ibid., 128, 156, 157, 256.

15. Mary E. Kirchen et al., "Effects of Microgravity on Bone Healing in a Rat Fibular Osteotomy Model," *Clinical Orthopaedics and Related Research*, no. 318 (1995): 231–242; "The Effects of Weightlessness in Spaceflight on the Healing of Bones (SSIP-29)," LSDA, Johnson Space Center, https://lsda.jsc.nasa.gov/Experiment/exper/273 (accessed May 26, 2020).

16. John Glisch, "Four Rats' Legs to Be Broken for Discovery Experiment," *Orlando Sentinel*, Jan. 12, 1989 (synopsis); UPI, "Breaking Rats' Legs Could Save Astronauts," *Antelope Valley Press*, Jan. 8, 1989; David Foy, "Chance of a Lifetime: Trade a Swine for Rat," *AVP*, Jan. 13, 1989, all in folder (hereafter f.) "Life Sci. Biological Payloads US/USSR (Animals in Space) to 1997," RN 9870, Life Sciences Coll., NASA Hist. Ref. Coll., Hist. Prog. Off., NASA HQ, Washington, DC.

17. Thora W. Halstead to AA for Space Sci. and Apps., Mar. 8, 1989, f. "Life Sci. Biological Payloads US/USSR (Animals in Space) to 1997," RN 9870, NASA Hist. Ref. Coll.

18. Fras's first experiment, "The Effect of Weightlessness on the Aging of Brain Cells," flew on STS-51D in 1985. The LSDA contains no data or results for this experiment.

19. George G. Gerondakis, *Get Away Special (GAS) Educational Applications of Space Flight*, TM-101267 (Washington, DC: NASA, 1989), 15. Gerondakis reported that "some students that have graduated attribute the association with the GAS program as a motivator for their careers in engineering and sciences."

20. See Thora W. Halstead and Patricia A. Dufour, *Biological and Medical Experiments on the Space Shuttle 1981–1985*, TM-108025 (Washington, DC: NASA-OSSA, 1986) for the first 24 SSIP and GAS experiments by US K-12 students. Subjects included insects, humans, sponges (*Microciona porifera*), plants, seeds, rats, brine shrimp (*Artemia*), planaria (*Dugesia tigrina*), bacteria, mold, algae (*Chlorella*), and yeast (*Kefir*). See also

Man/Systems Division, Shuttle Student Involvement Program (SSIP) Final Reports of Experiments Flown, NASA/JSC Internal Note, JSC 24005, Oct. 20, 1989, 1–2.

21. "We Are Professional Inventors: The Egg Came First," Techshot, https://techshot.com/aerospace/about-us/ (accessed June 8, 2020). Techshot is a NASA contractor.

22. "RTQ for STS-29, Space Shuttle Student Involvement Program (SSIP) Experiments, Final Revision 3/3/89," f. "Life Sci. Biological Payloads US/USSR (Animals in Space) to 1997," RN 9870, NASA Hist. Ref. Coll.

23. NASA Office of HR and Educ., "SSIP . . . Education at its Best!" (Arlington, VA: NASA, 1997), NAS 1.2:SCI 2/13/997.

24. Thomas O'Toole, "Space Ants Display No Signs of Life," *Washington Post*, July 7, 1983, A2; David Garrett et al., "STS-7 Experiments," Space Shuttle Mission STS-7 Press Kit, June 1983, NASA Release No. 83–87, edited by Richard W. Orloff (Jan. 2001), 49; Halstead and Dufour, *Biological and Medical Experiments on the Space Shuttle 1981–1985*, 30.

25. Gerondakis, *Get Away Special (GAS) Educational Applications of Space Flight*, 1–3, 10–11, 14–15.

26. [Notice: 05-045] RIN 2700-AC39: NASA Final Rule: Small Self-Contained Payloads (SSCPs) STATUS REPORT, NASA HQ, Mar. 18, 2005, in *Fed. Reg.*, Mar. 17, 2005, Vol. 70, No. 51, Rules and Regulations, 12966, *Federal Register Online* https://www.federalregister.gov/documents/2005/03/17/05-5089/small-self-contained-payloads-sscps (accessed Oct. 10, 2021).

27. E. Anderson, "Shuttle Student Involvement Program (SSIP): Benefits to Education," draft, Feb. 2, 1989, f. "Life Sci. Biological Payloads US/USSR (Animals in Space) to 1997," RN 9870, NASA Hist. Ref. Coll. Of the fifty-seven, "at least thirty-five" focused on the life sciences.

28. "Reach for the Stars Through NASA's Space Science Student Involvement Program," July 30, 1991, f. "Life Sci. Biological Payloads US/USSR (Animals in Space) to 1997," RN 9870, NASA Hist. Ref. Coll.

29. "RTQ for STS-29."

30. "SSIP . . . Education at its Best!"

31. Halstead and Dufour, *Biological and Medical Experiments on the Space Shuttle 1981–1985*, 2.

32. Special Payloads Division, GSFC, *Get Away Special . . . the First Ten Years*, NASA-TM-102921 (Greenbelt, MD: GSFC, 1989), 4 states: "An unusual feature of the GAS Program is that experimenters are not required to furnish postflight reports to NASA. NASA feels that GAS customers can best speak for their own experiments." See e.g., Halstead and Dufour, *Biological and Medical Experiments on the Space Shuttle 1981–1985*, 16–19, 23, 28–30.

33. Fras, as a coauthor of Kirchen et al., "Effects of Microgravity on Bone Healing in a Rat Fibular Osteotomy Model," *Clinical Orthopaedics and Related Research* 318 (Sep. 1995): 231–232, was cited in an article on traumatic injury in space, per the PubMed Central database.

34. Glynn Lunney interview, Houston, TX, by Carol Butler, Dec. 9, 1999, NASA JSC

Oral History Collection (OHC), https://historycollection.jsc.nasa.gov/JSCHistoryPortal/history/oral_histories/LunneyGS/lunneygs.htm (accessed May 26, 2020).

35. Tables 1, "Domestic R&D and R&D Abroad, PhRMA Member Companies, 1970–2005," and 2, "R&D as a Percentage of Sales, PhRMA Member Companies, 1970–2005," in Pharmaceutical Industry Profile, 2006 (Washington, DC: PhRMA, 2006), 44–45, http://www.phrma.org/profiles_%26_reports/ (accessed Sep. 29, 2006). By Mar. 21, 2011 this was no longer available online.

36. Henry R. Hertzfeld, "Space as an Investment in Economic Growth," in Exploring the Unknown, Vol. III: Using Space, edited by John M. Logsdon et al. (Washington, DC: NASA, 1998), 393–396, 475–476.

37. Thomas A. Sullivan, "Electrophoresis Demonstration, Apollo 14" and "Electrophoresis Demonstration, Apollo 16," in Catalog of Apollo Experiment Operations, NASA RP-1317 (Houston: NASA-JSC, 1993), history.nasa.gov/alsj/RP-1994-1317. pdf; Electrophoresis Technology (MA-011) and Electrophoresis Experiment (MA-014), LSDA, https://lsda.jsc.nasa.gov/Experiment/exper/399 and https://lsda.jsc.nasa. gov/Experiment/exper/403; R. S. Snyder, Electrophoresis Demonstration on Apollo 16, NASA TM X-64724 (Huntsville: NASA-MSFC, Nov. 1972), NASA Technical Reports Server (NTRS), ntrs.nasa.gov/archive/nasa/casi.ntrs.nasa.gov/19730009430.pdf (all accessed May 26, 2020).

38. A. V. Pertsov and E. A. Zaitseva (Baum), "Discovery of Electrokinetic Phenomenon in Moscow University," plenary address, III Int'l. Conf. on Colloid Chemistry and Physicochemical Mechanics, Moscow, June 24–28, 2008, http://www.icc2008.ru/en/conference/reuss.htm (accessed May 26, 2020). The process was based on the molecules' sizes and polarities.

39. "Space Processing Studies Underway for Use in '80s," Marshall Star, Apr. 3, 1974, NASA George C. Marshall Space Flight Center, 2, 4.

40. Hertzfeld, "Space as an Investment in Economic Growth," 395.

41. Judy A. Rumerman, "The McDonnell Douglas Corporation," Born of Dreams—Inspired by Freedom, US Centennial of Flight Commission, http://centennialofflight. net/essay/Aerospace/McDonnell/Aero31.htm (accessed June 8, 2020).

42. Charles D. Walker interview, Washington, DC, by Jennifer Ross-Nazzal and Rebecca Wright, Nov. 19, 2004, JSC OHC, https://historycollection.jsc.nasa.gov/JSCHistoryPortal/history/oral_histories/WalkerCD/WalkerCD_11-19-04.htm (accessed Oct. 17, 2021).

43. Hertzfeld, "Space as an Investment in Economic Growth," 396; "Feasibility of Commercial Space Manufacturing: Production of Pharmaceuticals, Final Report," Vol. I: Executive Summary, MDC E2104, MDAC, St. Louis Div., Contract NAS8-31353, Nov. 8, 1978, 26–30, 103, Doc. III-23 in Exploring the Unknown, Vol. III: Using Space, edited by Logsdon et al., 534–539.

44. Walker interview, Nov. 19, 2004; Percy H. Rhodes and Robert S. Snyder, Preparative Electrophoresis for Space, NASA TP-2777 (Huntsville, AL: NASA-MSFC, 1987), 1.

45. Walker interview, Nov. 19, 2004.

46. Hertzfeld, "Space as an Investment in Economic Growth," 394; Walker interview, Nov. 19, 2004; Lunney interview.

47. "Space Science," Nov. 5, 1980 and *Daily Activity Report*, Aug. 15, 1980, 1–2, f. "NASA Daily Activities Repts.—1980," b. 1, NASA Mgmt. Docs., NASA Daily Activity Repts. 1980–81, Ctr. Files, JSC Hist. Coll.; Walker interview, Nov. 19, 2004.

48. Walker interview, Nov. 19, 2004.

49. Walker interview, Nov. 19, 2004. Walker's first flight as Payload Specialist, aboard STS-41D in 1984, was delayed for over two months.

50. Walker interview, Nov. 19, 2004. Boeing acquired MDAC in 1997.

51. Robert S. Snyder et al., *Continuous Flow Electrophoresis System Experiments on Shuttle Flights STS-6 and STS-7*, NASA TP-2778 (Huntsville, AL: NASA-MSFC, 1987), 1.

52. Walker interview, Nov. 19, 2004; Elvin B. Pippert, *Flight Data File Crew Activity Plan STS-4 JSC-17878*, NAS 1.15: 85502 (Houston: NASA-JSC, FOD, Ops. Div., May 14, 1982), 4-15-4-21, 4-53-4-61, 5-17-5-21, 7-9. L+0 and L+3 refer to the number of days after launch, so L+0 means the launch day itself.

53. Snyder et al., *Continuous Flow Electrophoresis System Experiments on Shuttle Flights STS-6 and STS-7*, 1.

54. Ibid., 1–2, 4; Walker interview, Nov. 19, 2004.

55. Rhodes and Snyder, *Preparative Electrophoresis for Space*, 1, 9.

56. Snyder et al., *Continuous Flow Electrophoresis System Experiments on Shuttle Flights STS-6 and STS-7*, 10.

57. Percy H. Rhodes and Robert S. Snyder, MSFC, patent no. 4,752,372, "Moving Wall, Continuous Flow Electronphoresis [*sic*]," granted June 21, 1988, patent app. no. 904,128, filed Sep. 5, 1986, http://ntrs.nasa.gov/archive/nasa/casi.ntrs.nasa.gov/19880014461.pdf (accessed June 8, 2020).

58. Walker interview, Nov. 19, 2004; "McDonnell Engineer Carrying Shuttle Secret," *St. Louis Post-Dispatch*, May 23, 1984; "Industrial Astronaut to Fly with Biological Machine," *AWST*, June 18, 1984, pgs. 67, 69; "Countdown Starts for Shuttle's Flight," June 24, 1984, and Martha Shirk, "Area's History-Making Astronaut Still Flying High on 'Discovery,'" both *St. Louis Post-Dispatch*, Sep. 13, 1984, pgs. 1, 6.

59. Walker interview, Nov. 19, 2004.

60. Charles D. Walker interview, Washington, DC, by Jennifer Ross-Nazzal, Mar. 17, 2005, JSC OHC, https://historycollection.jsc.nasa.gov/JSCHistoryPortal/history/oral_histories/WalkerCD/WalkerCD_3-17-05.htm (accessed Oct. 10, 2021).

61. Walker interview, Nov. 19, 2004.

62. Glyn O. Roberts, *Analysis of Electrophoresis Performance*, Final Report, NASA-CR-171034, Apr. 6, 1984, 1, 1–1.

63. Charles D. Walker interview, Houston, TX, by Sandra Johnson, Apr. 14, 2005, JSC OHC, https://historycollection.jsc.nasa.gov/JSCHistoryPortal/history/oral_histories/WalkerCD/WalkerCD_4-14-05.htm (accessed Oct. 10, 2021).

64. Walker interview, Apr. 14, 2005; NASA, Space Shuttle Press Kit, STS-51D, Apr. 1985, 17.

65. Rumerman, *NASA Historical Data Books*, Vol. VI, 355; "Medicine Sales Forecast at $1 Billion," *AWST*, June 25, 1984, pgs. 52–56. James Rose of MDAC took over the job from Isaac Gillam of DFRC in Oct. 1987.

66. W. Henry Lambright, "The NASA-Industry-University Nexus: A Critical Alliance in the Development of Space Exploration," in *Exploring the Unknown*, Vol. II: *External Relationships*, edited by Logsdon et al., 432.

67. "NASA Commercial Space Policy, October 1984," Doc. III-27 in Lambright, "The NASA-Industry-University Nexus," 573.

68. Walker interview, Nov. 19, 2004; "Medicine Sales Forecast at $1 Billion," 52–53; Rick Stoff, "McDonnell Looks to Factories in Space," *St. Louis Globe-Democrat*, Apr. 16, 1984; "Space Industries Plans to Develop Orbiting Facility with Engineering Firm," *AWST*, Sep. 2, 1985, pg. 26.

69. "NASA Commercial Space Policy, Oct. 1984," Doc. III-27 in Lambright, "The NASA-Industry-University Nexus," 574.

70. "Medicine Sales Forecast at $1 Billion," 52–55 and "Lovelace Studies Cancer Research in Space," *AWST*, June 25, 1984, pg. 55. Lovelace was doing monoclonal antibody research on cancer and malaria.

71. NASA, Space Shuttle Mission STS-51D Press Kit, Apr. 1985, 16.

72. NASA, Space Shuttle Mission STS-61B Press Kit, Nov. 1985, 33.

73. Walker interview, Nov. 19, 2004 and Charles D. Walker interview, Springfield, VA, by Sandra Johnson, Nov. 7, 2006, JSC OHC, https://historycollection.jsc.nasa.gov/JSCHistoryPortal/history/oral_histories/WalkerCD/WalkerCD_11-7-06.htm (accessed Oct. 10, 2021).

74. Memorandum, Mar. 21, 1985, f. "March 1985," b. 3, Rdg. Files, July 1984–June 1985, Charlesworth Files, Ctr. Series, JSC Hist. Coll.

75. Reports from the Payload Specialist Liaison Officer (PSLO), Jan. 22 and Feb. 14, 1986, f. "Space Ops. Dir. Wkly. Activity Repts. Oct. 1985–Feb. 1986," b. 7, Charlesworth Files, Ctr. Series, JSC Hist. Coll.; Walker interview, Nov. 7, 2006. Five journalist-astronaut applicants had been scheduled the previous week for trips to JSC in Apr. for medical tests, tours, and a flight aboard a KC-135.

76. Ann Bradley to Robert Hoskins, in memorandum A. Ladwig to E. Johnson, July 2, 1986; draft and unsigned letter Albert T. Scroggins to "15 National Panelist[s]," July 11, 1986, Albert T. Scroggins Papers, The South Caroliniana Library, University of South Carolina, Columbia.

77. Astronaut Bio: Barbara R. Morgan, July 2010, NASA, https://www.nasa.gov/sites/default/files/atoms/files/morgan_barbara.pdf (accessed June 8, 2020).

78. Report from the PSLO, Feb. 19, 1986, f. "Space Ops. Dir. Wkly. Activity Repts. Oct. 1985–Feb. 1986," b. 7, Charlesworth Files, Ctr. Series, JSC Hist. Coll.

79. Fu-Kuen Lin et al., "Cloning and Expression of the Human Erythropoietin Gene," *Proceed. Natl. Acad. Sci.* 82 (Nov. 1985): 7580–7584. Amgen marketed Epoietin Alfa under the trade name Epogen. Ortho Biotech Products, the renamed Johnson & Johnson subsidiary, marketed it under license to Amgen as Procrit.

80. Walker interview, Nov. 19, 2004; Philip Culbertson interview, Arlington, VA, by Adam L. Gruen, Feb. 9, 1988, NASA History Program Office, Washington, DC; "Ortho Div. Drops Initial Effort to Develop Medicine in Space" and "Biological Processor Will Produce New Drugs on Shuttle," *AWST*, Sep. 16, 1985, 20–21.

81. Doc. I-31, Memorandum, Kenneth S. Pedersen, Washington, DC, n.d. [Aug.

1982], in *Exploring the Unknown*, Vol. II: *External Relationships*, edited by Logsdon et al., 93, 96.

82. Doc. I-34, Beggs to Kenneth Baker, Apr. 6, 1984, in *Exploring the Unknown*, Vol. II: *External Relationships*, edited by Logsdon et al., 109.

83. Doc. I-35, Article 2, Sec. 2.3, Article 3, Sec. 3.3, "MOU Between the U.S. NASA and the ESA on Cooperation in the Detailed Design, Development, Operation and Utilization of the Permanently Manned Civil Space Station," in *Exploring the Unknown*, Vol. II: *External Relationships*, edited by Logsdon et al., 112–113, 116.

84. R. P. Rauch, "Industrial Space Facility (ISF) Technical Status Briefing," Jan. 11, 1984, MDAC, rec. no. 19955, b. 2 "McDonnell Douglas Space Station Concepts," Maura Mackowski Life Sciences Collection, NASA Hist. Ref. Coll., Hist. Prog. Off., NASA HQ, Washington, DC.

85. Culbertson interview; Craig Covault, "NASA Approves Fly-Now, Pay-Later Plan for Orbiting Industrial Facility," *AWST*, Aug. 26, 1985, 16–17.

86. "SII/Westinghouse to Offer Materials Processing Facility in '90," *Defense Daily*, Sep. 30, 1986, 155; "Space Industries, Westinghouse Partnership Spurs New Marketing Campaign," *AWST*, Oct. 6, 1986, 25–26; Covault, "NASA Reaffirms Strong Support for Space Commercialization," *AWST*, Oct. 5, 1987, 31.

87. Donald G. James, "NASA Ames to Build Mockup of Proposed Private Sector Space Laboratory," press release (hereafter pr.) 87–19, Mar. 23, 1987, NASA-Ames History Office, Moffett Field, CA; Craig Covault, "NASA Approves Fly-Now, Pay-Later Plan," *AWST*, Aug. 26, 1985, 16–17; Wespace, "ISF Update," 4th qtr. 1987, 3.

88. Culbertson interview; Caldwell C. Johnson and Maxime A. Faget, pat. no. 4,903,919, "Apparatus and Method for Docking Spacecraft," granted Feb. 27, 1990, app. no. 07/125,993, Nov. 27, 1987, http://patft.uspto.gov (accessed June 8, 2020); "Fairchild Seeks Agreements on Leasecraft," *AWST*, June 25, 1984, 54; Stoff, "McDonnell Looks to Factories in Space"; Theresa M. Foley, "U.S. Opens Low Earth Orbit to Commercial Development," *AWST*, Feb. 22, 1988, 21–22 and "U.S. Space Platform Firms Aim for 1991 Service Start," *AWST*, Feb. 29, 1988, 36–38, 41; "Europe Offered Mini-Spacelab," *Flight Int'l.*, Nov. 22, 1986, 28.

89. "Program History, Space Station Freedom," f. 073562 "Redesign of the Space Station," b. 39, RG 255, Records of the National Aeronautics and Space Administration, Office of the Administrator, Records of NASA Administrator Daniel S. Goldin, NARA, College Park, MD, (hereafter Goldin Recs.); Covault, "U.S. Space Program Urged to Support Intl. Commercial Competitiveness" and "NASA Will Consider Use of Commercial Space Facility," *AWST*, Nov. 2, 1987, 55; Culbertson interview.

90. "NASA Use of Industrial Space Platform Ordered by WH Policy Group," 29; Foley, "Congress Directs NASA to Adapt Space Station to Shrinking Budget," *AWST*, Jan. 4, 1988, 32 and "U.S. Opens LEO to Commercial Development," 21.

91. Foley, "Space Station Faces Funding Cutoff in NASA Dispute with Congress," *AWST*, Feb. 1, 1988, 22.

92. Jim Van Nostrand, "Report Hits NASA's Microgravity Materials Program," *Elec. Engrg. Times*, Sep. 14, 1987, 48; Covault, "NASA Delays Space Station Contracts as Industrial Research Interest Grows," *AWST*, Nov. 9, 1987, 30; "WH Group Backs Funds

for ISF," *Defense Daily*, Jan. 12, 1988, 43–44; Faget, "Industrial Space Facility and Space Station," letter, *AWST*, Jan. 18, 1988, 76; Foley, "U.S. Opens LEO to Commercial Development," 21; Bob Davis, "NASA Required to Lease Space on Private Craft," *WSJ*, Feb. 12, 1988, 1; Paulette Thomas, "Lofty Project: Against Sizable Odds a Small Firm is Poised to Run a Space Station," *WSJ*, Mar. 11, 1988, 1, 10; Andrew Lawler, "Space for Rent?" *Space World*, Sep. 1988, 16.

93. Microgravity Science and Applications Division-A Program Overview: 1986–1987, 45, in U.S. Congress. House Committee on Science, Space. *NASA Reports Required by Congress, 1987–1988. Congressional Publications*, 1989 https://play.google.com/books/reader?id=pjnjdilDRSQC&hl=en&pg=GBS.PP3 (accessed June 17, 2020).

94. Van Nostrand, "Report Hits NASA's Microgravity Materials Program," 48; Culbertson interview. Culbertson stated that "scientists in NASA," whom he lumped with Agency engineers "who might use space as a place to verify system design and verify that things work properly" did like the free-flyer idea, but as the first of several steps to the station, not as an end in itself, and having "committed a lot of thinking" to it in that sense, did not want to "revector" their planning. Dunbar's scientists were experimentalists flying payloads and human "experiments."

95. "NASA Use of Industrial Space Platform Ordered by WH Policy Group," *AWST*, Jan. 11, 1988, 29, 31; Foley, "U.S. Opens LEO to Commercial Development," 21–22.

96. "NASA Use of Industrial Space Platform Ordered by WH Policy Group," 29, 31; Foley, "Space Station Faces Funding Cutoff in NASA Dispute with Congress," 22.

97. Foley, "Congress Blocks NASA Plan to Lease Private Platform," *AWST*, Mar. 28, 1988, 16–17.

98. "Presidential Directive on National Space Policy," Feb. 11, 1988, unclassified summary, archived at https://www.hq.nasa.gov/office/pao/History/policy88.html (accessed July 7, 2020).

99. Foley, "U.S. Space Platform Firms Aim for 1991 Service Start," *AWST*, Feb. 29, 1988, 36–38 and "Congress Blocks NASA Plan to Lease Private Platform," 16–17.

100. Foley, "Congress Blocks NASA Plan to Lease Private Platform," 16–17; Lawler, "Space for Rent?" 17.

101. "Congress Questions NASA's Ability to Fund Mix of Orbital Platforms," *AWST*, Apr. 4, 1988, 22–23; *Washington Post*, Mar. 1, 1988 in "MDC Stock Update" (an MDAC newsletter), Mar. 1, 1988, rec. no. 19955, b. 2 "McDonnell Douglas Space Station Concepts," Mackowski Life Sciences Coll.

102. Lawler, "Space for Rent?" 17, 19.

103. Foley, "NASA Advised to Drop Plan for Commercial Platform Lease," *AWST*, Apr. 17, 1989, 20–21.

104. Culbertson interview; Joe Allen, "In Memoriam: Recollections of Max Faget: Colleague, Mentor, and Friend," *Quest* 12, no. 2 (2005): 41–42.

105. James, "NASA Ames to Build Mockup of Proposed Private Sector Space Laboratory," pr. 87–19, Mar. 23, 1987.

106. "Johnson Space Center, Group 2 (Wilkening, Allen, Paine, Herres) Fact-Finding Session," Oct. 22, 1990, f. 5 "Government File: Advisory Committee on the Future of the U.S. Space Program: Agendas and Working Calendars," b. 13; Dick Malow to Paine,

Aug. 26, 1991, f. 4 "Government File, Advisory Committee on the Future of the U.S. Space Program, Correspondence and Memos, 1991," b. 14, Thomas O. Paine Papers, Manuscript Division, Library of Congress (LOC), Washington, DC.

107. Culbertson interview; "Industry Observer," *AWST*, June 22, 1987, 15; "Calls for DOD/NASA Study of ISF," *Def. Daily*, Nov. 2, 1987, 3. Allen blamed ISF's demise on "the glacial pace of progress in NASA space projects through the 1980s."

108. James Beggs interview, Washington, DC, by Ivelisse Rodriguez, Dec. 15, 1982. See also "Table 1, Docs. Reviewed, Chap. III, Attchmt. A, Size of Astronaut Corps in Space Station Era," f. "NASA HQ Ops. Med. Supp. to LDMM in LEO & Beyond, Feb. 1982," b. 7, Ctr. Series–S&LS, JSC Hist. Coll. This contractor-JSC report foresaw 10–11 flights annually through 1991, 20 per year by 1993, and 34 flights by 1998, mainly for station construction and expansion.

109. James Fletcher interview, Washington, DC, by Roger Launius, Sep. 13, 1991.

Chapter 2. Working in the Space Environment

1. Ulf Merbold, Aug. 2002 and Roberta Bondar, July 1997, "Payload Specialists," and Marc Garneau, June 2004, "Partner Astronauts," *Astronaut Biographies*, NASA-JSC, https://www.asc-csa.gc.ca/eng/astronauts/canadian/former/default.asp (accessed Oct. 11, 2021) and Kenneth Money, Feb. 2011, *Canadian Astronauts*, Canadian Space Agency, http://www.asc-csa.gc.ca/eng/astronauts/biomoney.asp (accessed May 27, 2020).

2. SB-3/Mgr., Operational Med, [Arnauld Nicogossian], memo for record, June 1, 1981 and SM-8/Dir., Spacelab Flight Div. [Jesse W. Moore] to multiple addressees, Mar. 18, 1981, f. "NASA/HQ S-Codes Corr OSS, 4-7/81," b. 17, Ctr. Series, HQ Corr., JSC Hist. Coll.

3. Table 164a Statistical Values for Stature and Table 200a Statistical Values for Weight, *Military Handbook, Anthropometry of U. S. Military Personnel (Metric)*, DOD-HDBK-743A (Washington, DC: DOD, 1991), 393, 465.

4. Loyd S. Swenson Jr., James M. Grimwood, and Charles C. Alexander, *This New Ocean: A History of Project Mercury*, SP-4201 (Washington, DC: NASA, 1966), 131.

5. Habitability and Human Factors Office, Space Life Sci. Dir., "Anthropometry and Biomechanics," Vol. I, Sect. 3, *Anthropometry and Biomechanics, Man-Systems Integration Standards*, NASA-STD-3000, Rev. B (Houston: NASA-JSC, 1995), http://msis. jsc.nasa.gov/sections/section03.htm (accessed May 27, 2020). Statistics were for 40-year-olds. No one would be statistically average in every measurable dimension, therefore there was no truly average-sized astronaut.

6. Habitability and Human Factors Office, *Anthropometry and Biomechanics*, Fig. 3.3.1.3-1, notes for figs. 9 and 10 of 12, Sects. 3.3.5.2, 3.3.3.1, 3.3.3.2.1, 3.3.3.2.2, 3.3.5.3, 3.3.3.3.2, 3.3.1.3, and 3.3.4.2. Sitting height was noted to be technically "buttock vertex" in microgravity unless the person was strapped or held tightly downward into the seat.

7. Habitability and Human Factors Office, *Anthropometry and Biomechanics*, Fig. 3.3.5.3-1 and Sects. 3.3.5.2, 3.3.5.3, 3.3.3.2.1, 3.3.6.2. This was among American males only. NASA had no data for Japanese females.

8. "Man in Space: Today and Tomorrow," *USA Today*, Feb. 6, 1984, 4–6, attached to Leslie Lenkowsky to Abrahamson, Feb. 10, 1984, USIA Wireless File, NASA Hist. Ref. Coll.

9. W. E. Hull, chair, and a committee of 15 that included D. S. Nachtwey and Paul Rambaut wrote this untitled report, in f. "Hull, W. E. Life Sciences Guideline Data for Long-Duration Missions, n.d.," b. 8, Ctr. Series—S&LS, JSC Hist. Coll. For scale, the five *hundredth* person in space flew in 2009.

10. "Table 1, Docs. Reviewed, Chap. III, Att. A, Size of Astronaut Corps in Space Station Era"; S. Furukawa, "Preliminary Assessment of Needs, NASA Op. Med. in Space Station," Mar. 5, 1982, f. "Furukawa," b. 7, Ctr. Series—S&LS, JSC Hist. Coll.

11. Sudhakar L. Rajulu and Glenn K. Klute, *Anthropometric Survey of the Astronaut Applicants and Astronauts from 1985 to 1991*, NASA-RP-1304 (Houston: NASA-JSC, 1993), viii, ix, 1, 2. The survey contained data from 473 individuals, 399 males and 74 females. There were 277 mission specialists, 47 payload specialists, 120 pilots, and 9 observers. Of the 473, NASA chose 82 to become astronauts, including 12 female and 42 male mission specialists, 1 female and 28 male pilots, and the 9 observers. The authors included no data regarding female pilots due to the sample size.

12. Larry Toups, interview with the author, May 17, 2006; Stephen Capps and Nathan Moore, NASA/JSC-23848 Internal Note, "Lunar Base 'Construction Shack,'" 17. M-S Div., Aug. 25, 1989, f. "Lunar Base 'Construction Shack,' JSC-23848 Aug. 25, 1989," b. 6, Ctr. Series—Hab. Studies, JSC Hist. Coll. This study spoke of a crew of nine, but a 1988 University of Texas study of the "shack" crew transporter was based on a crew of eight. See Wallace T. Fowler, "Final Design for a Lunar Construction Shack Vehicle," NASA-CR-184754 (Houston: NASA-JSC, 1988), 2, available on the NASA Technical Reports Server, ntrs.nasa.gov/archive/nasa/casi.ntrs.nasa.gov/19890006704.pdf (accessed May 27, 2020).

13. Capps and Moore, "Lunar Base 'Construction Shack,'" 13–15; Fowler, "Final Design for a Lunar Construction Shack Vehicle," 2, 10, 16.

14. Capps and Moore, "Lunar Base 'Construction Shack,'" 6–9; Nathan Moore, interview with the author, May 19, 2006.

15. Martha E. Evert to Moore and David J. Gutierrez, July 8, 1991, f. "Action Item 91-12, 7/2/81," b. 8, Ctr. Series—Hab. Studies, JSC Hist. Coll.; Moore interview.

16. "Initial Lunar Habitat Simulation: 1st Crew Shift Results," n.d. [1991], f. "Initial Lunar Habitat," b. 8, Ctr. Series—Hab. Studies, JSC Hist. Coll.

17. Handwritten, "Test Results, KC-135," in f. "Team Meeting Materials, Oct. 1, 1991," b. 8 Ctr. Series—Hab. Studies, JSC Hist. Coll.

18. H. Reimers and Evert, PowerPoint, M-S Div., Oct. 1991, summary of presentation by Judith Aylward, US Army, Natick, MA, Sep. 26, 1991, f. "Team Meeting Materials, Oct. 1, 1991," b. 8, Ctr. Series, Hab. Studies, JSC Hist. Coll.

19. Isabel S. Abrams, "Beyond Night and Day," *Space World*, Dec. 1986, 12–13; C. M. Winget et al., "Comparison of Circadian Rhythms in Male and Female Humans," *Waking Sleeping* 1 (1977): 359–363.

20. Frederick W. Rudge, "Relationship of Menstrual History to Altitude Chamber

Decompression Sickness," paper presented at 69th AGARD Aerospace Medical Panel Symposium, Tours, France, Apr. 4–5, 1990, reprinted in *ASEM* 61 (1990): 657–659.

21. Victor A. Convertino, "Gender Differences in Autonomic Functions Associated With Blood Pressure Regulation," *Am. Jnl. Regltry. Integrtv. Compar. Physiol.* 275 (1998): R1909–1920; Mary M. Connors, *Living Aloft: Human Requirements for Extended Spaceflight*, SP-483 (Moffett Field, CA: NASA-ARC, 1985), 29, http://history.nasa.gov/SP-483/ch2-2.htm (accessed May 27, 2020).

22. Convertino, "Gender Differences in Autonomic Functions Associated with Blood Pressure Regulation," R1909–1920. Studies did not result in a different suit for females.

23. C. J. Fenrick, pr. 84-17, "NASA to Test Athletic Males in Simulated Weightlessness," Mar. 6, 1984, NASA ARC News Releases 1984, NASA Ames History Office, Moffett Field, CA.

24. Peter Waller and Linda Blum, pr. 85-48, "NASA-Developed Analyzer May Aid Bone Treatments on Earth," Dec. 5, 1985, NASA ARC News Releases 1984, NASA Ames History Office, Moffett Field, CA.

25. John Glenn with Nick Taylor, *John Glenn: A Memoir* (New York: Bantam, 1999), 359–360, 364–366, 369. NIA was established as a component of the NIH in 1974.

26. Maura Phillips Mackowski, *Testing the Limits: Aviation Medicine and the Origins of Manned Space Flight* (College Station: Texas A&M University Press, 2006), 199–213; Margaret A. Weitekamp, *Right Stuff, Wrong Sex: America's First Women in Space Program* (Baltimore: Johns Hopkins University Press, 2004), 180–183; AP, "NOW Starts Petition Drive to Orbit Female Ex-Trainee," Oct. 30, 1998; Yvonne Latty, "First Female Astronaut Candidate Tries Again to Get into Space," *Houston Chronicle*, Nov. 15, 1998, A39; Michael Precker, "Women Who Passed Test Never Got Chance in Space; John Glenn Encore Stirs Old Longings," [New Orleans] *Times-Picayune*, Oct. 11, 1998, A20. Cobb and two other women also passed sensory deprivation tests at an FAA facility and Cobb succeeded at in-flight tests at Pensacola NAS, site of the Navy's aviation medicine facility. Officially the program ended when the Navy cancelled the flight tests because NASA was unable to state that it had a specific need or mission for female astronauts at that time.

27. *A Mission of Discovery*, Shuttle Press Kit STS-95, NASA, Oct. 27, 1998, Space Shuttle Press Kits, https://historycollection.jsc.nasa.gov/JSCHistoryPortal/history/shuttle_pk/1998.htm (accessed June 19, 2020).

28. Glenn and Taylor, *John Glenn: A Memoir*, 404. Glenn was to drink five 8-oz. (0.24 L) pouches of a salty lemon-lime solution during reentry and recalled drinking at least three. After landing, Glenn did not pass out, but he did get sick to his stomach and reported that it felt as though the drink had never been absorbed, but just sat there.

29. *A Mission of Discovery*, 127, 129, 135, 136, 138, 139, 140; Glenn and Taylor, *John Glenn: A Memoir*, 375, 377.

30. *A Mission of Discovery*, 110; Glenn and Taylor, *John Glenn: A Memoir*, 377.

31. *A Mission of Discovery*, 128, 130.

32. *A Mission of Discovery*, 79; Glenn and Taylor, *John Glenn: A Memoir*, 375, 377–378,

387; John Bluck, pr. 98-65AR, "Miniaturized Transmitter to Be Used in Efforts to Save Babies," Nov. 18, 1998, http://www.nasa.gov/home/hqnews/1998/98-208.txt (accessed June 19, 2020). ARC engineers developed the pill in cooperation with the Fetal Treatment Center at the University of California–San Francisco to monitor the health of unborn infants after in-utero surgery.

33. A. C. Rossum et al., "Effect of Spaceflight on Cardiovascular Responses to Upright Posture in a 77-Year-Old Astronaut," *Am. Jnl. Cardiol.* 88, no. 11 (Dec. 1, 2001): 1335–1337, and R. P. Stowe et al., "Immune Responses and Latent Herpes Virus Reactivation in Spaceflight," *ASEM* 72, no. 10 (Oct. 2001): 884–891.

34. Glenn and Taylor, *John Glenn: A Memoir*, 372, 374, 375.

35. Glenn and Taylor, *John Glenn: A Memoir*, 375, 376, 382, 395, 396–397, 398–399.

36. NIA, *Exercise: A Guide from the National Institute on Aging* (Washington, DC: NIA, 1998; repub. online Dec. 16, 2005), www.niapublications.org/exercisebook/ExerciseGuideComplete.pdf (accessed Mar. 7, 2010).

37. Lillian D. Kozloski, *U.S. Space Gear: Outfitting the Astronaut* (Washington, DC: Smithsonian, 1994), 156–158, 162; Mark Hess, *Space Station Freedom Media Handbook May 1992* (Washington, DC: NASA, 1992), 73.

38. Kozloski, *U.S. Space Gear*, 157–158; Waller and Blum, "New All-Hard Spacesuit Developed for U.S. Space Station," ARC pr. 88-1, Jan. 12, 1988, ARC News Releases 1988, NASA Ames History Office, Moffett Field, CA; Hess, *Space Station Freedom Media Handbook May 1992*, 73.

39. Waller and Blum, "New All-Hard Spacesuit Developed"; Hess, *Space Station Freedom Media Handbook May 1992*, 73.

40. Eng. and S&LS Directorates, Mar. 24, 1989 and S&LS Direc. Mar. 31 and Apr. 14, 1989, f. "JSC Wkly. Activity Repts. March–April 1989," and Eng. Dir., May 5, 1989 and FCOD, May 18, 1989, f. "JSC Wkly. Activity Repts. May–June 1989," and Eng. Direc., Dec. 8, 1989, f. "JSC Wkly. Activity Repts. Nov.–Dec. 1989," b. 8, MSC Wkly./ Qtly. Repts., JSC Series, JSC Hist. Coll; Waller and Blum, "New All-Hard Spacesuit Developed"; Kenneth S. Thomas and Harold J. McMann, *U.S. Spacesuits* (Berlin: Springer Praxis, 2005), 309–310, 314, 324. This detailed volume contains technical and historical details beyond the scope of this book.

41. Kozloski, *U. S. Space Gear*, 158–163.

42. Joseph Kosmo, "EVA Systems: PSS Final Rev. March 26–28, 1991," f. "EVA Systems Mar. 26–28, 1991," b. 6, Ctr. Series—Hab. Studies, JSC Hist. Coll.

43. Kosmo, "EVA Systems." Emphasis in the original. Apollo astronauts used an average of 930 BTU per hour, ranging from 412 to 1,393 for the most strenuous activities. The Skylab average was higher, probably because astronauts were not confined to a small cabin for a large part of their mission. The range was broader, however: 279 to 2,250 BTU/hour. Shuttle astronauts had the lowest figures (794 BTU/hour) and narrowest range, 558-1,113.

44. Office of Exploration, Exploration Studies Technical Report, FY 1988 Status, Vol. II: Study Approach and Results, TM-4075, December 1988, 5–30.

45. OAST, *1991 Integrated Tech. Plan for the Civil Space Program* (Washington, DC: NASA, 1991), 3–22, 4–13.

46. Andrew Chaikin, "The Farthest Place," in Michael Light, *Full Moon* (New York: Knopf, 1999).

47. Henry O. Pohl, Engineering Directorate, Weekly Activity Report, Sep. 9, 1988, f. "JSC Weekly Activity Reports, September-October 1988," box 7, MSC Weekly/Quarterly Reports, 1988, JSC Series, JSC Hist. Coll. Work was still under way at UWI on "grip-assisted gloves" in 2004.

48. Kozloski, *U. S. Space Gear,* 163–164.

49. James W. McBarron II et al., "Individual Systems for Crewmember Life Support and EVA," in *Space Biology and Medicine,* Vol. II: *Life Support and Habitability,* edited by Sulzman and Genin, 275–329. Cuffs provided 15 minutes for cosmonauts to return to station.

50. Kozloski, *U. S. Space Gear: Outfitting the Astronaut,* 164.

51. Chaikin, *A Man on the Moon: The Voyages of the Apollo Astronauts* (New York: Penguin, 1994), 218. The odor reminded Neil Armstrong of "wet ashes in a fireplace" and Buzz Aldrin thought the dust smelled like "spent gunpowder."

52. Kosmo, "EVA Systems."

53. Ibid.

54. "Recent Activities," "ISSA EVA Prebreathe Protocol," and "Recommendations," in "Hamilton Standard/Zvezda EVA and Spacesuit (EMU) Activities, A Status Update & Proposal for the Acceleration of Joint Work Between U.S. and Russian Industry," f. 073944 "Common Space Suit," b. 46, Goldin Recs. Differences in atmospheric pressure and prebreathe time made US suits significantly less useful for emergency rescue EVAs.

55. Eve Rasmussen, "Recycling Could Save Money for Future NASA Space Missions," pr. 83-09, Mar. 11, 1983, ARC News Releases 1983, NASA Ames History Office, Moffett Field, CA.

56. Wayne Hale et al., eds., *Wings in Orbit: Scientific and Engineering Legacies of the Space Shuttle, 1971–2010,* SP-2010-3409 (Washington, DC: NASA, 2011), 413.

57. David Suffia, "Background Information: Space Garden," Boeing Aerospace & Electronics A-713, Sep. 1989, rec. no. 19955, b. 2, "McDonnell Douglas Space Station Concepts," Mackowski Life Sciences Coll.

58. Suffia, "Background Information."

59. W. M. Knott, "Controlled Ecological Life Support System: Breadboard Project—1988," in *Controlled Ecological Life Support Systems: CELSS '89 Workshop,* TM-102277, edited by Robert D. MacElroy (Moffett Field, CA: NASA-ARC, Mar. 1990), 295, 298.

60. *Educational Briefs for the Middle and Secondary-Level Classroom,* EB-109, Sep. 1992, NASA, RN 9930, "Life Sciences (1989–1992)," Life Sciences Coll., NASA Hist. Ref. Coll.

61. OSSA Life Sciences Div., "CELSS: Supplying Humans in Space," *NASA Life Sciences Rept. 1987* (Washington, DC: NASA, 1987), 43; David L. Bubenheim, "The Crop Growth Research Chamber: A Ground-Based Facility for CELSS Research," in MacElroy, *Controlled Ecological Life Support Systems,* 304.

62. B. G. Bugbee and F. B. Salisbury, "Current and Potential Productivity of Wheat

for a Controlled Environment Life Support System," *Adv. in Sp. Res.* 9, no. 8 (1989): 5–15. Bugbee and Salisbury noted that a reduction from 23° to 17°C (73.4° to 62.6° F) increased yield by 20 percent but lengthened the life cycle from 62 to 89 days.

63. R. M. Wheeler et al., "NASA's Biomass Production Chamber: A Testbed for Biogenerative Life Support Studies," *Adv. in Sp. Res.* 18, no. 4/5 (1996): 215–224.

64. G. I. Meleshko et al., trans. Lydia Razran Stone, "Biological Life Support Systems," in *Space Biology and Medicine*, Vol. II: *Life Support and Habitability*, edited by Sulzman and Genin, 364.

65. Ibid., 364; Bugbee and Salisbury, "Wheat Production in the Controlled Environments of Space," *Utah Science* (Winter 1985): 145–151; Theodore W. Tibbitts et al., "Utilization of White Potatoes in CELSS," *Adv. in Sp. Res.* 9, no. 8, (8)53–(8)59. Articles referred to the "harvest index" when comparing crop types. This was the mass of the edible portion divided by total biomass of the plant. Field-grown wheat was typically 40–50% edible, potatoes 80%. Ames also funded some of this CELSS research.

66. OSSA Life Sciences Div., "CELSS: Supplying Humans in Space," *NASA Life Sciences Rept. 1987* (Washington, DC: NASA, 1987), 44–46; Meleshko et al., "Biological Life Support Systems," 357. The USAFSAM, Boeing, and the USSR had experimented on single-celled algae, *Chlorella*, in 1960–1961 with negative results.

67. Waller, "Fact Sheet: Role of ARC in the Space Station," pr. 89-38, May 3, 1989, ARC News Releases 1989, ARC History Archives, Moffett Field, CA; Bubenheim, "The Crop Growth Research Center," 303–312.

68. OSSA Life Sciences Div., "CELSS: Supplying Humans in Space," 46.

69. Mackowski, "Salad Bars in Space: The Controlled Ecological Life Support System (CELSS)," *Griffith Observer* 54, no. 9 (Sep. 1990): 5–6; Jane Hutchison, Del Harding, Paula Cleggett-Haleim, pr. 90-31, June 14, 1990, ARC News Releases 1990, NASA Ames History Office, Moffett Field, CA.

70. Michael Braukus and Pam Alloway, "Spring Planting and Crop Harvest Time Underway at NASA," HQ and JSC pr. 91-78, May 24, 1991, RN 9930, Life Sciences Coll., NASA Hist. Ref. Coll.; Mackowski, "Salad Bars in Space," 5–7. Boeing and KSC experimented in the late 1980s with vertically stacked accordion trays and a circulating misting system to achieve successful root growth.

71. After the VSE announcement in 2005, NASA organized the Advanced Environmental Monitoring and Control project at the Jet Propulsion Laboratory (JPL), under the Exploration Systems Technology Development Program within the Exploration Systems Mission Directorate. It incorporated the study of plant production and bioregeneration.

72. G. W. Easterwood et al., "Lunar Base CELSS—A Bioregenerative Approach," *The 2nd Conf. on Lunar Bases and Space Activities of the 21st Century*, NASA Conf. Pub. 3166, Vol. 2, ed. Wendell Mendell (Washington, DC: NASA, 1992), 519–523.

73. B. C. Wolverton and John D. Wolverton, *Growing Clean Water: Nature's Solution to Water Pollution* (Picayune, MS: Wolverton Environmental Systems, Inc., 2001), 42–44.

74. Ibid., 106–111, 113–121.

75. Hardware Information, "Growth Apparatus for the Regenerative Development

290 · Notes to Pages 65–71

of Edible Nourishment (GARDEN)," https://lsda.jsc.nasa.gov/Hardware/hardw/1141 and Variable Pressure Growth Chamber (VPGC), https://lsda.jsc.nasa.gov/Hardware/hardw/1142; Image Information, Lunar-Mars Life Support Systems Integration Facility (LSSIF), https://lsda.jsc.nasa.gov/PhotoGallery/detail_result/2002; and Lunar-Mars Life Support Test Project (LMLSTP), https://lsda.jsc.nasa.gov/Research/research_detail/24?researchtype=, LSDA (accessed June 23, 2020).

76. Advanced Astroculture TM, https://www.nasa.gov/mission_pages/station/research/experiments/explorer/Investigation.html?#id=178; Biomass Production System, https://www.nasa.gov/mission_pages/station/research/experiments/explorer/Investigation.html?#id=20; Photosynthesis Experiment and System Testing Operation, https://www.nasa.gov/mission_pages/station/research/experiments/explorer/Investigation.html?#id=215; Plant Generic Bioprocessing Apparatus, https://www.nasa.gov/mission_pages/station/research/experiments/explorer/Investigation.html?#id=266; Optimization for Root Zone Substrates, https://www.nasa.gov/mission_pages/station/research/experiments/explorer/Investigation.html?#id=99, ISS National Laboratory Space Station Research Explorer, (accessed June 23, 2020).

77. Toups interview; Mackowski, "Salad Bars in Space," 8–9; Matt Schudel, "Maurice M. 'Mel' Averner, 72: NASA Scientist Studied Possibility Of Living on Mars," *Washington Post*, Mar. 2, 2009, http://www.washingtonpost.com/wpdyn/content/article/2009/02/28/AR2009022801927.html (accessed Jan. 18, 2011).

78. John A. Pitts, *The Human Factor: Biomedicine in the Manned Space Program to 1980*, SP-4213 (Washington, DC: NASA, 1985), 175–192, 204–207, 276–277.

79. James M. Beggs, Special Announcement, NASA HQ, Sep. 29, 1981, f. "HQ Org 1959–1987," b. 6, Org. Series, JSC Hist. Coll.

80. Vernikos, "Life Sciences in Space," 280; NASA HQ Pictorial Chart, June 1982, f. "HQ Organization 1959–1987," b. 6, Org. Series, JSC Hist. Coll.

81. Vernikos, "Life Sciences in Space," 280, 695.

82. Lennard Fisk, interview, Ann Arbor, MI, by Rebecca Wright, Sep. 8, 2010, NASA JSC OHC, https://historycollection.jsc.nasa.gov/JSCHistoryPortal/history/oral_histories/NASA_HQ/Administrators/FiskLA/fiskla.htm (accessed Aug. 31, 2020); James Beggs, transcribed interview by Ivelisse Rodriguez, Dec. 15, 1982, 13, NASA HQ History Office. See also Judy A. Rumerman, *NASA Historical Data Book*, Vol. V, *NASA Launch Systems, Space Transportation, Human Spaceflight, and Space Science 1979–1988*, SP-4012 (Washington, DC: NASA, 1999), 373–374.

83. Vernikos, "Life Sciences in Space," 281.

84. "Medical Reqs. for an Extended Orbiting Space Station," 1982; handwritten memorandum "Bill Bishop 12-8-82 telephone call, Follow-up on presentation at JSC," attached to "Program Plan," June 1982, ARC, in f. "Space Station Medical Reqs." and "Program Plan . . . June 1982," b. 7, Ctr. Series—S&LS, JSC Hist. Coll.

85. Waller, ARC, "Fact Sheet: The NASA-Ames Role in Space Station," pr. 86-41, Nov. 13, 1986, NASA ARC News Releases 1986, NASA Ames History Office, Moffett Field, CA.

86. "Key Assignments, Lyndon B. Johnson Space Center," Feb. 3, 1987, blue folder (no title), b. 6, sec. 106, Org. Series, JSC Hist. Coll.

87. Thompson, a NASA engineer from 1963 to 1983, left for Princeton University, but in Mar. 1986 became vice chair of the Rogers Commission investigating the *Challenger* accident, then MSFC Director in Sep. of that year. He would become Deputy Administrator at HQ under Richard Truly in 1989.

88. Org. chart, MSFC, in Andrew J. Dunar and Stephen P. Waring, *Power to Explore: A History of MSFC 1960–1990*, SP-4313 (Washington, DC: NASA, 1999), 636 and "Directory Chart, NASA, George C. Marshall Space Flight Center," Apr. 1987, f. "MSFC—Organization 1959–1987," b. 7, Org. Series, JSC Hist. Coll. Launched in 1999, AXAF was renamed Chandra to honor Indian American Nobel astrophysicist Subrahmanyan Chandrasekhar.

89. L. A. Fisk to multiple addressees, Sep. 17, 1987, f. "HQ—OSSA org., 1967, 1987," b. 6, Org. Series, JSC Hist. Coll.

90. Douglas Isbell, "NASA Considers Moving Life Sciences Research to Office of Exploration," *Space News*, Jan. 21–Feb. 3, 1991, 29; William J. Broad, "Cut-Down Stations May Fall Short on Space Biology," *New York Times*, Mar. 29, 1991, C1+.

91. Vernikos, "Life Sciences in Space," 281.

Chapter 3. Safety, Science, and Operational Medicine: Shuttle and Station in the 1980s and 1990s

1. Ross-Nazzal, "An Interview with Joe Allen," Part II, *Quest* 14, no. 3 (2007): 32. STS-5 astronaut Allen revealed that mission commander Vance Brand ordered the two pilot ejection seats "pinned," i.e., disabled, because they had not been removed in time. Allen cited "historical evidence," UK studies Brand was apparently aware of, about survivor guilt among military pilots who ejected from planes not equipped with such seats for their passengers.

2. John Young to Dir., FCO, JSC, Jan. 4, 1984, f. "JY 1984–85," carton (hereafter c.) 19, Truly Collection, Regis University, Denver, CO.

3. Richard A. Colonna, Orbiter and GFE Projs. Off., Wkly. Rept., July 1, 1988, f. "JSC Weekly Activity Reports, July–Aug. 1988," b. 7, JSC Series—MSC Wkly/Qtly Repts. and "Orbiter Ditching Studies," May 1, 1985, f. "Orbiter Ditching Studies," b. 23, Shuttle Series, MPAD/RTG Docs. (Brown), JSC Hist. Coll.

4. K. Barickman and S. N. Chen, "OV-103 Orbiter Structural Analysis for Water Ditching Conditions," Feb. 1987, contract NAS 9-17650, attached to transmittal memo MDAC-Houston Div. to C. T. Modlin, f. "Summary of the Effects of Water Ditching on the OV-103 Orbiter Structure," b. 23, Shuttle Series, MPAD/RTG Docs. (Brown), JSC Hist. Coll.

5. A speed of 200 knots per hour is 230.16 mi/hr or 370.4 km/hr.

6. Henry O. Pohl, Eng. Direc., Wkly. Rept., Feb. 12, 1988; Colonna, Orbiter and GFE Projs. Of., and Pohl, Eng. Direc., Wkly. Rept., Mar. 4, 1988; Pohl, Eng. Direc. and Charles S. Harlan, Safety, Reliability, and Quality Assurance (SR&QA), Wkly. Repts., Mar. 18, 1988, f. "JSC Wkly. Activity Repts. Jan.—March 1988," b. 7, JSC Series—MSC Wkly./Qtly. Repts., JSC Hist. Coll.; James W. McBarron II et al., "Individual Systems for Crewmember Life Support and EVA," in *Space Biology and Medicine*, Vol. II, *Life*

Support and Habitability, edited by F. M. Sulzman and A. M. Genin (Washington, DC: AIAA, Moscow: Nauka Press, 1994), 281–282.

7. SR&QA, Wkly. Rept., June 10, 1988, f. "JSC Wkly. Activity Repts. April-June 1988"; Aug. 26, 1988, f. "JSC Wkly. Activity Repts. July-Aug. 1988"; and Sep. 2, 1988, f. "JSC Wkly. Activity Repts. Sep.-Oct. 1988," b. 7, JSC Series—MSC Wkly./Qtly. Repts., JSC Hist. Coll.

8. Ctr. Ops. Direc., Wkly. Rept., Aug. 12, 1988 and Eng. Direc., Wkly. Rept., Aug. 19, 1988, f. "JSC Wkly. Activity Repts. July-Aug. 1988" and Eng. Direc., Wkly. Repts., Sep. 9, 1988 and Oct. 7, 1988, f. "JSC Wkly. Activity Repts. Sep.-Oct. 1988," b. 7, JSC Series-MSC Wkly./Qtly. Repts., JSC Hist. Coll.

9. S&LS Direc., Wkly. Rept., Mar. 25, 1988, f. "JSC Wkly. Activity Repts. Jan.-March 1988"; FCOD and SR&QA, Wkly. Repts., Aug. 19, 1988, f. "JSC Wkly. Activity Repts. July-Aug. 1988"; and WSTF, Wkly. Rept., Oct. 28, 1988, f. "JSC Wkly. Activity Repts. Sep.-Oct. 1988," b. 7 and WSTF, Wkly. Rept., Jan. 6, 1989, f. "JSC Wkly. Activity Repts. Jan.-Feb. 1989," b. 8, JSC Series—MSC Wkly./Qtly. Repts., JSC Hist. Coll.; McBarron II et al., "Individual Systems for Crewmember Life Support and EVA," 281.

10. McBarron II et al., "Individual Systems for Crewmember Life Support and EVA," 280–283.

11. Carolyn L. Huntoon, S&LS Direc. Wkly. Activity Rept., Feb. 5 and Mar. 10, 1988, f. "JSC Wkly. Activity Repts. Jan.-March 1988," b. 7, JSC Series—MSC Wkly./Qtly. Repts., JSC Hist. Coll.

12. Huntoon, S&LS Direc. Wkly. Activity Rept., Mar. 3 and 18, 1988, f. "JSC Wkly. Activity Repts. Jan.-March 1988"; Huntoon, S&LS Direc., Apr. 15 and June 10, 1988 and William J. Huffstetler, NIO Wkly. Rept., June 10, 1988, f. "JSC Wkly. Activity Repts. April-June 1988"; Huntoon, S&LS Direc., July 29 and Aug. 26, 1988 and Donald R. Puddy, FCOD, Wkly. Rept. July 1, 1988, f. "JSC Wkly. Activity Repts. July-Aug. 1988"; Huntoon, S&LS Direc., Sep. 9 and Oct. 21, 1988, f. "JSC Wkly. Activity Repts. Sep.-Oct. 1988"; and Puddy, FCOD, Nov. 10 and S&LS Direc. Dec. 2, 1988, f. "JSC Wkly. Activity Repts. Nov.-Dec. 1988," b. 7, JSC Series—MSC Wkly./Qtly. Repts., JSC Hist. Coll.

13. "GFE" stood for "government furnished equipment," including the orbiters.

14. Huntoon, S&LS Direc. Wkly. Activity Repts., Mar. 10, 1988, f. "JSC Wkly. Repts. Jan.-March 1988"; Oct. 21, 1988, f. "JSC Wkly. Repts. Sep.-Oct. 1988"; Dec. 2 and 16, 1988, f. "JSC Wkly. Activity Repts. Nov.-Dec. 1988," b. 7; Colonna, O&GFE Projects Office, Wkly. Rept., Jan. 20, 1989, f. "JSC Weekly Reports January–February 1989," and Apr. 15, 1989, "JSC WR March-April 1988," and Huntoon, S&LS and Charles S. Harlan SR&QA, Wkly. Repts., Feb. 3, 1989, f. "JSC Wkly. Activity Repts. Jan.-Feb. 1989," b. 8, JSC Series-MSC Wkly./Qtly. Repts., JSC Hist. Coll.

15. Young to J. D. Weatherbee, July 26, 1985; Young to Dep. Dir. NSTS, Apr. 17, 1989; Young to Dep. Dir. Space Shuttle Prog., Aug. 14, 1990; Young to Dep Dir SS Prog., Sep. 4, 1991, f. "JY 1984–85," "JY 1989," "JY 1990," and "JY 1991," c. 19, Truly Coll.

16. Harlan, SR&QA, Wkly. Rept., Apr. 14, 1989, f. "JSC Wkly. Repts. March–April 1989," b. 8, JSC Series—MSC Wkly./Qtly. Repts., JSC Hist. Coll.

17. "Shuttle Orbiter Arresting System," Space Shuttle Transoceanic Abort Landing

(TAL) Sites FS-2001–05–012-KSC, NASA Facts Online, KSC, last revised Jan. 18, 2006, http://www-pao.ksc.nasa.gov/kscpao/nasafact/talsup.htm (accessed Mar. 21, 2011); Landing Safety, Orbiter Arresting System, *Implementation of the Recommendations of the Presidential Commission on the Space Shuttle* Challenger *Accident* (Washington, DC: NASA, June 1987), part 1, recc. VI., 58, http://www.hq.nasa.gov/pao/History/rogersrep/v6ch5.htm (accessed May 27, 2020); "History," *Engineered Arresting Systems Corporation: Who We Are,* Zodiac Aerospace, http://www.esco.zodiacaerospace.com/about-esco/history.php (accessed Mar. 21, 2011).

18. Young to Aaron Cohen, July 1, 1988, f. "JY 1986–88," c. 19, Truly Coll.

19. W. M. Hinckley et al., "Space Shuttle Range Safety Command Destruct System Analysis and Verification," Phase I–Destruct System Analysis and Verification, Feb. 2, 1976, and Phase II—Ordnance Options for a Space Shuttle Range Safety Command Destruction, Dec. 10, 1976, repub. with Phase III–Breakup of Space Shuttle Cluster via Range Safety Command Destruct System (Dahlgren, VA and Silver Spring, MD: NSWC, Mar. 1981), I-1-I-4, f. "Space Shuttle Range Safety"; J. H. Wiggins Co., "STS Range Safety Hazards Analysis," Apr. 26, 1979, f. "STS Range," b. 34, MPAD, Shuttle Series, JSC Hist. Coll.

20. This is a term from aerobatic flying, describing the path of an aircraft ascending at a very steep angle, then nosing over and plunging straight into the ground, with catastrophic results.

21. "Proposed Change to Section 3.2.3.1 of the NSTS-08116 Doc.," n.d. [after Oct. 1, 1987], f. "Proposed Change," b. 26, MPAD, Shuttle Series, JSC Hist. Coll.

22. Robert F. Schultz to Richard Kohrs, June 12, 1987, f. "Prevention of Shuttle Intact High Velocity Impact," b. 26, MPAD, Shuttle Series, JSC Hist. Coll.

23. Presidential Commission on the Space Shuttle *Challenger* Accident, *Report of the Presidential Commission on the Space Shuttle* Challenger *Accident* (Washington, DC: The Commission, 1986), 184–185.

24. William R. Marshall to NSWC, Dec. 5, 1986, f. "Revision to Shuttle Breakup Data," b. 28, MPAD, Shuttle Series, JSC Hist. Coll.

25. B. D. O'Connor to All Astronauts, July 17, 1989; "Workshop Timetable," June 7, 1989; "Ground Track for Buran, 1st Flight"; and "Russian Cities and Towns Essentially under the 51.6° Inclination Manned Launch Vehicle Ground Track," attached to memo Young to Depy. Dir., Natl. STS Prog., Sep. 5, 1989, f. "JY 1984," c. 19, Truly Coll.

26. Presidential Commission, *Report of the Presidential Commission on the Space Shuttle* Challenger *Accident,* p. 41. NASA recovered 20 percent of the ET, mostly interior bits.

27. Young to Depy. Dir., Space Shuttle Program, June 27, 1990, f. "JY 1990," c. 19, Truly Coll. By June 1990, thrust reduction had proven problematic, so in this letter Young backed an LaRC scheme for "case venting" in planned second generation SRBs. Underlining in original.

28. Young to multiple addressees, Aug. 30, 1991, f. "JY 1991," c. 19, Truly Coll.

29. Presidential Commission, *Report of the Presidential Commission on the Space Shuttle* Challenger *Accident,* 184.

30. "Space Shuttle External Tank," Wikipedia, http://en.wikipedia.org/wiki/External_tank (accessed Mar. 21, 2011); Allard Beutel and Bruce Buckingham, "Countdown Begins Aug. 24 for Space Shuttle Atlantis Launch," Media Advisory M06-132, Aug. 18, 2006, https://www.nasa.gov/home/hqnews/2006/aug/HQ_06132_115_launch_milestones.html; "STS-120 Launch Countdown Milestones and Times," Oct. 17, 2007, https://www.nasa.gov/mission_pages/shuttle/news/120_milestones.html; and "STS-123 Launch Countdown Milestones and Times," Mar. 4, 2008, https://www.nasa.gov/mission_pages/shuttle/news/123_milestones.html (all accessed June 24, 2020); also Joel Wells, online status report, KSC Nov. 9, 1999, http://www-pao.ksc.nasa.gov/kscpao/status/stsstat/1999/nov/11-09-99.htm (accessed Dec. 11, 2009).

31. Kenneth Souza et al., *Life into Space, 1965–1990*, RP-1372 (Moffatt Field, CA: NASA-Ames, 1996), 8–9, 42–45, 65.

32. President's letter; William Gilbreath, "LifeSat, a New Satellite for Biological Space Research"; "Comings and Goings"; "1986–87 ASGSB Membership," *ASGSB Newsletter* 3, no. 1 (Mar. 1987): 1, 6, 11, 14.

33. Linda Neuman Ezell, *NASA Historical Data Book*, Vol. II, *Programs and Projects 1958–1968*, SP-4012 (Washington, DC: NASA 1988), 202, 232–236, 254, 259–260, 263–264, 274, 288, 302–303, 311–312, 319–320, 326–327, 331–332; Ezell, *NASA Historical Data Book*, Vol. III, *Programs and Projects 1969–1978*, SP-4012 (Washington, DC: NASA, 1988), Appendix B.

34. Antha Adkins et al., *LifeSat Engineering In-House Vehicle Design*, TM-104752 (Houston: NASA-JSC, July 1992), 1; Lennard A. Fisk, interview, Ann Arbor, Sep. 9, 2010, NASA JSC OHC, https://historycollection.jsc.nasa.gov/JSCHistoryPortal/history/oral_histories/NASA_HQ/Administrators/FiskLA/fiskla.htm (accessed Aug. 31, 2020).

35. Gilbreath, "LifeSat, a New Satellite for Biological Space Research," 11.

36. Byron L. Swenson et al., *A Conceptual Design Study of the Reusable Reentry Satellite*, TM-101043 (Moffett Field: NASA-Ames, Oct 1988).

37. Adkins et al., *LifeSat Engineering In-House Vehicle Design*, 1; Swenson et al., *A Conceptual Design Study of the Reusable Reentry Satellite*, IV, V; OSSA, Life Sciences Div., *Life Sciences Div. Strategic Implementation Plan*, TM-102907 (Washington, DC: NASA, April 1989), 32, 34. LifeSat predated NASA's adoption of the metric system of measurement.

38. Adkins et al., *LifeSat Engineering In-House Vehicle Design*, 1–2, 18. "NASA Releases RFP for LifeSat," *ASGSB Newsletter* 5, no. 1 (Jan. 1989): 3; *Life Sciences Div. Strategic Implementation Plan*, p. 27; OSSA, Life Sciences Div., *Space Life Sciences: Programs and Projects*, TM-105459 (Washington, DC: NASA, April 1989), 9.

39. Adkins et al., *LifeSat Engineering In-House Vehicle Design*, 1–2, 10, 15; Swenson et al., *A Conceptual Design Study of the Reusable Reentry Satellite*, 4; "Two Companies Chosen for Work on LifeSat," *ASGB Newsletter* 5, no. 3 (Sep. 1989): 10.

40. See, e.g., John W. Wilson et al., *Preliminary Analysis of a Radiobiological Experiment for LifeSat*, TM-4236 (Washington, DC: NASA, 1991). This study examined the likelihood of inducing a certain type of tumor in rodents based on orbital inclination,

thickness of shielding, and time in orbit to prove that statistically useful oncogenesis data could be created with a LifeSat flight.

41. NASA, "Nasafacts: Biosatellite II," NF-3/10–68, n.d. [1968], 2–6, 8–10; Souza et al., *Life into Space, 1965–1990*, 47.

42. Francis A. Cucinotta et al., *Predictions of Cell Damage Rates for Lifesat [sic] Missions*, TM-102170 (Houston: NASA JSC, Nov. 1990), 3, 6; Wilson et al., *Preliminary Analysis of a Radiobiological Experiment for LifeSat*, 1, 2; Swenson et al., *A Conceptual Design Study of the Reusable Reentry Satellite*, 10–12, 16, 71, 73, 76.

43. Adkins et al., *LifeSat Engineering In-House Vehicle Design*, 2, 16, A-1, A-2. The landing parameters allowed for 20 Gs in the 1988 concept.

44. Barbara Selby, "NASA Selects Small Business Innovation Research Projects," HQ Release 91–15, Jan. 29, 1991, http://www.nasa.gov/home/hqnews/1991/91-015.txt (accessed Apr. 27, 2020); Micro-G Research Inc. "Variable-G Facility for LIFESAT," SBIR 1988–1, https://sbir.nasa.gov/SBIR/abstracts/88/sbir/phase1/SBIR-88-1-12.08-9339. html (accessed Dec. 30, 2020). In 1988 Micro-G's "centrifuge facility" was described as "at least two independently-controlled rotors capable of providing g-force environments within the range of zero to somewhat above one while supplying simultaneous 1 g control data."

45. Adkins et al., *LifeSat Engineering In-House Vehicle Design*, 3–4.

46. Adkins et al., *LifeSat Engineering In-House Vehicle Design*, 2, 5, 6, 10, 15, 114–117.

47. In other words, fly one-way through the hole in the doughnut and one-way through the Van Allen Belts themselves.

48. Adkins et al., *LifeSat Engineering In-House Vehicle Design*, 2, 3–4, 5–6, 10, 13, 15, 95–96; Swenson et al., *A Conceptual Design Study of the Reusable Reentry Satellite*, 62–64, 68–69. The 41 authors of the LifeSat Technical Memorandum 104752 are by-lined in simple alphabetical order but 11 were with Lockheed, 2 with MDCA, 1 with Computer Sciences Corporation, and 27 with JSC.

49. Adkins et al., *LifeSat Engineering In-House Vehicle Design*, 3; Title XIII: Budget Enforcement, Subtitle A: Amendments to the Balanced Budget and Emergency Deficit Control Act of 1985 and Related Amendments—Part I: Amendments to the Balanced Budget and Emergency Deficit Control Act of 1985, H.R.5835—Omnibus Budget Reconciliation Act of 1990, https://www.congress.gov/bill/101st-congress/house-bill/5835 (accessed Sep. 7, 2020).

50. *Cong. Rec.*, Nov. 22, 1991, S17743.

51. *Cong. Rec.*, Sep. 27, 1991, S13917. Emphasis added.

52. *Cong. Rec.*, Sep. 27, 1991, S13917; *Cong. Rec.*, Nov. 22, 1991, S17743.

53. David W. Garrett, "Space Science Dominates 1991 NASA Activities," HQ pr. 91–209, Dec. 17, 1991.

54. OSSA, Life Sciences Div., *Space Life Sciences Strategic Plan*, TM-107856, 12, 28; "NASA Mixed Fleet Manifest 8/91," Aug. 20, 1991, repr. Spacelink, http://www.textfiles.com/science/payload.txt (accessed Apr. 27, 2020).

55. "W. P. Bishop, etc." to "W. E. Rice, etc.," attached to memo Nachtwey to multiple addressees, Sep. 22, 1982, f. "TM-58248," b. 7, Ctr. Series–Space & Life Sciences, JSC Hist. Coll. "Not sent" is handwritten across memo by "B.," presumably Bill Bishop.

56. "Space Station MOF/Human Life Sciences Research Laboratory/HLSRL Concept and Development Plan," f. "Space Station Medical Reqs. 1982," b. 7, Ctr. Series—Space & Life Sciences, JSC Hist. Coll.

57. Ibid.

58. Ibid.

59. Ibid. All the documents in this folder used "men" when referring to astronauts. Writers also assumed the doctor to be male.

60. Space Station Program Description WG, "Space Station Mission Description Doc., Sci. and Apps., Sec. 3.1.4, Life Sciences," 6, 7–8; "7.1.4.14 Health Maintenance," Book 3, Post RID Markup (11/9/82); and "6.11 Health Maintenance," f. "SSP Dem Doc, LSS 1982," b. 7, Ctr. Series—Space & Life Sciences, JSC Hist. Coll.

61. S. Furukawa et al., "Medical Support and Technology for Long-Duration Space Missions," paper IAF-82-174, AIAA 1982, 4–6, f. "IAF 82–174," b. 7, Ctr. Series—Space & Life Sciences, JSC Hist. Coll.

62. An informative history and analysis of psychological and behavioral issues related to habitability is Jack Stuster, *Bold Endeavors: Lessons from Polar and Space Exploration* (Annapolis: Naval Institute Press, 1996). A behavioral sciences researcher specializing in extreme environments, Stuster consulted with NASA's Life Sciences Division on lunar and planetary habitats.

63. "Medical Reqs. Doc. for a Long Duration Manned Orbital Facility," 7, f. "Space Station Medical Reqs. 1982," b. 7, Ctr. Series—Space & Life Sciences, JSC Hist. Coll. This analog data 1969–1979 was exclusively on males, as the United States forbade females to serve aboard Navy submarines.

64. "3.4.2.3 Medical"; David H. Suddeth and Gary P. Barnhard, Advanced Miss. Analysis Office, GSFC, App. A "Medical Apps. Expts. Tech." and Nachtwey, Biomedical Apps. Branch, JSC, App. B, "Support Tech. Dev.," in "3.4.2.3 Medical," f. "SSP Desc Doc, LSS 1982," b. 7, Ctr. Series—Space & Life Sciences, JSC Hist. Coll.

65. Nachtwey, Tech. Dev. Review Form, Nov. 26, 1982, attached to Richard Kennedy, JSC, "Crew Systems—Emesis Station," in "3.4.2.2 Habitability," f. "SSP Desc Doc, LSS 1982," b. 7, Ctr. Series—Space & Life Sciences, JSC Hist. Coll. KSC was concerned also with "overall crew acceptance of the setup, operation, and cleanup aspects."

66. "Key Development Items," f. "SSP Desc Doc, LSS 1982"; Furukawa et al., "Medical Support and Tech. for Long-Duration Space Missions," 7–8; "Crew Health Hazards"; Space Station Office, "Space Station System Reqs. Doc.," Oct. 8, 1982, f. "Space Station Medical Reqs. 1982," b. 7, Ctr. Series—Space & Life Sciences, JSC Hist. Coll.

67. LMSC, for Man-Systems Division, JSC, Nos. 201M05, "Contamination/Odor Control" and 201M06, "Contamination Limits," 2070301, "Animal Payload Bioisolation," Mgmt. Plan Overview, *Space Station Human Productivity Study Final Rept.—v. 3*, contract no. NAS9–17272 DR SE-1093T, Nov. 1985, f. "Space Station Human Productivity Study Final Rept.—v. 3–Reqs.," b. 2, Habitability Studies, Ctr. Series, JSC Hist. Coll.

68. "Specimens Habitat Tech. Reqs. for Man-Visit Missions" and "Critical or Unique Operational Reqs., Additional Reqs. for Man-Visited Missions," Space Station Life Sciences Research Facility Tech. Assessment and Tech. Dev. Plan, Technical Progress

Rev. No. 1, McDonnell Douglas, NASA-ARC Contract NAS2-11539, Mar. 3, 1983, f. "MDAC," b. 8, Ctr. Series—Space & Life Sciences, JSC Hist. Coll.

69. "7.1.4.14 Health Maintenance," Book 3, Post RID Markup (11/9/82), f. "SSP Dem Doc, LSS 1982," b. 7, Ctr. Series—Space & Life Sciences, JSC Hist. Coll.

70. Numbers 201M01, "Atmosphere Specification" and 207M02, "Animal Payload Bioisolation," Mgmt. Plan Overview, *Space Station Human Productivity Study*.

71. MDAC, "Specimens Habitat Tech. Reqs. for Man-Visit Missions."

72. Numbers 205M03, "Long Duration 0-G Noise Exposure Limits," and 205M01, "Prediction of Low Frequency Noise," Mgmt. Plan Overview, *Space Station Human Productivity Study*; D. G. Stephens, "Space Station Acoustics Control Tech. Dev.," in "3.4.2.2 Habitability," f. "SSP Dem Doc, LSS 1982," b. 7, Ctr. Series—Space & Life Sciences, JSC Hist. Coll. LaRC suggested such tests, to be done on station, in 1982.

73. John F. Murphy to Rep. Dan Glickman, Apr. 3, 1986, f. "Life Sciences (1980–86)," RN 9923, Life Sciences Coll., NASA Hist. Ref. Coll.

74. "Medical Reqs. Doc. for a Long Duration Manned Orbital Facility," 7.

75. Maura J. Mackowski, "Lifeboat! The Crew Emergency Rescue Vehicle (CERV)," *Griffith Observer* 52, no. 9 (Sep. 1988): 9.

76. Kornel Nagy, *Contingency Return Vehicle Study for Space Station*, Engrg. Direc., JSC, Oct. 1986, photocopies of slide presentation, rec. no. 19955, b. 2 "McDonnell Douglas Space Station Concepts," Mackowski Life Sciences Coll.

77. Francis X. Kane, "A Thirty-Year Perspective on Manned Space Safety and Rescue: Where We've Been, Where We Are, Where We Are Going," *Space Safety and Rescue, 1984–1985, Proceeding. of the Symposia of the IAA*, Vol. 64 (San Diego: AAS, Univelt, 1986), 61–88.

78. Ron Gerlach, "Space Station Crew Emergency Return Vehicle Background 10/1/86," in Nagy, *Contingency Return Vehicle Study for Space Station*.

79. Nagy, *Contingency Return Vehicle Study for Space Station*.

80. JSC's Strategic Game Plan: Charting A Course to the Year 2000 and Beyond (Houston: JSC, 1987), 11, 19.

81. Gerlach, "Space Station Crew Emergency Return Vehicle Background, 10/1/86," Nagy, "Contingency Return Vehicle Requirements" and "Issues for Contingency Return Vehicle Design," and S. Nagel, "Basic CERV Objectives" and "Minimum CERV Design Ideas," in "CERV Crew Requirements," Oct. 1, 1986, all in Nagy, *Contingency Return Vehicle Study for Space Station*.

82. Patricia Santy, "Medical Concerns for a CRV," Sep. 29, 1986 and Nagel, "Basic CERV Objectives," Nagy, *Contingency Return Vehicle Study for Space Station*.

83. Santy, "Medical Concerns for a CRV."

84. Chris Cerimele, Adv. Progs. Office, "Aero Trades," Sep. 9, 1986 in Nagy, *Contingency Return Vehicle Study for Space Station*.

85. A. Steiner, Electromagnetic Systs. Branch, "CRV Comm and Tracking," Oct. 1986, in Nagy, *Contingency Return Vehicle Study for Space Station*; Douglas J. Mudgway, *Uplink-Downlink: A History of the Deep Space Network* (Washington, DC: NASA, 2001), 288. The third and final TDRSS would not be deployed until Feb. 1989.

86. John Kennedy, Structures and Mechanics Div., "Contingency Return Vehicle (CRV) Landing and Recovery," Oct. 1, 1986, in Nagy, *Contingency Return Vehicle Study for Space Station.*

87. R. McElya, Structures and Mechanics Div., "CRV Concept Development," Sep. 19, 1986, T. Moore, Avionics Syst. Div., "CRV Avionics Requirements, Bare Bones," Oct. 1, 1986, and Kent Joosten, Mission Ops. Direct., "Crew Monitoring & Backup Procedures," Oct. 1, 1986, in Nagy, *Contingency Return Vehicle Study for Space Station.*

88. Nagy, "CERV: Where to Next?" and Joan Baker, "Contingency Return Vehicle Cost Estimate," Oct. 1, 1986, in Nagy, *Contingency Return Vehicle Study for Space Station.*

89. Mackowski, "Lifeboat! The Crew Emergency Return Vehicle (CERV)," 9–11.

90. NIO, Wkly. Activity Rept., June 10, 1988, f. "JSC Wkly. Activity Repts. April-June 1988," b. 7, JSC Series—MSC Wkly./Qtly. Repts., JSC Hist. Coll.

91. NIO and Admin. Direct., Wkly. Activity Repts., Apr. 14, 1989, f. "JSC Wkly. Activity Repts. March–April 1989," b. 8, JSC Series—MSC Wkly./Qtly. Repts., JSC Hist. Coll.

92. NIO, Wkly. Activity Rept., Aug. 4, 1989, f. "JSC Wkly. Activity Repts. July-Aug. 1989," b. 8, JSC Series—MSC Wkly./Qtly. Repts., JSC Hist. Coll.

93. SR&QA, Wkly. Activity Rept., Aug. 25, 1989, f. "JSC Wkly. Activity Repts. July-Aug. 1989," b. 8, JSC Series—MSC Wkly./Qtly. Repts., JSC Hist. Coll. This might have been an early reference to the eventual strap-on Simplified Aid for EVA Rescue (SAFER). The STS-64 crew tested it in 1994.

94. Legal Dept., Wkly. Activity Rept., Dec. 15, 1989, f. "JSC Wkly. Activity Repts. Nov.-Dec. 1989," b. 8, JSC Series—MSC Wkly./Qtly. Repts., JSC Hist. Coll.; Cerimele et al., pat. no. 5,064,151, Assured crew return vehicle," approved Nov. 15, 1991, US PTO, DOC, http://patft.uspto.gov (accessed Oct. 10, 2020).

95. "HL-20 Model of Personnel Launch System Research," *NASA Facts On Line*, NF172, Apr. 1992, http://www.nasa.gov/centers/langley/news/factsheets/HL-20.html (accessed June 10, 2020).

96. Arnie Aldrich, "ACRV: Summary of ESA Proposal & Summary," 6–7, f. 073608 "Senior Mgmt. Mtg. 4/12/92," b. 40, Goldin Recs.

97. "Memorandum of Discussion . . . On Civil Space Cooperation," Moscow, July 1992, f. 073510 "9/30/92," b. 38, Goldin Recs.

98. "JSC Puts 'Life Boat' to Wave Test," *Station Break* 4, no. 5 (May 1992): 4, 10.

99. Aldrich, "ACRV: Summary of ESA Proposal & Summary," 6, 15–16, f. 073608, b. 40, Goldin Recs.

100. "The Manned Space and Microgravity Programmes," *The ESA Programmes*, BR-114 (Frascati: ESA-ESRIN, Aug. 1995), ESA, http://www.esa.int/esapub/br/br114/br114man.htm (accessed Apr. 24, 2020).

101. Office of the Press Secy., WH, "Joint Statement on Cooperation in Space," June 17, 1992 cited in Logsdon, "The Development of International Space Cooperation," in *Exploring the Unknown: Vol. II: External Relationships*, edited by Logsdon et al., 221; Rex D. Hall and David J. Shayler, *Soyuz: A Universal Spacecraft* (London: Springer-Praxis, 2003), 382; "Talking Points for Use in Moscow on Monday, Oct. 5, 1992," f. 073515 "Russia Trip Oct 4–6, 1992," b. 38, Goldin Recs.

102. Hall and Shayler, *Soyuz*, 382–383, 392.

103. Notes, Vest Committee mtg. June 7–8, 1993, f. 075980 "Vest Committee 6/7/93–6/8/93," b. 137, Goldin Recs.

104. Yuri Y. Karash, *The Superpower Odyssey: A Russian Perspective on Space Cooperation* (Reston, VA: AIAA, 1999), 175–176, 197–198.

105. Hall and Shayler, *Soyuz*, 384–386.

106. ACRV Proj. Office, JSC, "Phase B Stmt. of Work, Soyuz ACRV (Exercise Part of Contract Line Item 0007)," Doc. # JSC-34031, Mar. 31, 1993, f. "NASW-4727 Amndmt./Mod. #7," b. 1, Center Series: US/Russian Docs., JSC Hist. Coll.

107. Office of the VP, WH, "U.S.—Russian Joint Commiss. on Energy and Space—Joint Stmt. on Coop. in Space," Sep. 2, 1993, cited in Logsdon, "The Development of International Space Cooperation," 229.

108. Hall and Shayler, *Soyuz*, 364, 378–381, 387. The space station's official name was briefly Alpha in 1993, then International Space Station Alpha late in the year and in 1994.

109. Hall and Shayler, *Soyuz*, 377, 390. "A" was for "anthropometric," as the redesign was to accommodate NASA astronaut size ranges.

110. "European Participation in NASA's X-38 Program," attached to John D. Schumacher to Goldin, Feb. 9, 1998, f. 077518 "2/10/98 mtg. with the Director General of CNES . . . Gérard Brachet," b. 179, Goldin Recs.

111. "X-38 Technology," DFRC News, n.d., http://www.nasa.gov/centers/dryden/research/X38_Save/index.html and "X-38 Fact Sheet," Fact Sheets, DFRC News, n.d., http://www.nasa.gov/centers/dryden/news/FactSheets/FS-038-DFRC.html (both accessed May 27, 2020).

112. "Feature," DFRC; "X-38 Fact Sheet," Fact Sheets, DFRC News; Mark Carreau, "NASA Sinks Space Station's Lifeboat Plan; Budget Move Irks European Partnership," *Houston Chronicle*, July 6, 2002; Marcia S. Smith, "Issue Brief for Congress: Space Stations," Feb. 3, 2003, CRS, LOC, 6–10, 13; Schumacher to Goldin and Talking Points Tab D, Sep. 9, 1997, f. 076147 "9/10/97 Luncheon with Ambassador Bujon de L'estang," b. 139 and "Europe's Part in NASA's X-38 Program," Dec. 10, 1997, f. 077518, b. 179 "Talking Points Tab B," f. 076176 "12/10/97 mtg. with Mr. Uchida, NASDA president," b. 140, Goldin Recs.

113. Hall and Shayler, *Soyuz*, 394.

Chapter 4. Science and Scientists: Peer Review, the Extended Duration Orbiter Medical Project (EDOMP), Neurolab, and a Station Centrifuge

1. Minutes of LSAC Mtg., Mar. 14, 1981, 1, 3, 5, 6, 7–8, 9, 10, 12, f. "NASA HQ S-Codes OSS April–July 1981," b. 17, HQ Corr., Ctr. Files, JSC Hist. Coll.

2. Minutes of LSAC Mtg., Mar. 14, 1981, 8, 9, 10, 11.

3. John B. Charles interview, JSC, by Wright et al., Aug. 28, 1998, "People: Oral Histories," *History of Shuttle–Mir* (accessed Mar. 21, 2011); Charles A. Berry, "The Beginnings of Space Medicine," *Aviation, Space, and Environmental Medicine* 57 (Oct. 1986): A62. Charles stated that the Soviets published results but the wait was "months or years" for a translation. Typically it was data the United States already had and the

Soviets had followed protocols not felt to be "meaningful" or which changed repeatedly. Berry gave the example of information suppression regarding the deaths of the Soyuz 11 crew on return from the Salyut 1 station when their craft depressurized. It took him two years to gain access to the data on that event, alone and "under very controlled circumstances."

4. Minutes of LSAC Mtg., Mar. 14, 1981, 2–4, 12.

5. John Charles, interview with the author, Jan. 25, 2006.

6. See, for example, OSSA, "Management Plan, Spacelab 1 Payload," 1–3, 10, App. B, C, and D, f. 17.3.17 "SL-1 & 2 Project Plan," sh. 1 & 2, cb. 2, Lawrence Chambers Files, NASA Hist. Ref. Coll., Hist. Prog. Off., NASA HQ, Washington, DC.

7. E. Anderson, "Shuttle Student Involvement Program (SSIP): Benefits to Education"; Terri Sindelar, "RTQ for STS-29, Space Shuttle Student Involvement Program (SSIP) Experiments" Mar. 3, 1989; and "Reach for the Stars Through NASA's Space Science Student Involvement Program," July 30, 1991, RN 9870, NASA Hist. Ref. Coll.

8. Ray A. Williamson's essay "Developing the Space Shuttle" in *Exploring the Unknown*, Vol. IV, *Accessing Space*, edited by Logsdon et al. (Washington, DC: NASA 1999), 164–167, 179–182, 185–187 reviews the design and operation of the shuttle as a truck. The shift in attitudes toward the shuttle's role is reflected in the contrast between Jimmy Carter's Oct. 10, 1978 Presidential Directive/NSC 42, "Civil and Further National Space Policy," in *Exploring the Unknown*, Vol. I, *Organizing for Exploration*, edited by Logsdon et al., 575–578 and Ronald Reagan's "Presidential Directive on National Space Policy" of Feb. 11, 1988, http://www.hq.nasa.gov/office/pao/History/policy88.html (accessed May 27, 2020).

9. Henry R. Hertzfeld, "Space as an Investment in Economic Growth," in *Exploring the Unknown*, Vol. III, *Using Space*, edited by Logsdon et al., 392–396 describes the expanding policy emphasis on commercialization, including space manufacturing, and US economic growth in the 1980s.

10. "Space Science & Applications Notice," Oct. 1992. See for example Thora W. Halstead and Patricia A. Dufour, eds., *Biological and Manned Experiments on the Space Shuttle 1981–1985* (Washington, DC: NASA OSSA, 1986). The preface noted that many were "tests or preliminary observations, rather than true experiments." Half of the student data was stated to be unavailable for publication years later.

11. Halstead, "Guidelines for Participation in the Space Biology Program," 10, *Space Science & Applications Notice, Emerging Opportunities in the Space Biology Program*, f. 41.2 "Animal Welfare Organizations," sh. 2/3, cb. 2, Chambers Files.

12. S&LS Dir. Weekly Activity Reports (WAR), Nov. 10, 1988, f. "JSC Weekly Activity Reports, November–December 1988," b. 7, MSC W/QR, JSC Series, JSC Hist. Coll.

13. "Microgravity Sciences on Freedom: Selecting the Principal Investigators," *Station Break* 4, no. 1 (Jan. 1992): 2.

14. Guy Fogleman, interview with the author, Apr. 19, 2006.

15. Holloway, introduction and Earl W. Ferguson, remarks, NASA-NIH Joint Advisory Committee on Biomedical and Behavioral Research, mtg. mins., Jan. 9–10, 1994, 2, 4; Fogleman interview.

16. Holloway to multiple addressees, May 18, 1993, Holloway to Dir., Life and Biomedical Sciences and Apps., Sep. 7, 1993, Vernikos to Holloway Sep. 20, Holloway to multiple addressees, n.d. [Nov. 3, 1993], Goldin to Barbara A. Mikulski, Oct. 26, 1993, and Mikulski, "Report [To accompany H.R. 2491]," Sep. 7, 1993, 1, in Vernikos, "Peer Review of Research Proposals and Programs," Nov. 5, 1993, f. 27.1 "Peer Reviews," sh. 2/3, cb. 2, Chambers Files. Vernikos commented on this briefly and included the Sep. 7, 1993 memo in Logsdon et al., *Exploring the Unknown*, Vol. VI: *Space and Earth Science*, edited by Logsdon et al., 416–417.

17. Fogleman interview.

18. Life and Biomedical Sciences and Apps. Div., OLMSA, "Science Mgmt. Plan," rev. Feb. 1994, i, 2–3, 6–7, 8–11, 13–14, f. 41.5 "Science Management Plan," sh. 2/3, cb. 2, Chambers Files.

19. Ferguson, 4, Dan Golden [sic] remarks, 5, and Panel Disc./Recs., 10–11, NASA-NIH APBBR minutes.

20. Charles A. LeMaistre to Goldin, Apr. 12, 1994, RN 9930, Life Sciences Coll., NASA Hist. Ref. Coll.

21. Charles H. Evans Jr. and Suzanne T. Ildstad, eds., *Small Clinical Trials: Issues and Challenges* (Washington, DC: National Academies Press, 2001), 4, 15, 75.

22. Evans and Ildstad, eds., *Small Clinical Trials*, 8–10, 89–90.

23. Charles Sawin, interview with the author, May 16, 2006; S&LS Dir., JSC Committee for the Protection of Human Subjects, *Guidelines for Investigators Proposing Human Research for Space Flight and Related Investigations*, JSC 20483, Rev. C (Houston: NASA-JSC, Feb. 2004). The IMBP is sometimes translated as the "Institute for Biomedical Problems" with the acronym IBPM.

24. Sawin interview.

25. Pool, "Space Medicine Branch Report: Extended Duration Orbiter Medical Project," *ASEM*, July 1990, 679; Sawin interview; Charles interview. The official report for the EDOMP, which Sawin edited, stated that Congress allocated money in 1988 and Pool's article said that "NASA elected to develop" the EDO orbiter in 1989. However, Sawin said, the four-week mission "required a lot of equipment changes that the program couldn't afford, so it never happened." An example of early plans for shuttle missions up to 90 days or with 10 crewmembers is Hamilton Standard, "Concept Definition for an EDO ECLSS," Contract NAS 9–14782, Sep. 1977, https://ntrs.nasa.gov/archive/nasa/casi.ntrs.nasa.gov/19790015495.pdf (accessed May 27, 2020).

26. Sawin, "Introduction to the Extended Duration Orbiter Medical Project," in *Extended Duration Orbiter Medical Project: Final Report, 1989–1995*, SP-1999-534, edited by Sawin et al. (Houston: JSC, n.d.), xxiii; Charles interview.

27. Charles et al., "Cardiovascular Conditioning," Frances Mount and Tico Foley, "Assessment of Human Factors," and n.a., "Hardware," in *Extended Duration Orbiter Medical Project: Final Report*, 1–1, 6–11, 6–13, 8–1.

28. "Shuttle Extended Duration Orbiter Medical Project (EDOMP)," Hist. Res., LSDA, https://lsda.jsc.nasa.gov/books/ground/SP-1999-534.pdf (accessed June 25, 2020).

29. Pool, "Space Medicine Branch Report: Extended Duration Orbiter Medical Project," 679; Sawin, "Biomedical Investigations Conducted in Support of the Extended Duration Orbiter Medical Project," *ASEM* 70, no. 2 (Feb. 1999): 169–180.

30. Charles et al., "Cardiovascular Conditioning," 1–1,1–4.

31. Ibid., 1–6, 1–7, 1–10, 1–17, 1–18.

32. Ibid., 1–4, 1–6, 1–10.

33. Janice Meck Fritsch-Yelle, "In-Flight Use of Florinef to Improve Orthostatic Intolerance Postflight (DSO 621)," https://lsda.jsc.nasa.gov/Experiment/exper/1255 (accessed June 25, 2020). This experiment was done on STS-45, -49, -50, -52, -55, -56, -64, -66, -67, -68, -70, -73, and -74. See also Janice Meck Fritsch-Yelle, "Test of Midodrine as a Countermeasure Against Postflight Orthostatic Hypotension, (SMO 006)," https://lsda.jsc.nasa.gov/Experiment/exper/931 (accessed June 25, 2020) for the disappointing 2006–2008 results of a countermeasure approved by the FDA in 1998.

34. Charles et al., "Cardiovascular Conditioning" and "Hardware," 1–2, 1–13, 1–16, 8.4–6. Astronauts drank 1 L (1.06 qt.) of water or artificially sweetened fruit drink containing 8 gm. (0.28 oz.) sodium chloride. Flight crews used several evolutions of the complicated LBNPS and the Re-Entry Anti-gravity Suit (REAGS), the result of a 30-month study by the USAF's Armstrong Lab. Worn with the Advanced Crew Escape Suit (ACES), it had fuller leg coverage and deleted the abdominal bladder. The REAGS provided greater gravity protection but at lower pressure and with less discomfort, with the abdominal bladder gone. The STS-71 crew used it first, in July 1995.

35. Ibid., 1–2–1–4. Disrhythmia could vary from palpitations to cardiac arrest.

36. Helen W. Lane et al., "Regulatory Physiology," in *Extended Duration Orbiter Medical Project,* edited by Sawin et al., 2–1, 2–2, 2–3, 2–5, 2–8.

37. Lane et al., "Regulatory Physiology," 2–2, 2–5, 2–6.

38. Greenisen et al., "Functional Performance Evaluation," in *Extended Duration Orbiter Medical Project,* edited by Sawin et al., 3–1.

39. Greenisen et al., "Functional Performance Evaluation," 3–8–3–10, 3–20–3–21.

40. This difference came into question, however, because of NASA's requirement that all astronauts on missions of 10 or more days exercise to some extent.

41. Greenisen et al., "Functional Performance Evaluation," 3–8–3–10, 3–11, 3–14–3–16, 3–17, 3–21. Investigators ran up against mission extensions, NASA regulations concerning duty-day work-hour limits and the 10-day mission exercise mandate, fatigue and soreness causing crew to switch exercise machines or forego an exercise or test, and one individual's insistence on designing his own new fitness regime in-flight.

42. Wayne Hale et al., eds., *Wings in Orbit: Scientific and Engineering Legacies of the Space Shuttle, 1971–2010,* SP-2010-3409 (Washington, DC: NASA, 2011), 372–375 describes these sensations and illusions and their impact on performance.

43. Reschke, "Neuroscience Investigations, an Overview of Studies Conducted," Deborah L. Harm et al., "Visual-Vestibular Integration Motion Perception Reporting," Reschke et al., "Visual-Vestibular Integration as a Function of Adaptation to Space Flight and Return to Earth," William H. Paloski et al., "Recovery of Postural Equilibrium Control Following Space Flight," and Jacob J. Bloomberg, "Effects of Space Flight

on Locomotor Control," in *Extended Duration Orbiter Medical Project*, edited by Sawin et al., 5.1–1, 5.2–1, 5.2–4–7, 5.3–5, 5.4–2–3, 5.4–9–10, 5.5–1–4. Paloski noted that even after nominal landings orbiters rested at a 6° forward pitch.

44. Mount and Foley, "Assessment of Human Factors," 6–1–2.

45. Mount and Foley, "Assessment of Human Factors," 6–3–6–5.

46. Duane Pierson et al., "Environmental Health," in *Extended Duration Orbiter Medical Project*, edited by Sawin et al., 4–1, 4–4, 4–7.

47. "Facilities," in *Extended Duration Orbiter Medical Project*, edited by Sawin et al., 7.1–7.3; Sawin interview. DFRC built the Postflight Science Support Facility (PSSF) for EDOMP. Beginning with STS-40 in June 1991, crew exited the orbiter into a medical exam facility on wheels, the Crew Transport Vehicle (CTV). Adapted from airline passenger transporters, it let doctors gather data right away, before postflight recovery started. CTVs also offered astronauts some privacy in which to recover their "Earth legs."

48. Charles interview.

49. Arnauld Nicogossian, interview with the author, Apr. 18, 2006; J. Travis Brown and Sawin, "Preface," *Extended Duration Orbiter Medical Project*, edited by Sawin et al., iii.

50. Nicogossian interview.

51. Named for the post–World War II Nuremberg, Germany trials of medical personnel who conducted dangerous, painful, and often lethal experiments on prisoners during the Hitler regime. It gave absolute preeminence to an individual's right to decline participation.

52. Astronauts could be, and were, *prevented* from doing what they wanted, even in their private lives, though. See, e.g., Dir. FCO to multiple addressees, June 21, 1988 and Chief, Astronaut Office to Robert L. Gibson, Jan. 2, 1991, f. "Astros Misc," c. 31, Truly Coll. The memo ruled out "high risk recreational activities," including vehicle racing, by all assigned flight crew and no "nominal risk" sports for eight months prior to launch. NASA had grounded Gibson, a two-time mission commander, in July 1990 and then put him on six months' probation on T-38s in Jan. 1991 for competing in an air race. Management (and the public) found out when another plane struck his, killing the other pilot.

53. James Logan, interview with the author, May 19, 2006.

54. Sawin, "Project Summary and Conclusions," *Extended Duration Orbiter Medical Project*, edited by Sawin et al., 9–3–9–5, 9–7.

55. Logsdon, "The Development of International Space Cooperation," 7–9 and Doc. I-29, Memorandum, Arnold Frutkin to Administrator, May 9, 1973, in *Exploring the Unknown*, Vol. II: *External Relationships*, edited by Logsdon et al., 85–87.

56. Vernikos, "Life Sciences in Space," 285–286, 298 and Doc. III-29, SSB, NAS, "A Strategy for Research in Space Biology and Medicine in the New Century," Nov. 28, 2009, 437 in *Exploring the Unknown*, Vol. VI: *Space and Earth Sciences*, edited by Logsdon et al. The cancelled rhesus experiments became the Bion 11 and 12 payloads.

57. Vernikos, "Life Sciences in Space," 408.

58. George H. W. Bush, Pres. Proc. 6158, July 17, 1990, online at *Project on the Decade of the Brain,* LOC, http://www.loc.gov/loc/brain/proclaim.html (accessed May 27, 2020).

59. Dave Williams, "Foreword," in *The Neurolab Spacelab Mission: Neuroscience Research in Space. Results from the STS-90 Neurolab Space Mission,* NASA SP-2003-535, edited by Jay C. Buckey and Jerry L. Homick (Houston: JSC, 2003), iii.

60. Williams, "Foreword," Buckey and Homick, "Preface," and Buckey et al., "Surgery and Recovery in Space," in *Neurolab Spacelab Mission,* edited by Buckey and Homick, iii, v–vi, 275.

61. Buckey et al., "Animal Care on Neurolab," in *Neurolab Spacelab Mission,* edited by Buckey and Homick, 295.

62. Buckey et al., "Animal Care on Neurolab," in *Neurolab Spacelab Mission,* edited by Buckey and Homick, 295–298. Kathie Olsen and Ann Carlson, interview with the author, Apr. 19, 2006. A reproductive neuroendocrinologist at the NSF, Olsen had faulted the protocol two years earlier for using new rat mothers.

63. "The Balance System"; "Sensory Integration and Navigation"; Giles Clément et al., "Perception of the Spatial Vertical During Centrifugation and Static Tilt"; Steven Moore et al., "Ocular Counter-Rolling During Centrifugation and Static Tilt"; Gay R. Holstein and Giorgio P. Martinelli, "The Effect of Spaceflight on the Ultrastructure of the Cerebellum"; Ottaviano Pompeiano, "Gene Expression in the Rat Brain During Spaceflight"; Muriel D. Ross and Joseph Varelas, "Ribbon Synaptic Plasticity in Gravity Sensors of Rats Flown on Neurolab"; James J. Knierim et al., "Ensemble Neural Coding of Place in Zero-G"; Meredith D. Temple et al., "Neural Development Under Conditions of Spaceflight," in *Neurolab Spacelab Mission,* edited by Buckey and Homick, 1–2, 5–10, 11–17, 19–25, 27–37, 39–44, 52, 63–68, 161–168.

64. "Nervous System Development in Weightlessness"; Robert Kalb et al., "Motor System Development Depends on Experience: A Microgravity Study of Rats"; Gregory R. Adams et al., "Gravity Plays an Important Role in Muscle Development and the Differentiation of Contractile Protein Phenotype"; Michael L. Wiederhold et al., "Early Development of Gravity-Sensing Organs in Microgravity"; Eberhard R. Horn, "The Development of an Insect Gravity Sensory System in Space (Crickets in Space)"; Jacqueline Raymond et al., "Development of the Vestibular System in Microgravity"; Tsuyoshi Shimizu et al., "Development of the Aortic Baroreflex in Microgravity," in *Neurolab Spacelab Mission,* edited by Buckey and Homick, 91–92, 95–103, 113, 123–132, 133–142, 143–149, 151–159.

65. "Blood Pressure Control," Benjamin D. Levine et al., "Neural Control of the Cardiovascular System in Space"; Andrew C. Ertl et al., "The Human Sympathetic Nervous System Response to Spaceflight," in *Neurolab Spacelab Mission,* edited by Buckey and Homick, 171–172, 175–185, 197–202.

66. "Circadian Rhythms, Sleep, and Respiration"; Derk-Jan Dijk et al., "Sleep, Circadian Rhythms, and Performance During Space Shuttle Missions; G. Kim Prisk et al., "Sleep and Respiration in Microgravity," in *Neurolab Spacelab Mission,* edited by Buckey and Homick, 207–208, 211–221, 223–232.

67. Horn et al., "Crickets in Space"; Shimizu et al., "Development of the Aortic

Baroreflex in Microgravity"; Charles M. Oman, "Neurolab Virtual Environment Generator"; Buckey, "Surgery and Recovery in Space," in *Neurolab Space Mission*, edited by Buckey and Homick, 140, 154, 158, 253–257, 275–277.

68. Goldin, talking points, 19, Dec. 2, 1997, f. 076175 "FASEB Board luncheon," b. 140, Goldin Recs.

69. Olsen and Carlson interview.

70. Neurolab flew a short-arm centrifuge of 0.5 m (1.6 ft) radius and producing 0.5 or 1.0 g's.

71. Broad, "Cut-Down Station May Fall Short on Space Biology," C1, C8; ARC, "Research Centrifuge Requirements for the Space Station," June 1986, 1, 5, 7, attached to Kenneth Souza to multiple addressees, July 9, 1986, f. "Space . . . by ARC, June 1986," b. 10, SSPO Reports & Presentations; Robert H. Moser to Burton I. Edelson, Aug. 15, 1986, attached to M. Devirian to R. Hook, Sep. 4, 1986, f. "Consideration," b. 4, Critical Evaluation Task Force subseries, Space Station series, JSC Hist. Coll.

72. ARC, "Vestibular Research Facility for Spacelab," brochure, 1980 and LMSC, "Vestibular/Variable-Gravity Research Facility: A Versatile In-Flight System for Life Science Investigations," brochure, Sunnyvale, CA, Sep. 1980, RN 9923, "Life Sciences (1980–86)," Life Sciences Coll., NASA Hist. Ref. Coll.

73. MDAC, Space Station Life Sciences Research Facility Technology Assessment and Technology Development Plan, Technical Progress Rev. No. 1.

74. Committee on Human Exploration of Space, NRC, *Human Exploration of Space: A Review of NASA's 90-Day Study and Alternatives* (Washington, DC: National Academies Press, 1990), 19.

75. "NASA Advisory Council: Recommendations and NASA Responses, Life Sciences Responses," n.d., 1, attached to memorandum Lynn D. Griffiths to Exec. Secy., NASA Advisory Council, Aug. 4, 1988, RN 9922, Life Sciences Coll., NASA Hist. Ref. Coll.

76. Waller, "Fact Sheet: Role of ARC in the Space Station," pr. 89–38, May 3, 1989, ARC News Releases 1989, ARC History Office, Moffett Field, CA.

77. SSPO, Wkly. Activity Repts., Oct. 20, 1989, f. "JSC Weekly Activity Reports Sep.–Oct. 1989," b. 8, JSC Series, MSC W/QR, JSC Hist. Coll.

78. Hess, *Space Station Freedom Media Handbook May 1992*, 69, 96; Clément et al., "Perception of the Spatial Vertical During Centrifugation and Static Tilt," and Bernard Cohen et al., "Adaptation to Linear Acceleration in Space (ATLAS) Experiments: Equipment and Procedures," in *Neurolab Spacelab Mission*, edited by Buckey and Homick, 6, 280–281; Lynn Harper, interview with the author, Sep. 16, 2005. Harper pointed out that a scientist might be selected to fly one shuttle experiment and not get a chance to repeat it for several years, if ever.

79. William J. Broad, "Cut-Down Station May Fall Short on Space Biology," *New York Times*, Mar. 26, 1991, C1, C8.

80. Broad, "Cut-Down Station May Fall Short on Space Biology," C1, C8; "Space Station Freedom Chief Scientist Speaks Out," *Station Break* 4, no. 6 (June 1992): 5; Hess, *Space Station Freedom Media Handbook May 1992*, 69–70, 73.

81. "NASA, ISA Sign MOU," *Station Break* 4, no. 1 (Jan. 1992): 1, 8.

82. John J. McCarthy, meeting notes, 18, f. 075980, b. 137; Goldin to Officials-in-Charge, Mar. 9, 1993, f. 073594 "Senior Mgmt. Mtg. on Space Station 3/10/93," b. 40, Goldin Recs.

Chapter 5. Organizing in the 1980s–1990s: Ethics, Institutes, and Biological Modeling

1. Mary B. Kerwin, memo for record, Aug. 28, 1981, RN 9923, Life Sciences Coll., NASA Hist. Ref. Coll.

2. Howard Rosenberg, "Informed Consent: How the Space Program Experimented with Dwayne Sexton's Life," *Mother Jones*, Sep.-Oct. 1981, 31–44.

3. Joanne Omang, "U.S. Scientists Accused of Testing Radiation on Leukemia Patients," *Washington Post*, Aug. 20, 1981, A4. During the same period, the AEC gave the clinic $25.4 million, meaning that NASA money accounted for 8.3 percent of ORINS funds.

4. Kerwin memo; Paul Rambaut to Kerwin and Larry Medway, Aug. 25, 1981, f. "Life Sciences (1980–86)," RN 9923, Life Sciences Coll., NASA Hist. Ref. Coll.

5. Lynne Murphy, memo for the record, Sep. 25, 1981, f. "Life Sciences (1980–86)," RN 9923, Life Sciences Coll., NASA Hist. Ref. Coll.

6. Statement of Andrew J. Stofan before the Subcommittee on Investigations and Oversight, Committee on Science and Technology, US House of Representatives, Sep. 23, 1981, f. "Life Sciences (1980–86)," RN 9923, Life Sciences Coll., NASA Hist. Ref. Coll.

7. Statement of Andrew J. Stofan. This was the "Retrospective Study of Radiation Effects."

8. Kerwin memo.

9. See *Statement by Dan Goldin to the Independent Advisory Committee on Human Radiation Experiments, April 21, 1994* and related internal documents in f. 073794 "April 21, 1994 Radiation Advisory Committee," b. 43, Goldin Recs.

10. See "Appendix: Markey Report 'Experiment List,'" in Appendices, Roadmap to the Project: The DOE Roadmap, US Department of Energy, https://ehss.energy.gov/ohre/roadmap/roadmap/abbrev.html (accessed June 29, 2020). See also Subcommittee Staff Report, Subcommittee on Energy Conservation and Power, Committee on Energy and Commerce, US House of Representatives, "American Nuclear Guinea Pigs: Three Decades of Radiation Experiments on U.S. Citizens," Oct. 1986, 4, 27, 34–35, f. "Life Sciences (1980–86)," RN 9923, Life Sciences Coll., NASA Hist. Ref. Coll.

11. David Tomko, interview with the author, June 15, 2006.

12. One consequence was the passage of the Animal Enterprise Protection Act of 1992, Public Law 102–346, outlawing "animal terrorism," the destruction of property by activist groups.

13. Stofan to Anne Higgins, note, July 22, 1981; Eleanor Seiling to Ronald Reagan, July 2, 1981; Rosemarie R. Vitagliano, memo to unspecified person, Apr. 21, 1981; Leslie Sorg to NASA, WH referral doc., July 15, 1981, f. 41.2, sh. 2/3, cb. 2, Chambers Files.

14. John F. Murphy to Rep. Christopher H. Smith, Apr. 21, 1986 and to Sen. Lloyd Bentsen, May 19, 1986, f. "Life Sciences (1980–86)," RN 9923, Life Sciences Coll.,

NASA Hist. Ref. Coll.; Murphy, memo to Legislative Affairs Staff, Jan. 28, prob. 1985 and Terence T. Finn to Rep. Samuel S. Stratton, Aug. 21, 1981, f. "Life Sci. Biological Payloads US/USSR (Animals in Space) to 1997," RN 9870, NASA Hist. Ref. Coll. Smith and Bentsen's constituents expressed concern about alleged cat tests. Murphy assured them the Agency had none underway or planned. He did not mention that NASA was conducting such studies on rats. Another couple had written Stratton to complain, apparently citing a brochure from United Action for Animals on NASA activities, written two decades before.

15. Dianne C. Lambert, memo for record, Sep. 19, 1984, f. "Life Sci. Biological Payloads US/USSR (Animals in Space) to 1997," RN 9870, NASA Hist. Ref. Coll.

16. "Public Law 99–198, Food Security Act of 1985, Subtitle F—Animal Welfare," National Agricultural Library, US Department of Agriculture, https://www.nal.usda.gov/awic/public-law-99–198-food-security-act-1985-subtitle-f-animal-welfare (accessed June 29, 2020).

17. J. Davis, "Space Motion Sickness," 1, f. "Astros Misc," c. 31, Truly Coll. Through STS-32 in Jan. 1990 approximately two-thirds of astronauts incurred what is called "Space Adaptation Syndrome" or "Space Motion Sickness." One out of six cases was "severe," with repeated vomiting over days.

18. Victor J. Wilson, Vestibular Reflexes of Otolith Origin, Final Technical Report Apr. 1, 1979–Dec. 31, 1987, NASA Res. Grant NSG-2380 (Washington, DC: NASA, 1988).

19. Barbara Stagno, "Rockefeller Relents: Anatomy of a Victory," Satya, Apr. 1998, archived at http://www.satyamag.com/apr98/rockefeller.html (accessed Dec. 7, 2020); Wilson, "Laboratory of Neurophysiology," Rockefeller University: Scientists and Research: Emeritus Faculty: Victor J. Wilson, http://www.rockefeller.edu/research/abstract.php?id=191&status=eme (accessed Jan. 5, 2007). URL changed to https://www.rockefeller.edu/our-scientists/emeritus-faculty/926-victor-wilson/ (accessed Apr. 14, 2020). No reference was then made to cat experiments or to NASA.

20. Tomko interview; Victor Stolc, interview with the author, Sep. 16, 2005.

21. Homer E. Newell, Beyond the Atmosphere: Early Years of Space Science, NASA SP-4211 (Washington, DC: NASA, 1980), 238, 241; "LPI Celebrates 40 years" timeline, LPI, 2008, http://www.lpi.usra.edu/lpi_40th/ (accessed May 27, 2020). The name was changed to the Lunar and Planetary Institute in Jan. 1978.

22. List of members in USRA Researcher, Summer 2005, 11, "Publications" and "Member Institutions," About USRA, Andrew Bradley, ed., 6–7, http://www.usra.edu/ (accessed Mar. 17, 2008); M. H. Davis and A. Singy, eds., USRA 1994: Introduction to USRA—Universities Space Research Association, NASA-CR 195980 (Boulder, CO: USRA, 1994). Nonmember institutions could also participate in USRA activities.

23. USRA, Ann. Progr. Analysis of the NASA Space Life Sciences Res. and Education Support Progr., Dec. 1, 1993–Nov. 30, 1994 (Columbia, MD: USRA, 1994), 42.

24. OIG, Quality Control Review Report: Ernst & Young LLP Audit of the Universities Space Research Association (USRA) for Fiscal Year Ended June 30, 1999, and Follow-up of Audit of USRA for Fiscal Year Ended June 30, 1998, IG-00-001

(Washington, DC: NASA, 2000), 1, "Audits–Issued Reports," NASA OIG, https://oig.nasa.gov/audits/reports/FY00/pdfs/ig-00-001.pdf (accessed June 29, 2020).

25. "USRA Restructures Corporate Management," *USRA Researcher,* Fall 2000, 9.

26. Davis and Singy, eds., *Introduction to USRA,* 12, 13, 14, 15, 17; USRA, *Ann. Progr. Analysis of the NASA Space Life Sciences Res. and Education Support Progr.,* Dec. 1, 1993– Nov. 30, 1994, 15–16, 38–39.

27. Davis and Singy, eds., *Introduction to USRA,* 9, 12–13, 14–15, 24–25.

28. "NASA Space Radiation Summer School at the Brookhaven National Laboratory"; "The Program and its Objectives"; "Aerospace Medicine Grand Rounds"; "DSLS Brown Bag Seminar Series"; and "DSLS Student Research Program," DSLS, USRA, http://www.dsls.usra.edu (accessed Mar. 17, 2008 and Sep. 24, 2010). The Postdoctoral program ran through 2006, training eight early career researchers.

29. *Technology Development and Aerospace Environments (TDAE) 1999–2000 Annual Rept.,* TDAE Program, Huntsville Program Office, USRA, http://space.hsv.usra.edu/tdae/ (accessed Sep. 22, 2010).

30. Robert A. Cassanova and Diana E. Jennings, "NIAC, A Legacy of Revolutionary Creativity," 9 and "Short Report: Long-term Success of NIAC-Funded Concepts," 10–13 in *NASA Institute for Advanced Concepts, 9th Annual & Final Rept., 2006–2007,* archived at http://www.niac.usra.edu/library/annual_report.html (accessed June 29, 2020).

31. Amy M. Grunden and Wendy F. Boss, NIAC Phase I Final Rept.: Redesigning Living Organisms to Survive on Mars; Edward Hodgson et al., A Chameleon Suit to Liberate Human Exploration of Space Environments, NIAC Contract No. 07600–082, June 16, 2004; Pamela A. Menges, Artificial Neural Membrane Flapping Wing, USRA Contract No. NAS5–03110, May 3, 2006; Paul Todd, Lunar Robotic Ecopoiesis Test Bed, "NIAC Funded Studies," NIAC, USRA, http://www.niac.usra.edu/studies/studies.html (accessed June 29, 2020). These and other NIAC-backed research were archived here. See also "NIAC Director Receives NASA Award," *The Researcher,* Fall 2003, 7 and "NIAC Nominated for World Technology Award," *The Researcher,* Summer 2004, 8. The NIAC idea was resurrected in 2011 as the "NASA Innovative Advanced Concepts."

32. Huntoon, "NASA Biomedical and Bioengineering Research Institute (NBBRI) Preliminary Planning Concepts," PPT, Oct. 30, 1992, NSBRI—Rummel/Fogleman Notebook, rec. no. 19954, b. 1 "National Space Biomedical Research Institute (NSBRI)," Mackowski Life Sciences Coll. The NASA Office of Inspector General's Feb. 1, 2018 "Audit of the National Space Biomedical Research Institute" (IG-18–192) describes the transition from one Institute to another and provides insight into ongoing tensions over the Institute concept.

33. Huntoon interview, Barrington, RI, by Wright, June 5, 2002, JSC OHP, https://historycollection.jsc.nasa.gov/JSCHistoryPortal/history/oral_histories/HuntoonCL/huntooncl.htm (accessed June 29, 2020), 29, 33–34.

34. Huntoon, "NASA Biomedical and Bioengineering Research Institute (NBBRI) Preliminary Planning Concepts"; Fogleman interview; Vernikos, "Life Sciences in Space," 299; Huntoon interview, 32–33.

35. Aaron Cohen, "NASA Technology Research Institute–JSC Institute for Bio-

medicine and Biotechnology–Planning Status," PPT, Apr. 14, 1993, NSBRI—Rummel/ Fogleman Notebook, Mackowski Life Sciences Coll.; Huntoon interview.

36. Fogleman interview; Hans Mark to Truly, May 6, 1989, attached to an unsigned letter ("a group of concerned NASA employees," JSC) to Mark, May 3, 1989; IG Office, Report of Investigation: Unauthorized Disclosure of Astronaut Medical Records I-JS-89–104, 1989; Pool, "NASA Program for the Medical Evaluation and Certification of Astronauts"; Nicholas C. Chriss, "NASA Accused of Failing to Ground Medically Unfit," "Disqualification Urged for Colorblind Astronaut," and "NASA Chief Mum on Astronauts' Waivers," Feb. 12, 13, and 14, 1989, *Houston Chronicle*, 1A+, f. "Astro Medical Issue '89," c. 31, Truly Coll.

37. Huntoon, "NASA Techn. Res. Inst. (NTRI)—JSC Inst. for Biomed. and Biotech., Planning Status," PPT, Apr. 26, 1993, slides 5, 8, NSBRI—Rummel/Fogleman Notebook, Mackowski Life Sciences Coll.

38. JSC, draft "Estab. of an NTRI, Inst. for Biomed. and Biotech., NASA Coop. R. A. Soliciting Res. Proposals for the Period Ending _____," June 2, 1993, 2, 11, 14, 15, NSBRI—Rummel/Fogleman Notebook, Mackowski Life Sciences Coll.

39. JSC, draft "Estab. of an NTRI," 2–4, 6.

40. Lambright, "Transforming Government: Dan Goldin and the Remaking of NASA" (Arlington, VA: PricewaterhouseCoopers Endowment for The Business of Government, Mar. 2001), 18, 20–22; Wesley T. Huntress Jr. to multiple addressees, May 30, 1995, Solar Data Analysis Center, GSFC, http://umbra.nascom.nasa.gov/spd/ZBR. html (May 27, 2020); "Background and Talking Points, Mr. Dan S. Goldin," Nov. 4, 1993, f. 073672 "NRC Space Studies Board Joint Meeting 11/4/93," b. 41, Goldin Recs.

41. Huntress to multiple addressees; Laurie Boeder and Dwayne Brown, "Review Team Proposes Sweeping Management, Organizational Changes at NASA," HQ pr. 95–73, May 19, 1995, 1–2, NSBRI—Rummel/Fogleman Notebook, Mackowski Life Sciences Coll.

42. "ZBR Status Dec. 1, 1994," 2, f. 073938 "Sr. Mgmt. Briefing," b. 46, Goldin Recs.

43. Boeder and Brown, "Review Team," 1–2; Lambright, "Transforming Government," 21; Wisniewski et al., "Pre-Meeting Briefing," 41, f. 073999 "4-27-95 [ZBR]" ZBR Team Status Report, Apr. 27, 1995, and Goldin, "NASA Employees/ZBR" talk, May 18, 1995, f. 074019 "3-18-95 [ZBR]," b. 48, Goldin Recs. Lambright wrote that Mikulski, who paid close attention to NASA matters on the Appropriations Committee, was so furious that she placed explicit language in the legislation preventing the cutbacks, and summoned Goldin to Capitol Hill to explain himself before her committee. No layoffs took place; the downsizing was handled through attrition.

44. "ZBR Status Report, Pre-decisional Briefing," 42, 48; "JSC Response to Zero Base Team Recs. 4/18/95"; John M. Klineberg, "GSFC Response to HZ Zero Base Study Team Position"; Code A, "A/Mr. Goldin 'Eyes Only,'" 10, f. 073998 "4-18-95 [ZBR]," b. 47; Wisniewski et al., "Pre-Meeting Briefing," 23, 27, 29–31, f. 073999 and Goldin, notes, May 1995, f. 074019, b. 48, Goldin Recs.

45. Goldin, notes, f. 074019, b. 48, Goldin Recs. In this seven-page thought piece he remarked, "We assume NASA people are free," meaning the Agency had not incorporated the "people costs" in projects. He also believed that "We have no method of

comparing which NASA centers manage resources best," i.e., against each other or as "best in class" among similar outside organizations.

46. Huntress to multiple addressees; Boeder and Brown, "Review Team," 2.

47. Lambright, "Transforming Government," 21.

48. NASA Science Institutes Plan, *Report of the NASA Science Institutes Team: Final Publication (Incorporating Public Comments and Revisions)* (Washington, DC: NASA, Feb. 1996).

49. NASA OIG, "Executive Summary," *Audit Report: NASA Science Research Institutes, Sep. 30, 1998,* IG-98-037 (Washington, DC: NASA, 1998), 1.

50. Vernikos, "Life Sciences in Space," 299; NASA OIG, "Executive Summary," 1; Boeder and Brown, "Review Team," 5–6. Steven Dick and James Strick go into some detail on the work of the Astrobiology Institute in *The Living Universe: NASA and the Development of Astrobiology.*

51. Douglas Isbell, "Diaz Named to Lead Planning Effort for Science Institutes," HQ pr. 95-110, July 12, 1995, "1995 News Releases," NASA News, HQ, http://www.nasa.gov/home/hqnews/1995/95-110.txt (accessed May 27, 2020).

52. Boeder (HQ) and Jeff Carr Johnson (JSC), "Huntoon to Lead Planning Effort for Life Sciences Institute," pr. 95-132, Aug. 4, 1995, "1995 News Releases," NASA News, HQ, http://www.nasa.gov/home/hqnews/1995/95-132.txt, (accessed May 27, 2020).

53. SIP Visits–All Notes, Aug. 27, 1995, NSBRI—Rummel/Fogleman Notebook, Mackowski Life Sciences Coll.

54. Ibid.

55. John [A.] Rummel to Abbey, Sep. 8, 1995, and Rummel, "DRAFT-4/24/02, NASA Reference Materials, Planning and Implementation of the National Space Biomedical Research Institute (NSBR)," 4, NSBRI—Rummel/Fogleman Notebook, Mackowski Life Sciences Coll. The 2002 Draft stated that "the most important mission of NASA scientists is to bind NASA's immense engineering and technical capabilities to the still larger and more diverse industrial and academic research communities across the country and around the world."

56. Ibid. Writing retrospectively in 2002, Rummel commented that "the scope and funding levels for the NSBRI were highly debated within NASA."

57. Diaz, "Science Institute Planning"; Don Savage, "NASA Science Institutes Report Available for Public Comment," Notice to Editors N95-79, HQ, Dec. 1, 1995, NSBRI—Rummel/Fogleman Notebook, Mackowski Life Sciences Coll.

58. Diaz, "Science Institute Planning"; *NASA Science Institutes Plan, Report of the NASA Science Institutes Team,* Final Publication (Incorporating Public Comments and Revisions) (Washington, DC: NASA, Feb. 1996), 12–13, NSBRI—Rummel/Fogleman Notebook, Mackowski Life Sciences Coll.

59. "Draft 1/20/96, Planning Implementation Status, Natl. Inst. for Space Life Sci. or NASA Biomed. Res. Inst.," 1/23/96, JSC, PPT, 4–6, 15; Thomas S. Luedtke, Deputy AA Procurement, to JSC Procurement Office, Feb. 13, 1996 and attached "Minutes of the Acquisition Strategy Meeting for the NASA Biomed. Res. Inst. at JSC," 1–2, 5, NSBRI—Rummel/Fogleman Notebook, Mackowski Life Sciences Coll.

60. "Draft 1/20/96, Planning Implementation Status," 6, 19, 25; "Minutes of the Acquisition Strategy Meeting," 3–4.

61. *NASA Science Institutes Plan*, 1, 5. The Microgravity Institute, at Lewis, would specialize in fluid and combustion science.

62. Ibid., 11–12; Cindy Buck, Executive Secretary. NASA Science Institutes Planning Team to Planning Team Members, Mar. 1, 1996, with attached "NASA Science Institute Report Foreword," 2, NSBRI—Rummel/Fogleman Notebook, Mackowski Life Sciences Coll.

63. "Draft 1/20/96, Planning Implementation Status," 30–31, 43; "Minutes of the Acquisition Strategy Meeting," 4.

64. Ibid., 6–8, 12, 13; "NASA Science Institutes Report Q&A," attached to Buck memo, 6–7.

65. *NASA Science Institutes Plan*, 4, 9, 10, 12–13; "NASA Science Institutes Report Q&A," 8.

66. *NASA Science Institutes Plan*, i.

67. Graph, "Draft–JSC Life Sci. Budget Profile," n.d. (Oct. 1995); Sulzman, chart "Inst. for Space Biomed. Res.," Feb. 20, 1996; "NSBRI Budget Discussions, 3/20/96, JSC/HQ," PPT, 2–6; "NSBRI, Planning Status, 4/5/96, JSC, NASA HQ," PPT, 4; "NSBRI, Planning Status, Presented to Gen. John Dailey, 4/9/96, JSC," PPT, 1, NSBRI—Rummel/Fogleman Notebook, Mackowski Life Sciences Coll.

68. NSBRI SOL 9-CAN-96–01, posted Apr. 19, 1996 and attached memorandum, "TO: PROSPECTIVE OFFERORS," n.d.; [John A. Rummel?], "NSBRI, Coop. Agrmt. Notice (CAN), Summ. Presentation to JSC Senior Staff, 5/6/96," PPT, 2, 7, 10, NSBRI—Rummel/Fogleman Notebook, Mackowski Life Sciences Coll.

69. CAN, Soliciting Proposals for the Establishment of the NSBRI, 9-CAN-96–01, June 10, 1996, 12–15, 24, NSBRI—Rummel/Fogleman Notebook, Mackowski Life Sciences Coll.; Ron White, interview with the author, May 18, 2006.

70. Braukus, "NASA Names a New National Biomedical Research Institute," pr. 97–43, HQ, Mar. 14, 1997, "1997 News Releases," NASA News, HQ, http://www.nasa.gov/home/hqnews/1997/97-043.txt (accessed May 27, 2020).

71. White interview.

72. "Cooperative Agreement Management Plan (CAMP) for the Establishment of the NSBRI," May 29, 1997, NASA-JSC, 4, 7, NSBRI—Rummel/Fogleman Notebook, Mackowski Life Sciences Coll.

73. Ibid., 5, 6, 12.

74. Ibid., 7; Abstr., Proceedgs. of the First Bienn. Space Biomed. Investigators' Wkshp. (Houston: NASA-JSC, 1999), NTRS, http://ntrs.nasa.gov/search.jsp, (accessed Sep. 21, 2010). The CAMP did not specify who would attend the workshop and the NTRS has only the one report, from 1999. An NSBRI website search shows that after 1999 some individual disciplines held or took part in symposia, but did not always publish proceedings.

75. "Coop. Agrmt. Mgmt. Plan," 9–10, 12, 13.

76. Ibid., 11–12.

77. Rummel, "NSBRI Overview," JSC, May 6, 1998, NSBRI—Rummel/Fogleman

Notebook, Mackowski Life Sciences Coll. By May 1998, the NSBRI had enlisted 10 "active institute associates": BNL, Brooklyn College, Dartmouth College, Loma Linda Univ. Med. Ctr., UC-Irvine, Univ. of Florida, Univ. of Houston, Univ. of PA, Univ. of TX Health Sci. Ctr., and York Univ. in Toronto, ON.

78. Goldin to Bobby Alford, July 9, 1998, NSBRI—Rummel/Fogleman Notebook, Mackowski Life Sciences Coll.

79. Alford to Barbara J. Kirkland, BD/Procurement Systems Office, JSC, Nov. 18, 1998 and attached NSBRI, "Vision 2005," 1, NSBRI—Rummel/Fogleman Notebook, Mackowski Life Sciences Coll.

80. NSBRI, "Vision 2005," 1, 8.

81. Ibid., 4, 8, 9, 10.

82. Wilbur C. Trafton, "HEDS: A Summary Report on HEDS Res. & Tech. to the NASA Sr. Mgmt. Council," rev., Oct. 30, 1997, f. 076162 "Sr. Mgmt. Mtg.," b. 140, Goldin Recs.

83. Mtg. Notes, LMSA Advisory Committee, NASA HQ, Oct. 19–20, 2000, 3–4, 7, App. D, 1, www.nasa.gov/pdf/189494main_LMSAAC_minutes_102000.pdf (accessed May 27, 2020); Malcolm L. Peterson, "FY 1999 Budget Passback Presentation to the CIC," Dec. 17, 1997, f. 076180 "CIC Mtg.," b. 140, Goldin Recs.

84. NSBRI Annual Review, First Year Ops., FY 1997–1998, 1–3, NSBRI—Rummel/Fogleman Notebook, Mackowski Life Sciences Coll.

85. Carrie Ash to White, Mar. 11, 1999, NSBRI—Rummel/Fogleman Notebook, Mackowski Life Sciences Coll.

86. "NSBRI FY 99 Accomplishments," Dec. 31, 1999; "Bioastronautics Program: NASA's Goals," "Bioastronautics Facility Concept of Ops.," and NSBRI, "FY 2001 Augmentation Plan," all marked "Draft," Feb. 10, 2000; Nicogossian to Abbey, Oct. 22, 1999; Olsen to multiple addressees, Dec. 8, 2000 and attached "Site Visit Review Report of the NSBRI," Nov. 28–Dec. 1, 2000; Rummel, "DRAFT-4/24/02, NASA Reference Materials, 9–10; Olsen to Alford, Mar. 16, 2001; Alford to Olsen, Mar. 16, 2001; Olsen to Alford, Apr. 11, 2001; Alford to Olsen Apr. 12, 2001; "Goldin Briefing–Radiation Health and JSC Reorganization," Dec. 7, 1999; e-mail, White to Lillie Coney, June 5, 2001; e-mails Rummel to Olsen Sep. 6, 2001, Olsen to Joe Rothenberg, Sep. 7, 2001, and Olsen to Rummel, Sep. 19, 2001; "Impact of President's Budget Plan for FY 2003 on the NSBRI," Feb. 20, 2002, all in NSBRI—Rummel/Fogleman Notebook, Mackowski Life Sciences Coll.; OBPR, "BPRAC Mtg. Rept.," NASA HQ, Feb. 19–20, 2002, 2, 7, http://www.nasa.gov/pdf/189500main_BPRAC_minutes_022002.pdf (accessed May 27, 2020); OBPR, "BPRAC Mtg. Rept.," May 20–21, 2004, 14, http://www.nasa.gov/pdf/189486main_BPRAC_minutes_052004.pdf (accessed May 27, 2020).

87. Olsen to JSC Director, Mar. 4, 2002, Rummel/Fogleman Notebook, Mackowski Life Sciences Coll.

88. NSBRI, Rev. Strategic Plan, July 17, 2003, 2, 4, 10–11, 16; NSBRI, Annual Sci. and Tech. Rept., Oct. 1, 2004–Sep. 30, 2005, 1–2, 4–5, 18–19, 20–22.

89. Alford to Roy S. Estess, and "Impact of President's Budget Plan for FY 2003 on the NSBRI," Feb. 20, 2002; Rummel, "DRAFT-4/24/02, NASA Reference Materials, 11," NSBRI—Rummel/Fogleman Notebook, Mackowski Life Sciences Coll. The Apr.

24, 2002 Rummel draft document noted that "A level of $50–100M was constantly identified as the level that would be required to successfully establish and implement the NSBRI." He added that the idea had been to "support beyond low Earth orbit human space flight in the 2010–2020 time frame" but acknowledged that such had "not been officially approved."

90. See, e.g., OIG, "Barters on the International Space Station Program," IG-02-024, Sep. 5, 2002, 1–3, 6–7, obtained via OIG FOIA request 2008–37. Agency auditors found that NASA had bartered with foreign partners to lessen cash outflow to contractor Boeing, but "did not maintain adequate documentation to support its estimates of bartered item values" totaling $1.5B or delete from NASA's accounts payable ledger the full $.5B in work taken from Boeing. Items included MPLMs traded for Italian astronaut training and flight time. Other reports at "OIG Audit Reports," https://oig.nasa.gov/ (accessed June 29, 2020) detailed poor contract and accounting practices 1998–2001, potentially or actually overpaying Boeing hundreds of millions of dollars and providing inadequate oversight. Regarding Abbey's removal, see Goldin, Memo for the Record, Aug. 5, 1999, addressed to AA for Space Flight Joseph H. Rothenberg, signed by all three, in f. 077613 "Memos/Notes from the Code M AA," b. 180, Goldin Recs.

91. Lambright, "Leadership and Large-scale Technology: The Case of the International Space Station," *Space Policy* 21 (2005): 200 and "Leading Change at NASA: The Case of Dan Goldin," *Space Policy* 23 (2007): 41. Goldin officially resigned in Oct., effective Nov. 17, citing the 9/11 attacks as an impetus for spending more time with his family.

92. White interview; R. Srini Srinivasan, Joel I. Leonard, and Ronald J. White, "Mathematical Modeling of Physiological States," in *Space Biology and Medicine*, Vol. III, Book 2, edited by Nicogossian et al., 559–594.

93. It later became the "Human Research Roadmap," within the Human Research Program (HRP) of the Space Life Sciences Program at JSC. See https://humanresearchroadmap.nasa.gov/ (accessed Sep. 19, 2011 and Apr. 14, 2020).

94. NASA, *Bioastronautics Roadmap: A Risk Reduction Strategy for Human Space Exploration*, SP-2004-6113 (Washington, DC: NASA, Feb. 2005, previously published under JSC 62577), 7–8; David E. Longnecker and Ricardo A. Molins, eds., *A Risk Reduction Strategy for Human Exploration of Space: A Review of NASA's Bioastronautic Roadmap* (Washington, DC: National Academies Press, 2006), 20, 22, A-5.

95. NASA, *Bioastronautics Roadmap*, 7–8; Longnecker and Molins, eds., *A Risk Reduction Strategy for Human Exploration of Space*, 19; Charles interview and Charles Sawin, interview with the author, May 16, 2006.

96. NASA, *Bioastronautics Roadmap*, 8, A-4, A-6, A-7, A-9, A-11, A-12, A-14, A-17, A-21–22, A-26–27, A-31, A-34, A-37, A-39–40, A-42, A-44, A-66–67, A-69–70, A-72–73, A-74–75, A-77, A-78, A-81, A-83, A-85, A-87–88, A-89, A-90, A-92, A-98.

97. Longnecker and Molins, eds., *A Risk Reduction Strategy*, 19.

98. See Richard P. Feynman, "App. F–Personal Observations on the Reliability of the Shuttle," *Presidential Commission on the Space Shuttle Challenger Accident*, Vol. II (Washington, DC: Presidential Commission, 1986).

99. GAO, *Space Shuttle: Need to Sustain Launch Risk Assessment Process Improvements*, GAO/NSIAD-96–73 (Washington, DC: GAO, Mar. 1996), 2.

100. NASA, *Bioastronautics Roadmap*, 3.

101. Charles interview.

102. Ibid.

103. Ibid. See also P. A. Whitson, *Renal-Stone Risk Assessment during Space Shuttle Flights*, NASA TM-11752 (Houston: NASA-JSC, 1996), 3.

104. Charles interview; Radiation Health Program, Life Sciences Div., OSSA, *Radiation Health Research, 1986–1990*, TM-4270 (Washington, DC: NASA, 1991), 68–71.

105. Srinivasan et al., "Mathematical Modeling of Physiological States"; Hale et al., eds., *Wings in Orbit: Scientific and Engineering Legacies of the Space Shuttle, 1971–2010*, 434–435.

Chapter 6. Radiation and the Science of Risk Reduction

1. Richard S. Johnston, "Introduction," in *Biomedical Results of Apollo*, SP-368, edited by Richard S. Johnston, Lawrence F. Dietlein, and Charles A. Berry (Washington, DC: NASA, 1975), 5–6.

2. Barton C. Hacker and James M. Grimwood, *On the Shoulders of Titans: A History of Project Gemini*, SP-4203 (Washington, DC: NASA, 1977), 229–230, 537, 539, 541, 546–547, 549, 552–553, 558, 563–564, 567–568, 573–577; Swenson Jr., Grimwood, and Alexander, *This New Ocean: A History of Project Mercury*, SP-4201 (Washington, DC: NASA, 1989), 385, 399, 467, 497, 500.

3. W. Royce Hawkins and John F. Zieglschmid, "Clinical Aspects of Crew Health," in *Biomedical Results of Apollo*, 80. "Biostack" was a small apparatus layering biosamples and particle tracking devices.

4. J. Vernon Bailey, "Radiation Protection and Instrumentation," 107, 109–113; Hawkins and Zieglschmid, "Clinical Aspects," *Biomedical Results of Apollo*, 51; Nimi Rao, D. A. Hagemeyer, and D. B. Holcomb, US Dept. of Energy, Occupational Radiation Exposure Report for CY 2018, 1–1, 2–2, https://energy.gov/ehss/occupational-radiation-exposure (accessed Mar. 31, 2020).

5. HZEs are so named because they possess high atomic number (Z) and high energy (E).

6. Horst Bücker, "Biostack: A Study of the Biological Effects of HZE Galactic Cosmic Radiation," in *Biomedical Results of Apollo*, edited by Johnston et al., 341, 343–344, 346–351, 353.

7. W. Zachary Osborne, Lawrence S. Pinsky, and J. Vernon Bailey, "Apollo Light Flash Investigations," in *Biomedical Results of Apollo*, edited by Johnston et al., 355–365. As early as 1952 a researcher from the Division of Medical Physics at the UC Berkeley Donner Lab predicted the phenomenon. He reported it to the USAF's Wright Field Aero Medical Laboratory, which evaluated the Mercury astronauts seven years later. See Cornelius A. Tobias, "Radiation Hazards in High Altitude Aviation," *Av. Med.* (Aug. 1952): 345–372.

8. Webb Haymaker et al., "The Apollo 17 Pocket Mouse Experiment (Biocore)," in *Biomedical Results of Apollo*, edited by Johnston et al., 381–403.

9. W. David Compton and Charles D. Benson, *Living and Working in Space: A History of Skylab*, SP-4208 (Washington, DC: NASA, 1983), 168; Expt. Info.: Radiological Protection and Medical Dosimetry for the Skylab Crew (SKYRAD), LSDA, https://lsda.jsc.nasa.gov/Experiment/exper/840 (accessed June 10, 2020).

10. Expt. Info.: Cytogenetic Studies of Blood (M111), LSDA, https://lsda.jsc.nasa.gov/Experiment/exper/422 and Visual Light Flash Phenomena (M106), LSDA, https://lsda.jsc.nasa.gov/Experiment/exper/415, (accessed June 10, 2020).

11. "Risk of Cancer Mortality Among LSAH Participants Compared with the General Population," *LSAH Newsletter* (Feb. 1997), v. 6, iss. 1, https://lsda.jsc.nasa.gov/LSAH/LSAH_nav (accessed July 17, 2020). The LSAH compared current and former astronauts with a control population of several thousand civil service volunteers receiving the same annual medical tests. This 1997 cancer study also included statistics on the general population of the Texas Gulf Coast. It found that astronauts had a lower rate of cancer in comparison. LSAH was later renamed the Lifetime Surveillance of Astronaut Health.

12. D. E. Philpott et al., "Cosmic Ray Effects on the Eyes of Rats Flown on Cosmos No. 782, Expt. K-007" and Philpott et al., "Retinal Changes in Rats Flown on Cosmos 936: A Cosmic Ray Expt.," *ASEM* 49, no. 1 (Jan. 1978): 19–28 and 51, no. 6 (June 1980): 556–562; Expt. Info.: "Space Radiation Dosimetry Aboard Cosmos 1129: U.S. Portion of the Expt. (COS 1129-14)," LSDA, https://lsda.jsc.nasa.gov/Experiment/exper/70; "Radiation Dosimetry and Spectrometry (COS 1887-30)," LSDA, https://lsda.jsc.nasa.gov/Experiment/exper/102; "Radiation Dosimetry and Spectrometry: Passive Systems (8809A18)," LSDA, https://lsda.jsc.nasa.gov/Experiment/exper/171 (all accessed June 10, 2020). The eye tissue studies were done only on 2 missions in the 1970s.

13. Press kit for STS-41-G, 5, 25.

14. Press kits for STS-9, 43; STS-61-A, 19; and STS-42, 21–22, 32–33.

15. G. Reitz et al., "Influence of Cosmic Radiation and/or Microgravity on Development of *Carausius morosus*," *Adv. Sp. Res.* 9, no. 10 (1989): 161–173; H. Bücker et al., "Embryogenesis and Organogenesis of *Carausius morosus* Under Spaceflight Conditions," *Adv. Sp. Res.* 6, no. 12 (1986): 115–124. "Biorack" was a floor-to-ceiling reconfigurable workstation, typically outfitted with glovebox, storage area, microscope, incubator, refrigerator, centrifuge, camera, and laptop computer.

16. G. V. Dalrymple et al., "A Review of the USAF/NASA Proton Bioeffects Project: Rationale and Acute Effects," 117–119 and D. H. Wood, "Long-Term Mortality and Cancer Risk in Irradiated Rhesus Monkeys," 132–140, both in *Rad Res* 126 no. 2 (May 1991).

17. Lawrence W. Townsend, John W. Wilson, and John E. Nealy, *Prelim. Estim. of GCR Shielding Reqs. for Manned Interplanetary Missions*, TM-101516 (Hampton, VA: NASA-LaRC, Oct. 1988); A. Konradi et al., "Low Earth Orbit Radiation Dose Distribution in a Phantom Head," *Nucl. Tracks Radiat. Meas., Intl. J. Radiat. Appl. Instrum. Part D* 20, No. 1 (1992): 52. The NRL updated CREME in 1986 and 1996.

18. Courtney G. Brooks, James M. Grimwood, and Loyd S. Swenson Jr., *Chariots*

for Apollo: A History of Manned Lunar Spacecraft, SP-4205 (Washington, DC: NASA, 1979), 37.

19. M. P. Billings and W. R. Yucker, *Summ. Final Rept.: The Computerized Anatomical Man (CAM) Model,* MDC G4655, for NASA-JSC, Contr. NAS9–13228, Sep. 1973, 1, 3, 5, 20–23, 31, 92, 112, https://ntrs.nasa.gov/search (accessed July 19, 2020).

20. M. G. Yochmowitz, D. H. Wood, and Y. L. Salmon, "Seventeen-Year Mortality Experience of Proton Radiation in *Macaca mulatta*," *Rad Res* 102 (1985): 14–34; Dennis D. Leavitt, "Analysis of Primate Head Irradiation with 55-MeV Protons" 127–131; Wood, "Long-Term Mortality and Cancer Risk in Irradiated Rhesus Monkeys," 132–140; Dalrymple et al., "A Review of the USAF/NASA Proton Bioeffects Project," 117–119, all in *Rad Res* 126 no. 2 (May 1991). Similar to the CAM project, for a 1963–1969 radiation study researchers had to first compile an atlas of cross-sectional monkey anatomy, then develop a monkey phantom containing a rhesus skeleton to measure radiation dose distributions. By 1991, analysis based on computerized tomography scans suggested that the 1980s test protocols had unwittingly caused 59 percent of the monkey brain to receive excess radiation, as much as 3 times the intended dose in some areas, impacting the number of cancers.

21. Expt. Descrip.: "Inflight Radiation Dose Distribution (DSO 469)," LSDA, https://lsda.jsc.nasa.gov/Experiment/exper/200 (accessed July 17, 2020); Konradi et al., "LEO Radiation Dose Distribution in a Phantom Head," 49–54 and Konradi et al., "DSO 469A: LEO Radiation Dose Distribution in a Phantom Head," both in *Results of Life Sciences DSOs Conducted Aboard the Space Shuttle 1988–1990* (NASA: Jan. 1991), 59–64.

22. BEVALAC would shut down in 1993, and reportedly "most of the facility's unique heavy-ion data had never been published in any form." See Stephen M. Maurer, Richard B. Firestone, and Charles R. Schriver, "Science's Neglected Legacy," *Nature* 405 (May 11, 2000): 117–120.

23. See for example, Lisa C. Simonsen et al., *Radiation Dose to Critical Body Organs for Oct. 1989 Proton Event,* NASA TP-3237 (Hampton, VA: NASA-LaRC, 1992), 4. GOES observed increased x-ray fluxes and a solar particle "event" during STS-28 in Oct. 1989.

24. CIRRCP, *Fifth Ann. Rept.,* June 30, 1989, 1–2, 5, 10, "Information Bridge," OSTI, DOE, http://www.osti.gov/bridge/product.biblio.jsp?osti_id=204992 (accessed Apr. 27, 2020).

25. Huntoon, "Medical Science," WAR, S&LS Dir., Jan. 22, 1988, f. "JSC Weekly Activity Reports, Jan.–March 1988," b. 7, JSC Series, JSC Hist. Coll. Based on CIRRCP annual reports and the final report when it dissolved in 1995, this does not appear to have taken place. Reports are at US DOE, http://www.osti.gov/bridge/basicsearch.jsp (accessed June 11, 2020). Search by full name of the CIRRCP.

26. Comm. on Adv. Sp. Tech., Aero. and Sp. Engrg. Bd., Commission on Engrg. and Tech. Syst., NRC, *Space Technology to Meet Future Needs* (Washington, DC: National Academies Press, 1987), 55–64; Barney B. Roberts, "Assumptions for Jointly Funded Program," Interim Rev., Dec. 11–13, 1990, f. "Welcome, Expectations, and Long Range Plans," b. 5, Hab. Studies, Ctr. Series, JSC Hist. Coll.

27. J. Winter et al., *The NASA CSTI High Capacity Power Project*, TM 105813 (Cleveland: NASA Lewis, Aug. 1992), 1.

28. Mark D. Bowles and Robert S. Arrighi, *NASA's Nuclear Frontier: The Plum Brook Reactor Facility*, SP-2004-4533 (Washington, DC: NASA, 2004), 73, 75–76, 138, 140.

29. Wesley E. Bolch et al., *A Radiological Assessment of Nuclear Power and Propulsion Operations near Space Station Freedom*, Contract Rept. 185185 (Cleveland, OH: NASA-Lewis, Mar. 1990).

30. Michael Braukus, "NASA Selects Institution for Life Sciences Research," HQ, pr. 97–37, Mar. 12, 1997. The NSCORT program ended in 2002.

31. J. F. Weiss et al., "Radioprotection by Metals: Selenium" and Myra L. Patchen et al., "Radioprotection by Polysaccharides Alone and in Combination with Aminothiols," *Adv. Sp. Res.* 12, no. 2–3 (1992): 223–231, 233–248.

32. R. Meehan et al., "Human mononuclear cell function after 4 degrees C storage during 1-G and microgravity conditions of spaceflight," *ASEM* 60, no. 7 (July 1989): 644–648; D. M. Surgenor et al., "Human Blood Cells at Microgravity: The NASA Initial Blood Storage Experiment," *Transfusion* 30, no. 7 (1990): 605–616.

33. P. E. Segall, H. Sternberg, and H. D. Waitz, "Low Temp. Preservation and Space Medicine," *Jnl. Grav. Physiol.* 2, Vol. 1 (1995): 143–144. Abstract on PubMed, accessed Nov. 25, 2020.

34. See for example Townsend et al., *Prelim. Est. of Rad. Exposures for Manned Interplanetary Missions from Anomalously Large Solar Flare Events*, TM-100620 (Hampton, VA: NASA-LaRC, May 1988), 2, 7–14 and Townsend et al., "Human Exposure to Large Solar Particle Events in Space," *Adv. Sp. Res.* 12, no. 2–3 (1990): 39. These combined OSSA Interplanetary Monitoring Platform (IMP) satellite data on a 1972 solar event with 1956 and 1960 flares and extrapolated to compare expected radiation doses for different types and thicknesses of shielding. Tests were for structural aluminum, not tile foam. See also G. D. Badhwar et al., "Active Dosimetric Measurements on Shuttle Flights," *Nucl. Tracks Radiat. Meas., Int. J. Radiat. Appl. Instrum. Part D* 20, no. 1 (1992): 17. The SAA had been drifting to the west at 0.27 degrees per year, and data this team of JSC and Batelle scientists had was 20 years old.

35. See for example Francis A. Cucinotta et al., *Cellular Track Model of Biological Damage to Mammalian Cell Cultures from Galactic Cosmic Rays*, NASA TP 3055 (NASA-LaRC, 1991), 1; J. E. Keith, G. D. Badhwar, and D. J. Lindstrom, "Neutron Spectrum and Dose-Equivalent in Shuttle Flights during Solar Maximum," *Nucl. Tracks Radiat. Meas., Int. J. Radiat. Appl. Instrum. Part D* 20, no. 1 (1992): 41–42.

36. Andrew Lawler, "Space Radiation Hazard Threatens Mars Mission," *Space News*, Feb. 26–Mar. 4, 1990, 8; "New Solar Storm Detector Sending Real-Time Images Used to Warn of Sun's Damaging Storms," Jan. 30, 2003, NOAA Satellite and Info. Services, DOC, http://www.noaanews.noaa.gov/stories/s1087.htm (accessed Sep. 24, 2010). *Space News* also cited hopes of NASA radiation researchers for a three-satellite system around the sun by 2010 to warn of coming storms on Mars.

37. See Townsend et al., "Interplan. Crew Exposure Estim. for Galactic Cosmic Rays, *Rad Res* 129, no. 1 (Jan. 1992): 48–52; Townsend et al., "Interplan. Crew Exposure Estim. for the Aug. 1972 and Oct. 1989 Solar Particle Events," *Rad Res* 126, no. 1

(April 1991): 108–110; Townsend et al., "Risk Analyses for the Solar Particle Events of Aug. through Dec. 1989," *Rad Res* 130, No. 1 (April 1992): 1–6; Badhwar et al., "Active Dosimetric Measurements on Shuttle Flights," 18–20.

38. Donald E. Robbins et al., "Ionizing Radiation," in *Space Biology and Medicine III: Humans in Spaceflight, Book 2*, edited by Carolyn S. Leach Huntoon, Vsevolod V. Antipov, and Anatoliy I. Grigoriev (Washington, DC: AIAA, 1996), 388.

39. "Life Sciences Div. Strategic Implementation Plan," 3, 8, 10–11.

40. An accessible history of SEI is Thor Hogan's in *Mars Wars: The Rise and Fall of the Space Exploration Initiative*, NASA SP-2007-4410 (Washington, DC: NASA, 2007). Dan Quayle's experience, including his frustration with former astronaut and NASA Administrator Richard Truly, are covered in a chapter of his *Standing Firm: A Vice-Presidential Memoir* (New York: HarperCollins, 1994).

41. Marc S. Allen, "The Space Exploration Initiative: Perspectives from the US Space Science Community," *Space Policy*, Nov. 1992, 307–13.

42. "White House Considers Inquiry Into NASA," *Washington Post*, July 15, 1990, A18; Ann Devroy, "In Wake of NASA's Problems, Administration Orders Review of Space Program," *Washington Post*, July 17, 1990; David Lauter, "Outside Experts to Review NASA Status," *Los Angeles Times*, July 17, 1990; John M. Broder, "Bush Lets Space Agency Remain in Orbit, Scrubs External Review," *Los Angeles Times*, July 26, 1990; Lee Dye, "Countdown for NASA: Restructuring Movement Launched as Time Runs out for Ailing Space Agency," *Los Angeles Times*, Aug. 20, 1990. Lauter's article quoted Quayle as denying "reports published over the weekend," likely a mix of speculation and leaks, that the NSC "is planning a wholesale restructuring of the agency." Dye's local sources in Los Angeles, possibly Paine or Bruce Murray of the CIT, both at the Quayle meeting that spawned AC-FUSSP's formation, said that "fierce opposition from NASA's leadership," probably Truly and J. R. Thompson, persuaded the president to back off the "major external investigation" idea.

43. Truly, NMI 1156.35, Aug. 10, 1990, f. 7, b. 13, Paine Papers. AC-FUSSP reported to Truly, who had it do a 120-day study.

44. Memorandum, Augustine to Members of AC-FUSSP, Dec. 17, 1990, f. 2, "Adv Comm . . . Corr & Memoranda," b. 14, Paine Papers.

45. "Rept. of the AC-FUSSP" (Washington, DC: NASA, 1990), https://history.nasa.gov/augustine/racfup1.htm (accessed June 11, 2020).

46. Roberts, "Early High Visibility Accomplishments of FY91 Constrained," "Augustine Report and Relationship to SEI," "FY 1991 Approps. Conf. Rept. Language," and "Summary," PSS Interim Review; "The SEI," Mar. 4, 1991, attached to Mark K. Craig to Roberts, Mar. 14, 1991, f. "SEI Promotional Materials," b. 6, Ctr. Series–Hab. Studies, JSC Hist. Coll.

47. "Life Sciences Div. Strategic Implementation Plan," 1989, 8. The connection is stated in Arnauld E. Nicogossian, "Foreword," *Space Life Sciences Strategic Plan 1992*, TM-107954 (Washington, DC: NASA, 1992), i: "The disapproval of the LifeSat series of biosatellites and the associated radiation initiative funding to support the ground-based research to address HZE biological effects has jeopardized one of the

major areas of scientific investigation. Re-planning was initiated to adjust to this decision."

48. Life Support Branch, Life Sciences Div., OSSA, "Space Radiation Health Program Plan," TM-108036 (Washington, DC: NASA, Nov. 1991), 20–21, 25–29. LET is "Linear-Energy Transfer," the amount of E transferred to an object, such as tissue, when an ionizing particle (e.g., an alpha-, beta-, gamma- or x-ray or a high-energy photon) travels through it, expressed as E per unit of path length.

49. The National Cancer Institute was sponsoring research there at the same time. See Joseph R. Castro, Stanley B. Curtis, and William T. Chu, "Bevalac Funding," *Science* 257, no. 5066 (July 3, 1992): 10–11. Some NASA money had been non–life sciences dollars, too, for example to calibrate particle detectors carried on board research satellites and balloons. See David P. Hamilton, "NASA Researchers Protest DOE Turnoff," *Science* 256, no. 5062 (June 5, 1992): 1388.

50. LSB, LSD, OSSA, "Space Radiation Health Program Plan," 10, 12, 21, 29–30; Roberts, "Assumptions for Jointly Funded Program," PSS Interim Review; Ken Frankel and Jack Miller, "Radiation Effects," LBNL, n.d. (before Sep. 1991), f. "Action Item 91–14. Sep. 16, 1991," b. 8, Ctr. Series—HQ Series, JSC Hist. Coll. As of Dec. 1990, the Houston document had shown the DOE as providing "the ion source, particle acceleration, and operating infrastructure at Brookhaven" in 1991. The NSRL opened in 2003.

51. NASA Advisory Council, AMAC, *Strategic Considerations for Support of Humans in Space and Moon/Mars Exploration Missions: Life Sciences Research and Technology Programs*, Vol. 1, TM-107983, June 1992.

52. Check the Life Sciences Data Archive (LSDA), which allows searching by year, mission, and other parameters: https://lsda.jsc.nasa.gov/.

53. NAC, AMAC, *Strategic Considerations*, Vol. 1, Caleb Hurtt, NAC Chair, to Dan Goldin, Washington, DC, June 3, 1992 in front matter, 4–5, 15, 21–22, 30–31, 36, 48–50, 52, 54, 56, 74–75, 78–79.

54. NAC, AMAC, *Strategic Considerations*, Vol. 1, 1–2, 36.

55. Ruth P. Liebowitz, *History of the Space Radiation Effects (SPACERAD) Program for the Joint USAF/NASA CRRES Mission. Part 1. From the Origins through the Launch, 1981–1990* (Hanscomb AFB, MA: Phillips Lab, Mar. 16, 1992), 42, 46, 47.

56. M. A. Green, *A Comparison of the Space Station Version of ASTROMAG with Two Free-Flyer Versions*, LBL-32977, UC-412, a paper presented at the 1992 Space Cryogenics Workshop, Ottobrunn, Germany, June 15–16, 1992.

57. NAC, AMAC, *Strategic Considerations*, Vol. 1, 10, 13.

58. NAC, AMAC, *Strategic Considerations*, Vol. 1, 36, 54. For example, John W. Wilson, John E. Nealy, and Walter Schimmerling, *Effects of Radiobiological Uncertainty on Shield Design for a 60-Day Lunar Mission*, TM-4422 (Hampton, VA: NASA-LaRC, 1993).

59. Lyn Ragsdale, "Politics Not Science: The U.S. Space Program in the Reagan and Bush Years," in *Spaceflight and the Myth of Presidential Leadership*, edited by Roger D. Launius and Howard E. McCurdy (Urbana: University of Illinois Press, 1997), 135–139, 140–144, 162–163. Emphasis in the original.

60. Ragsdale, "Politics Not Science," 162–164.

61. A. Steve Johnson et al., *Spaceflight Radiation Health Program at the Lyndon B. Johnson Space Center*, TM-104782 (Houston: NASA-JSC, 1993), 12.

62. See, for example, A. B. Cox and J. T. Lett, "The Quantification of Wound Healing as a Method to Assess Late Radiation Damage in Primate Skin Exposed to High-Energy Protons," *Adv. Sp. Res.* 9, no. 10 (1989): 125–130. This was a USAFSAM and CSU study of the irradiated colony. Rabbits, specifically their ears and eyes, were also common test subjects.

63. "Experiment Description for: Chromosome Analyses of Gemini Astronauts (GEMCYTO)," LSDA, https://lsda.jsc.nasa.gov/Experiment/exper/894 (accessed July 19, 2020).

64. Johnson et al., *Spaceflight Radiation Health Program*, 14.

65. "Space Radiation Analysis Group, Johnson Space Center," https://srag.jsc.nasa.gov/; Mark Weyland, "Operational Aspects of Space Radiation Analysis," Oct. 18, 2005, Earth-Sun System Division, https://hesperia.gsfc.nasa.gov/sspvse/oral/Mark_Weyland/presentation.ppt (all accessed July 19, 2020).

66. D. Robbins, "HEDS Radiation Roadmap," Feb. 7, 1995, f. 073967, "2/07/95 Briefing to Goldin Code U Radiation Road Map," b. 47, Goldin Recs.

67. Experiment Information: "Cytogenetic Effects of Space Radiation in Human Lymphocytes," https://lsda.jsc.nasa.gov/Experiment/exper/339, and "Inflight Radiation Measurements (5.2.1)," https://lsda.jsc.nasa.gov/Experiment/exper/340, LSDA (accessed July 19, 2020); T. C. Yang et al., "Biodosimetry Results from Space Flight Mir-18," *Rad. Res.* 148 (5 Suppl., Nov. 1997): S17–23.

68. An ISS "increment" includes the main Expedition team and any rotating crewmembers launched separately but cohabiting and working with the Expedition group.

69. "Fundamental Biology," "Human Life Sciences," and "ISS Risk Mitigation" Science, *History of Shuttle-Mir*, https://spaceflight.nasa.gov/history/shuttle-mir/science/sc-fb.htm (accessed June 11, 2020); Experiment Information: "Real-Time Radiation Monitoring Device (RRMD)," https://lsda.jsc.nasa.gov/Experiment/exper/756 (accessed Jan. 5, 2021), "Effective Dose Measurement during EVA (9401704)," https://lsda.jsc.nasa.gov/Experiment/exper/750; "Radiation Monitoring Equipment III (RME-3)," and "Cosmic Radiation and Effects Activation Monitor (CREAM)," LSDA, https://lsda.jsc.nasa.gov/Experiment/exper/757 (accessed July 19, 2020). CREAM was on Increments 6 and 7, RME on 5, 6, and 7.

70. Experiment Information: "Environmental Radiation Measurements on Mir Station (9401620)," LSDA, https://lsda.jsc.nasa.gov/Experiment/exper/762 (accessed July 19, 2020); E. V. Benton, A. L. Frank, and E. R. Benton, *Environmental Radiation Measurements on the Mir Station, Program 1–Internal Experiment, Program 2–External Experiment, Year 2 Progress Report*, CR-205067, Contract No. NCC2–893, NASA-ARC, Apr. 15, 1997, 2–3.

71. Experiment Information and Experiment Data Elements, "Active Dosimetry of Charged Particles (9401681)," LSDA, https://lsda.jsc.nasa.gov/Experiment/exper/763 (accessed July 12, 2020).

72. Aerospace Safety Advisory Panel (ASAP), Annual Report for 1997, 37 and handwritten note by Goldin, f. 077522 "2/12/98 Meeting with the ASAP," b. 179, Goldin

Recs. NASA axed TransHab, including Moon and Mars, in 2000. See "Transhab Concept," *International Space Station History*, June 27, 2003, archived at NASA Hist. Prog. Off. https://spaceflight.nasa.gov/history/station/transhab/ (accessed June 11, 2020).

73. Benton, Frank, and Benton, *Environmental Radiation Measurements on the Mir Station*, 2–4.

74. K. Walter, "Biological Research Evolves at Livermore," *Sci. & Tech. Rev.* (Nov. 2002), LLNL, DOE/University of CA, Dec. 12, 2002, https://str.llnl.gov/content/pages/past-issues-pdfs/2002.11.pdf (accessed July 12, 2020); T. C. Yang et al., "Biodosimetry Results from Space Flight Mir-18," J. N. Lucas, "Dose Reconstruction for Individuals Exposed to Ionizing Radiation Using Chromosome Painting," A. A. Edwards, "The Use of Chromosomal Aberrations in Human Lymphocytes for Biological Dosimetry," A. M. Chen et al., "Computer Simulation of Data on Chromosome Aberrations Produced by X Rays or Alpha Particles and Detected by Fluorescence *In Situ* Hybridization," *Rad. Res.* 148 (1997): S17-S23, S33-S38, S39-S44, S93-S101. Chromosome painting was invented at and patented by LLNL in the mid-1980s.

75. Benton, Frank, and Benton, *Environmental Radiation Measurements on the Mir Station*, 6; "The Christmas Radiation Brick," *Space Research* 1, no. 1 (Fall 2001): 11, MSFC and Hampton University, Hampton, VA; "ISS Expedition Crew and Shielding Augmentation: 'Christmas Bricks for a Safer Heaven,'" *Sp. Rad. Health Newsletter* 1, no. 2, Apr. 2001, https://nasa.gov/centers/johnson/pdf/513049main_V1–2.pdf (accessed July 12, 2020). For the Expedition 2 crew in 2001, for example, JSC used 10 years of data from LaRC, JSC, and LBNL to fulfill a two-year push by the SRHP to create an ad-hoc radiation-protected sleep station for the empty rack position in Destiny. Clean room workers fabricated CH_2 "POLY-Bricks," wrapped them in fire-retarding Nomex, and made a lightweight, inexpensive assembly that could be unfolded in orbit.

76. NASA, *Bioastronautics Roadmap*, 7–10, 12, 22, 28–29; Longnecker and Molins, eds., *A Risk Reduction Strategy for Human Exploration of Space*, 18–19.

77. Goldin to Alford, July 9, 1998; Fogelman notes, "Goldin Briefing–Radiation Health and JSC Reorganization," Dec. 7, 1999; and *Cooperative Agreement Management Plan for the Establishment of the NSBRI*, May 29, 1997, NASA-JSC S&LS Dir., 7., NSBRI—Rummel/Fogelman Notebook, Mackowski Life Sciences Coll.

78. Radiation Effects Team, NSBRI, "Previous Team Projects," n.d. (ca. 2000–2004), http://www.nsbri.org/Research/Projects/listprojects.epl?team=radiation&status=prev (accessed Sep. 24, 2010).

79. Allan Tobin et al., Site Visit Review Report of the NSBRI, Nov. 28–Dec. 1, 2000, Chief Scientist, Office of the Administrator, NASA HQ, 10–11; "Project Technical Summary" and "NASA Task Book Entry" for "Quantitation of Radiation Induced Deletion and Recombination Events Associated with Repeated DNA Sequences"; "Radiation-Induced Cytogenetic Damage as a Predictor of Cancer Risk for Protons and Fe Ions," "Countermeasures for Space Radiation Biological Effects," and "Countermeasures for Space Radiation-Induced Myeloid Leukemia," Radiation Effects Team, NSBRI, "Previous Team Projects." n.d. (ca. 2000–2004).

80. "Revised Strategic Plan," NSBRI, July 17, 2003, 2, 4–5.

81. "BNL's Marcelo Vazquez Selected as Space Radiation Liaison for the NSBRI,"

July 14, 2004 and "Neutron Detector under Development to Monitor Spacecraft Radiation," Nov. 11, 2003, *News Releases*, NSBRI, http://www.nsbri.org/NewsPublicOut/ (accessed Sep. 24, 2010).

82. LSD, OLMSA, *Strategic Program Plan for Space Radiation Health Research*, 1998, 13, 14, 19, 22. The plan cited breakthroughs in the ability to genetically predict cancer susceptibility, later sparking debate about medical privacy rights in the workplace.

83. LSD, OLMSA, *Strategic Program Plan*, 13, 14, 17, 23, 28.

84. LSD, OLMSA, *Strategic Program Plan*, 21, 23–24, 26. The document stated an assumption that the ethical issues involving acceptable risk and astronaut selection would have been resolved during Phase I, 1998–2002.

85. LSD, OLMSA, *Strategic Program Plan*, 21, 22.

86. LSD, OLMSA, *Strategic Program Plan*, 30–31.

87. "Visualization of Radiation Doses Estimated on the Martian Surface" and "Biography of Gautam D. Badhwar," accessed via "SRHP Featured Investigator: Dr. Gautam Badhwar (1940–2001)," *Space Rad. Health Newsletter* 1, no. 2 (Apr. 2001), Human Adaptation and Countermeasures Division, JSC, http://hacd.jsc.nasa.gov/web_docs/ radiation/newsletters/V1–2.htm; "MARIE: The Martian Radiation Environment Experiment," 2001 Mars Odyssey, JPL, http://mars.jpl.nasa.gov/odyssey/technology/ marie.html (both accessed Sep. 24, 2010). MARIE was launched in 2001, went into orbit around Mars in 2002, but malfunctioned Oct. 28, 2003, maybe ironically a victim of solar radiation flares. It found that in its orbital plane around Mars, radiation levels were 2.5 times that of the ISS orbit, perhaps because its orbit was higher than the station's. MARIE also observed some SPEs not picked up on Earth, confirming that they are directional.

88. "Organ Dose Measurement Using the Phantom Torso (93-E039)," https:// lsda.jsc.nasa.gov/Experiment/exper/906; "Dosimetric Mapping (96-E094)," https:// lsda.jsc.nasa.gov/Experiment/exper/900; "Bonner Ball Neutron Detector (BBND)," LSDA, https://lsda.jsc.nasa.gov/Experiment/exper/901; "A Study of Radiation Doses Experienced by Astronauts During EVA (96-E011)," https://lsda.jsc.nasa.gov/Experiment/exper/905; "Chromosomal Aberrations in Blood Lymphocytes of Astronauts (99-E010)," https://lsda.jsc.nasa.gov/Experiment/exper/933; "Matroshka-1," https:// lsda.jsc.nasa.gov/Experiment/exper/13397, all on LSDA, (accessed July 20, 2020). See also Karen Miller, "The Phantom Torso," May 3, 2001, NASA Science, https://science. nasa.gov/science-news/science-at-nasa/2001/ast04may_1/ (accessed Apr. 29, 2020). The Phantom Torso was a 95-pound, 3-foot-high model head and torso made from human bones, with plastic skin and organs that mimicked the density of human flesh. Sliced into 35 sections, it had 5 radiation detectors placed at the head, neck, chest, and abdomen.

89. "NASA Space Radiobiology Research Takes Off at New Brookhaven Facility," *Discover Brookhaven* 1, no. 3 (Fall 2003), http://www.bnl.gov/discover/Fall_03/ NSRL_1.asp; BNL, Associated Universities, Inc., Upton, L.I., NY, *Conceptual Design of the Booster Applications Facility (BAF) for the U.S. DOE*, Oct. 1997, http://server.c-ad. bnl.gov/esfd/nsrl/operations/index.html; "Space Radiobiology," BNL, http://www. bnl.gov/medical/NASA/LTSF.asp (all accessed Sep. 24, 2010). The word "booster" is

misleading, as scientists need a *lower* energy beam to simulate space radiation, less than 1 GeV. The BNL website described the Booster as a "lower energy pre-accelerator serving the AGS."

90. The calculated "lens equivalent dose" is the numerical product of the absorbed dose for the tissue, in this case the lens of the eye at 0.3 cm, and the biological impact of ionizing radiation measured in Sieverts (Sv).

91. F. A. Cucinotta et al., "Space Radiation and Cataracts in Astronauts," *Radiation Research* 156, no. 5 (2001): 460–466. Lens dose was determined using dosimeter badge data, modified by mission-specific readings, the type of protective shielding used, and the inefficiency of the badge in detecting high-LET particles.

Chapter 7. Design and Redesign: The Many Space Stations of NASA

1. For more history of very early LaRC space station studies see Chap. 9 of James R. Hansen, *Spaceflight Revolution: NASA Langley Research Center from Sputnik to Apollo*, SP-4308 (Washington, DC: NASA, 1995); For early MSFC work see Andrew J. Dunar and Stephen P. Waring, *Power to Explore: A History of Marshall Space Flight Center, 1960–1990* (Washington, DC: NASA, 1999), 529–532. See also Henry C. Dethloff, *Suddenly Tomorrow Came: A History of the Johnson Space Center* (Houston: NASA-JSC, 1993), 188–189.

2. Hansen, *Spaceflight Revolution*, 294–296; "Background and Viewpoints on Space Station," by MDAC, internal doc., Nov. 1981, rec. no. 19955, b. 2 "McDonnell Douglas Space Station Concepts," Mackowski Life Sciences Coll. MORL also was referred to as the Manned Orbital Research Laboratory.

3. *Big G*, McDonnell Douglas Corporation, internal pub., 97 bound pages, Dec. 1967, 2, 4–5, 7, 26, 32, b. 2 "McDonnell Douglas Space Station Concepts," Mackowski Life Sciences Coll.

4. US/USSR Coop. Space Laboratory (Skylab/Salyut), June 23, 1972, MDAC Eastern Division; Intl. Skylab Space Station Technical Considerations, April 1973, MDAC–West, both in Skylab Collection, Western Historical Manuscript Collection, University of Missouri–St. Louis, MO.

5. *Space Station: Exec. Summary*. MSFC-DPD-235/DR No. MA-04, Contract NAS8-25140, April 1972, MDC G2727. MDAC–West, Huntington Beach, CA, iii, 1–2, 38; "Background and Viewpoints on Space Station."

6. *Space Station: Exec. Summary*, MDAC–West, 3, 5, 10, 18–19, 37, 53; "Background and Viewpoints on Space Station."

7. Biomedical Expts. Scientific Satellite, 2nd Program Rev., Dec. 8, 1975, GE Space Div.; BESS Preliminary Design Study, Vol. 2, System Design, n.d., GE Space Div.; W. E. Berry, J. W. Tremor, and T. C. Aepli, "Biomedical Experiments Scientific Satellite (BESS)," Aerospace Div., American Society of Mechanical Engineers (ASME), presented at Intersociety Conf. on Environmental Systems, San Diego, CA, July 12–15, 1976, rec. no. 19955, b. 2 "McDonnell Douglas Space Station Concepts," Mackowski Life Sciences Coll.

8. "Background and Viewpoints on Space Station"; Manned Orbital Systems Concepts Study, Book 1—Executive Summary, MDC G5919 DPD 433 MA-04; Sep. 30, 1975, MDAC-West, Huntington Beach, CA, 2, 7–8, 14, 21–23, 35.

9. "Background and Viewpoints on Space Station"; Dunar and Waring, *Power to Explore*, 543–549.

10. "Background and Viewpoints on Space Station."

11. John Logsdon, *Together in Orbit: The Origins of International Participation in the Space Station* (Washington, DC: NASA, Nov. 1998), 8, 9–10, 17.

12. "Background and Viewpoints on Space Station."

13. White House, draft directive, "Presidential Directive—Space Transportation Policy," item "Space Policy Review, Terms of Reference," Secs. I. "Future Launch Vehicle Needs" and II. "Shuttle Organizational Responsibilities and Capabilities," 1–3, attached Allen J. Lenz to Martin Anderson, Edwin Harper, Verne Orr, Richard Darman, George A. Keyworth II, James Beggs, Hans Mark, and William Schneider, July 17, 1981, f. "NSDD 144," "NSDD—National Security Decision Directives, Reagan Administration," *Intelligence Resource Program*, Federation of American Scientists, https://fas.org/irp/offdocs/nsdd/nsdd-144.htm (accessed July 20, 2020). This was declassified in 1996.

14. *Soviet Space Programs: 1981–87. Piloted Space Activities, Launch Vehicles, Launch Sites, and Tracking Support*, prepared at the request of Hon. Ernest F. Hollings, Chair, Committee on Commerce, Science and Transportation, US Senate, Part 1, May 1988, 26–29.

15. David S. F. Portree, *Mir Hardware Heritage*, JSC 26770 (Houston: NASA-JSC, October 1994), 23, 26–32, 47–57.

16. Between 1978 and 1988 the Soviet Intercosmos program hosted 14 pilots from Czechoslovakia, Poland, East Germany, Bulgaria, Hungary, Vietnam, Cuba, Mongolia, Romania, France, India, Syria, and Afghanistan.

17. Geoffrey E. Perry, "Soviet Space Stations: Achievements, Trends, and Outlook," Nov. 1982, App. B in *Salyut: Soviet Steps Toward Permanent Human Presence in Space*, OTA-TM-STI-14 (Washington, DC: US Congress, OTA, Dec. 1983), 67. Perry cited several articles from various Soviet news sources.

18. "Experiments on this Mission," LSDA, Cosmos 782, https://lsda.jsc.nasa.gov/Mission/miss/57; Cosmos 936 https://lsda.jsc.nasa.gov/Mission/miss/58; Cosmos 1129, https://lsda.jsc.nasa.gov/Mission/miss/59 (accessed July 26, 2020).

19. D. Wensley to multiple addressees, n.d. [1982]; "Proposal Outline," and NASA RFP W 10–28647/HWC-2, June 28, 1982; Herbert S. Snyder to "All Offerors," Att. A, 4, Encl. 1, 17 and Encl. 3, 47, all in R. A. Husman, *Proposal Preparation Guidelines: Study of Space Station Needs, Attributes and Architectural Options*, 82W-152Q MDAC-West, rec. no. 19955, b. 2 "McDonnell Douglas Space Station Concepts," Mackowski Life Sciences Coll. Because of its EOS program using the shuttle, MDAC made biomedical electrophoresis a model in its RFP response. For international collaboration at this point, see also Logsdon, *Together in Orbit*, 11.

20. MDAC, "RFP Summary Data"; NASA, RFP "Space Station Definition and Prelim. Design," Solicitation No. 9-BF-10-4-01P, Aug. 20, 1984, B–6, B–7. rec. no. 19955, b. 2 "McDonnell Douglas Space Station Concepts," Mackowski Life Sciences Coll. In 1985 MDAC had already selected a second employee, Robert Wood, to fly the first production EOS equipment on a planned STS-61-M in June 1986 (Walker interview,

Nov. 7, 2006, JSC OHC, https://historycollection.jsc.nasa.gov/JSCHistoryPortal/history/oral_histories/WalkerCD/WalkerCD_11-7-06.htm [accessed Oct. 10, 2021], 2–3, 7–8, 41–42).

21. NASA, RFP "Space Station Definition and Prelim. Design," B-14, B-16; Rumerman, *NASA Historical Data Book*, Vol. V., 244; Sherman Shrock, *Space Station Needs, Attributes, and Architectural Options Study–Final Report*, v. 2, *Mission Definition* (Denver: Martin Marietta, 1983), 2–7 and 4–1 and v. 4 *Mission Implementation Concepts*, 7–1 and 7–3. Volume 6 was *DOD Mission Considerations (Classified)*.

22. NASA, RFP "Space Station Definition and Prelim. Design," B-11, B-12, C-20.

23. Ibid., C-2-7, C-2-8.

24. Ibid., C-4-47. As defined in NASA RP-1024 (J8400008) "Anthropometric Source book," Vol. I.

25. Astronauts were subject to OSHA regulations in space. Exposure to decibel (dB) levels greater than 85 over an 8-hour weighted average (launch produced 118 dB in the crew cabin) would have mandated hearing tests by an audiologist, hearing protection devices, etc. See "Occupational Noise Exposure–1910.95," OSHA, US DOL, https://www.osha.gov/laws-regs/regulations/standardnumber/1910/1910.95 (accessed July 26, 2020); Timothy Castellano, "An Investigation of Acoustic Noise Requirements for the Space Station Centrifuge Facility," TM-108811 (Moffett Field, CA: NARA-ARC, Feb. 1994), 21.

26. NASA, RFP "Space Station Definition and Prelim. Design," C-4-51, C-4-53, C-4-57, C-4-58.

27. Ibid., C-4-55.

28. Ibid., C-4-46, C-4-55, C-4-56.

29. Ibid., C-4-58, C-4-59.

30. Report of the Committee on the Space Station of the National Research Council (Washington, DC: National Academies Press, Sep. 1987), xi, 6–10, draft copy.

31. CBO, *The NASA Program in the 1990s and Beyond* (Washington, DC: Congress, 1988), 57–60.

32. Roger D. Bourke to Laurel Wilkening, Sep. 8, 1990, f. 10, "AC-FUSSP, Correspondence and Memoranda, 1990," b. 13; Wilkening to the Program and Building Blocks Group, Sep. 10, 1990, f. 9, "AC-FUSSP, Correspondence and Memorabilia 1989–90," b. 13; Jim Bain, typed notes, "Congressional Reception for Norm Augustine Oct. 2, 1990," and Wilkening to Paine, Bob Herres, and Joe Allen, Oct. 29, 1990, f. 10, b. 13; Anonymous to Paine, (1990), f. 4 "AC-FUSSP Reading File" [2 of 3] 1990," b. 19; Paine to Augustine, July 9, 1991, f. 4 "Govt. File, AC-FUSSP, 1991," b. 14, Thomas O. Paine Papers, MS Div., LOC, Washington, DC. Bourke, who had worked at HQ in 1985–86 and with JPL, used the term "blood feuds," comparing conditions at NASA to cinematic "Bedouin tribes" in *Lawrence of Arabia*, who "could not set aside their individual differences . . . to work toward a higher collective good." In an October 1 reply Wilkening called his "perspective on the warring factions . . . somewhat different from concerns that have been expressed by others." Bourke thought it was a bottom-up issue, not one of Truly's leadership but of "workers" who had "lost touch with the reason they are

there." At MSFC and HQ, the most competent people had been promoted away or left, leaving the "second or third rate" to do the work.

33. Paine to Augustine, Santa Monica, CA, Oct. 19, 1990, f. 10, b. 13, Paine Papers. Paine wrote, "The meat axe Congress swung at Space Station Freedom this week should ease our concern that our review might worsen the plight of NASA's manned space-flight program. We won't be accused now of destroying the village to save it. The timing of our December report looks better than ever."

34. Memorandum "Space Station," attached to Jim Beale to Paine, Oct. 19, 1990, f. 10, b. 13, Paine Papers; Stephen Daggett, "A Peace Dividend in 1990–91?" Feb. 14, 1990, Congressional Research Service.

35. See Julie Hatch and Angela Clinton, US Dept. of Labor, "Job Growth in the 1990s: A Retrospective," *Monthly Labor Review* (Dec. 2000): 8; Daniel H. Else, "The U. S. Defense Industrial Base: Trends and Current Issues," Oct. 27, 2000, Congressional Research Service, 3–5.

36. *Space Station Freedom Media Handbook* (Washington, DC: NASA, April 1989), 5; *Space Station Freedom Media Handbook* (Washington, DC: NASA, May 1992), 4; "Space Station," f. 10, b. 13, Paine Papers. The Sep. 29, 1986 MOU had called for the United States to pay 71.4 percent of an anticipated $7 billion price tag. ESA and Japan would each pay 12.8 percent and Canada 3 percent.

37. *Space Station Freedom Media Handbook* (1992), ii, 4.

38. John Cox, "Space Station Freedom Status," *Beyond the Baseline 1991*, Vol. 1: *Space Station Freedom, Part 1*, NASA Conf. Pub. 10083, Aug. 6–8, 1991.

39. K. Leath, R. J. Saucillo, and B. D. Meredith, "Evolution [of] User Requirements for the Restructured Space Station," William M. Cirillo, "SSF Growth Concepts and Configurations," Ken Flemming, "STV Fueling Options," Granville Paules, "Space Station Freedom Baseline Operations Concept," *Beyond the Baseline 1991*, Vol. 1: *Space Station Freedom, Part 1*, Vol. 2; Cox, "Space Station Freedom Status" and Donald W. Monell, "Evolution Design Requirements and Design Strategy," IAF, 43rd International Astronautical Congress, Aug. 28–Sep. 5, 1992. Leath and Flemming were from McDonnell Douglas; the other lead authors were from various NASA offices and centers.

40. *Space Station Redesign Team Final Rept. to the Adv. Comm. on the Redesign of the Space Station*, NASA-TM-109241 (Washington, DC: NASA, June 1993), 1, 263–264.

41. *Space Station Redesign Team Final Rept.*, 20, 259, 267; Goldin, notes, 1–5, f. 073594, b. 40, Goldin Recs. (emphasis in original).

42. See "Current Partner Participation in Phase 2," f. 073659, "Briefing Book NASA Delegation to Moscow, Oct. 2–6, 1993," b. 41, Goldin Recs. Also see Appendix A of the June 1993 *Space Station Redesign Team Final Rept.*, for the list of redesign team members. Of sixty-four names on the list, three represented Canada, four Japan, one the Italian Space Agency, and three ESA.

43. Memorandum, Margaret G. Finarelli to Daniel Goldin, "Visit of Ambassador Kuriyama April 19, 3:30–4:00," f. 073610 "April 19, 1993–Background Ambassador Kuriyama," b. 40, Goldin Recs. Finarelli referred to Sam Jameson's "Japan Unhappy with Space Station Cutback," *Los Angeles Times*, Apr. 14, 1993, which quoted Junji

Yoshihara, director of the Office of Space Utilization at the Japanese space agency as the source for the prime minister's discussions with the US president, and that Japanese displeasure might further impact the planned Superconducting Supercollider project. See also Takashi Otsuka, Akira Ozeki, and Norio Shimojima, "The Wandering Space Development Project at the Mercy of U.S. Convenience," *Asahi Shimbun*, Mar. 31, 1993.

44. "Concerns of the International Partners," f. 073631 "6/4/93 Briefing to DB [*sic*, DG], Int'l. Partner Meeting," b. 40, Goldin Recs.

45. Dennis Browne to John Boright, May 26, 1993, f. 073631, b. 40, Goldin Recs.

46. Report, "Intergovernmental Consultative Session, May 13, 1993," Dept. of State, f. 073631, b. 40, Goldin Recs.

47. *Space Station Redesign Team Final Rept.*, 20.

48. "Intergovernmental Consultative Session, May 13, 1993," DOS, f. 073631, b. 40, Goldin Recs.

49. *Space Station Redesign Team Final Rept.*, 6, 14, 267; "ISSA Assembly Sequence," July 12, 1994, for July 18, 1994 senior station mgmt. briefing, JSC, f. 073869 "Sr. Mgmt. Mtg.," b. 45; press release, Apr. 6, 1993, f. 073562, b. 39, Goldin Recs. The press release termed this "redesign guidance."

50. Goldin, typed notes, "Station Mgmt. Mtg.," Feb. 7, 8, 11, and 16, 1993 and handwritten notes n.d. [Feb. 8, 1993], f. 073562, b. 39, Goldin Recs.

51. McCarthy, typed notes, "Advisory Committee on the Redesign of the Space Station, May 4, 1993," 13–14, 16–18, f. 075980, b. 137, Goldin Recs. Regarding the lack of data, Goldin said, "I went to NASA's archives to find the CDR requirement for detailed drawings; it was removed 10 years ago."

52. *Space Station Redesign Team Final Rept.*, 259, 261, 295. The $9 billion was in addition to more than $8 billion already spent.

53. Hess and Jim Cast, "Space Station Host Center and Prime Contractor Announced," p.r. 93–148, HQ, Aug. 17, 1993, f. 073562, b. 39, Goldin Recs.

54. Per John Logsdon, *Together in Orbit*, 42, all the partners did not sign the agreement until Jan. 29, 1998.

55. USAF, *History of Research in Space Biology and Biodynamics at the U.S. Air Force Missile Development Center Holloman AFB, NM 1946–1958*, 33–41, https://permanent. fdlp.gov/lps70487/lps70487/www.hq.nasa.gov/office/pao/History/afspbio/top.htm (accessed Oct. 15, 2020); Harry G. Armstrong, ed., *Aerospace Medicine* (Baltimore: Williams & Wilkins, 1961), 609–612; Pitts, *The Human Factor*, 5–10.

56. Benjamin J. Loret, "Optimization of Manned Orbital Satellite Vehicle Design with Respect to Artificial Gravity" (master's thesis, Air University, 1961), 33–34, 76, 78 https://apps.dtic.mil/dtic/tr/fulltext/u2/268249.pdf (accessed Oct 18, 2020). See also Hale et al., *Wings in Orbit: Scientific and Engineering Legacies of the Space Shuttle, 1971–2010*, 413–414 about a 1992 shuttle centrifuge experiment that showed a threshold for plants of 15–20 "g-seconds," i.e., 1 g for 15 to 20 seconds for gravitropic response.

57. Dethloff, *Suddenly Tomorrow Came*, 144–145.

58. Hansen, *Spaceflight Revolution*, 269–270, 277–288, 292–297.

59. Long-arm centrifuges had been thought preferable for producing less dizziness and nausea but later US and Soviet/Russian ground and flight research demonstrated this was not so. See James B. Lackner and Ashton Graybiel, "The Effective Intensity of Coriolis, Cross-coupling Stimulation is Gravitoinertial Force Dependent: Implications for Space Motion Sickness," *ASEM* (March 1986): 229–235 and Aleksey A. Shipov, "Artificial Gravity," in *Space Biology and Medicine*, Vol. III: *Humans in Spaceflight, Book 2*, edited by Nicogossian et al., 349–363.

60. Arthur H. Smith et al., *Space Station Centrifuge: A Requirement for Life Science Research*, NASA TM-102873 (Moffett Field, CA: NASA-ARC, Feb. 1992), 1.

61. Charles interview.

62. Joseph F. Shea, "Preface," *Space Technology to Meet Future Needs*, Comm. on Adv. Space Tech., Aero. and Space Eng. Board, Commission on Eng. and Tech. Systems, NRC (Washington, DC: National Academies Press, 1987), vii-x. The committee's critique had in mind NASA's mandate in the Space Act of 1958 to pursue research of use to the military as well as a civilian space program. See pages 43, 47–48 125–126.

63. NRC, *Space Technology to Meet Future Needs*, 50, 65–66.

64. For example, R. A. J. Asher, "The Dangers of Going to Bed," *Brit. Med. Jnl.* 2, no. 4536 (Dec. 13, 1947): 967–968; Norman L. Browse, *Physiology and Pathology of Bed Rest* (Springfield, IL: Charles C. Thomas, 1965); R. Issekutz Jr. et al., "Effect of prolonged bed rest on urinary calcium output," *Jnl. Appl. Physiol.* 21, no. 3 (1966): 1013–1020 by a team under USAF contract; Stephen B. Hulley et al., "The Effect of Supplemental Oral Phosphate on the Bone Mineral Changes during Prolonged Bed Rest," *Jnl. Clin. Inv.* 50 (1971): 2506–2518; C. L. Giannetta and H. B. Castleberry, "Influence of Bed-rest and Hypercapnia Upon Urinary Mineral Excretion in Man," *Aerosp. Med.*, 1974, 45(7): 750–754 (partially NASA funded); Theodore N. Lynch et al., "Metabolic Effects of Prolonged Bed Rest: Their Modification by Simulated Altitude," *Aerosp. Med*, Jan. 1967, 10–20. The latter two were USAFSAM studies. See also "Poliomyelitis," https://www.cdc.gov/vaccines/pubs/pinkbook/downloads/polio.pdf (accessed July 26, 2020), which said US paralysis cases from polio peaked at 21,000 in 1952.

65. M. Hinsenkamp et al., "In Vivo Bone Strain Measurements: Clinical Results, Animal Experiments, and a Proposal for a Study of Bone Demineralization in Weightlessness," *ASEM* 52, no. 2 (1981): 95–103. Berry's hormone idea was closer to the truth, but Hinsenkamp noted it "cannot explain the local and preferential demineralization of the weightbearing bones."

66. Souza and Ilyin, "Major Results from Biomedical Experiments in Space," in *Space Biology and Medicine*, Vol. III: *Humans in Spaceflight, Book I, Effects of Microgravity*, edited by Nicogossian et al., 37–39.

67. John M. Vogel and Michael W. Whittle, "Bone Mineral Changes: The Second Manned Skylab Mission," *ASEM* 47, no. 4 (1976): 396–400.

68. Malcolm C. Smith Jr. et al. "Bone Mineral Measurement Experiment M078," in *Biomedical Results from Skylab*, edited by Johnston and Dietlein (Washington, DC: NASA, 1977), https://lsda.jsc.nasa.gov/books/skylab/biomedical_result_of_skylab.pdf (accessed July 26, 2020).

69. Frederick Elmore Tilton, "Long-term Follow-up of Skylab Bone Demineralization" (master's thesis, Univ. of TX Health Sci. Ctr. at Houston, June 1979), 1–2, 4–6, 10, 11, https://apps.dtic.mil/dtic/tr/fulltext/u2/a107965.pdf (accessed Oct. 18, 2020).

70. R. Srini Srinivasan, Joel I. Leonard, and Ronald J. White, "Mathematical Modeling of Physiological States," in *Space Biology and Medicine*, Vol. III, *Book II*, edited by Nicogossian et al., 559–562, 583–585.

71. N. C. Birkhead et al., Abstract of "Effect of Exercise, Standing, Negative Trunk and Positive Skeletal Pressure on Bed Rest-Induced Orthostatis and Hypercalciuria," Final Rept., Feb. 1964-Jan. 1965, Contract No. 615, AF336151538, Rept. No. 0129036, Jan. 1966, http://www.dtic.mil/docs/citations/AD0630921 (accessed Mar. 23, 2011).

72. TRW Defense and Space Systems Group, "Vestibular Research Facility for Spacelab," 1980 and LMSC, Adv. Syst. Div., Biotech. Dept., "Vestibular/Variable-Gravity Research Facility," Sep. 1980, both in RN 9923, Life Sciences Coll., NASA Hist. Ref. Coll.; Tomko interview. "Cooperation Between Space Agencies," *Life into Space, 1965–1990*, edited by Kenneth Souza, Robert Hogan, and Rodney Ballard, 8–9.

73. Smith et al., *Space Station Centrifuge*, vii.

74. Ibid., 1, 3, 4, 5, 8–9. The USAFSAM had done the red blood cell research in the late 1960s.

75. Larry Lemke to Carl Gustaferro, July 2, 1990 and attached ARC Space Proj. Div., "Executive Summary & Conclusions" and "Human-Rated Centrifuge External Rotor Concept," n.d. (FY 1989), f. 56.14.17, "Human Rated Centrifuge," sh. 1, cb. 3, Chambers Files.

76. Smith, "NASA's Space Station Program: Evolution of its Rationale and Expected Uses," Test. before Subcomm. on Science and Space, Comm. on Commerce, Science, and Transportation, US Senate (Washington, DC: CRS, Apr. 20, 2005), CRS-6 and "Space Stations" (Washington, DC: CRS, Aug. 9, 2005), CRS-3. Smith testified "the redesign excluded plans for a centrifuge," but in her later essay wrote that agency management lacked "firm plans" for orbiting one.

77. National Commission on Space, *Pioneering the Space Frontier* (New York: Bantam Books, 1986; repr. NASA History Program Office, updated Mar. 31, 2009), https://history.nasa.gov/painerep/cover.htm (accessed July 26, 2020), sections "Building the Technology Base" and "Highway to Space," 95–130. Reagan established the NCS, aka the Paine Commission, in October 1984 with a 20-year agenda.

78. L. G. Lemke, A. F. Mascy, and B. L. Swenson, "The Use of Tethers for an Artificial Gravity Facility," in *Space Tethers for Science in the Space Station Era*, Conf. Procdgs. Vol. 14, Venice, Oct. 4–8, 1987, edited by Luciano Guerriero and Ivan Bekey (Bologna: Società Italiana di Fisica, 1988), 385.

79. David N. Schultz et al., "A Manned Mars Artificial Gravity Vehicle," in *Space Tethers for Science in the Space Station Era*, edited by Guerriero and Bekey, 320, 324–327.

80. The effect of the Coriolis force, named for French engineer-mathematician Gustave-Gaspard Coriolis, is an apparent deflection of the path of an object that moves within a rotating coordinate system. The object does not actually deviate from its path, but it *appears* to do so. From the *Encyclopedia Britannica Online*, https://www.britannica.com/science/Coriolis-force (accessed July 27, 2020).

81. Schultz et al., "A Manned Mars Artificial Gravity Vehicle" and Lemke, Mascy, and Swenson, "The Use of Tethers for an Artificial Gravity Facility," 320–335, 379–387.

Chapter 8. The Cold War and Its Aftermath: Scientific Exchange, Social Change

1. Basic information about ASTP, Cosmos, and other missions can be found on NASA's Life Sciences Data Archive, https://lsda.jsc.nasa.gov, searchable by program, mission, year, species, field of research, and other parameters. See also the two-volume *Life into Space: Space Life Sciences Experiments* for the Cosmos/Bion project: Kenneth Souza, Robert Hogan, Rodney Ballard eds., *Ames Research Center 1965–1990*, RP-1372 and Souza, Guy Etheridge, and Paul X. Callahan, *Ames Research Center, Kennedy Space Center 1991–1998*, SP-2000-534. "Cosmos" was the program and "Bion" the biosatellite series.

2. Logsdon, "The Development of International Space Cooperation," in *Exploring the Unknown:* Vol. II: *External Relationships*, edited by Logsdon et al., 14–15. Information on the 1977 agreement to fly the shuttle to a Soviet Salyut station can be found in the same volume, 215–217, Doc. I-50, "Agreement Between the U.S.S.R. Academy of Sciences and the National Aeronautics and Space Administration."

3. WH, NSDD-75, U.S. Relations with the Soviet Union, Jan. 17, 1983, 5, 6, "NSDD—Reagan Administration," *Intelligence Resource Program*, FAS, https://fas.org/irp/offdocs/nsdd/nsdd-75.pdf (accessed July 20, 2020). This secret directive was declassified in 1994.

4. Souza to Robert Dunning, ARC, Jan. 31, 1980, f. 13.2.1.9 "Cosmos 1514 (1983)," sh. 1 / 2, cb. 1, Chambers Files. Cosmos 1514 did fly with NASA experiments in Dec. 1983.

5. Nicogossian to Harold P. Klein, Mar. 9, 1984, f. 13.2.1.9, sh. 1 / 2, cb. 1, Chambers Files.

6. Nicogossian, report to multiple addressees, Jan. 3, 1984, f. 13.2.1.9, sh. 1 / 2, cb. 1, Chambers Files.

7. Keefe, report, Mar. 12, 1984, f. 13.2.1.9.2 "60 Day Success Reports" and "Biosatellite Embryological Experiment with Mammals," trans. N. Timacheff et al., f. 13.2.1.90 "Paper on 'Cosmos' Biosatellite XII JWG," both sh. 1 / 2, cb. 1, Chambers Files. Capitalization in original. Ontogenesis is the study of the development of a particular feature or entire organism from conception until maturity.

8. J. Alberts, report, Mar. 1, 1984, f. 13.2.1.9.2 "60 Day Success Reports," sh. 1 / 2, cb. 1, Chambers Files.

9. Harold Sandler, H. L. Stone, John W. Hines, report, Mar. 13, 1984 and Christopher E. Cann, report, n.d. (Mar. 1984), f. 13.2.1.9.2 "60 Day Success Reports," sh. 1/2, cb. 1, Chambers Files.

10. Charles R. Doarn et al., "A Summary of Activities of the US/Soviet-Russian Joint Working Group on Space Biology and Medicine," *Acta Astronautica* 67 (2010): 649–658.

11. "Negotiations and Joint Planning" and "How Phase 1 Started," https://historycollection.jsc.nasa.gov/history/shuttle-mir/history/to-h-b-negotiations.htm and *History of Shuttle-Mir*, https://historycollection.jsc.nasa.gov/history/shuttle-mir/

history/h-b-organizational-frame.htm (accessed Jan. 16, 2020); "Implementing Agreement between NASA of the USA and RSA of the Russian Federation on Human Space Flight Cooperation," Oct. 5, 1992, f. 073515, "Russia Trip Oct 4–6, 1992," b. 38, Goldin Recs.

12. "Graphic Timeline," *History of Shuttle-Mir,* https://spaceflight.nasa.gov/history/shuttle-mir/references/r-timeline.htm (accessed Oct. 21, 2020). Vladimir Titov flew twice, on STS-63 and STS-86.

13. Guy Gardner and Bruce Luna, "Shuttle-Mir Program Customer Requirements," Nov. 13, 1992, f. 073542 "Shuttle Mir Issues," b. 39, Goldin Recs.

14. JSC, "Proposal for Enhancement of S. S. Freedom Capabilities through Utilization of the Russian Mir Space Station for US Biomedical Studies," Apr. 6, 1992 and attached US/USSR Biomedical Program reports: Gautam A. Badhwar (Radiation), Peggy A. Whitson (Biochemistry and Metabolism), Victor S. Schneider (Bone), John B. Charles (Cardiovascular), James M. Waligora (Barophysiology), Albert W. Holland (Human Behavior and Performance), Michael C. Greenisen (Exercise), and Millard F. Reschke (Neuroscience), n.d., f. 073492, "Station 7/17/92," b. 38, Goldin Recs.

15. George C. Nield and Pavel Vorobiev, eds., *Phase 1 Program Joint Report,* NASA SP-1999-6108 (Washington, DC: NASA, 1999), 2; John Uri interview, JSC, by Rebecca Wright and Mark Davison, May 15, 1998, and John B. Charles interview, JSC, by Wright, P. Rollins, and Glen Swanson, Aug. 28, 1998, "People: Oral Histories," *History of Shuttle-Mir,* https://spaceflight.nasa.gov/history/shuttle-mir/people/oral-histories.htm (accessed Oct. 24, 2020).

16. John Krige, Angelina Long Callahan, and Ashok Maharaj, *NASA in the World: Fifty Years of International Collaboration in Space* (New York: Palgrave Macmillan, 2013), 6–18, 147–151, 154–158, 160–162, 166–169, 171–182, 249–265. Krige contrasts the policymaking of Arnold J. Frutkin (1959–1979), who came from a space and science background, with that of his successors, who had civilian nuclear and arms control experience. The latter arrived with Reagan's SDI and stayed for Bush's Gulf War.

17. See for example, f. 077523, "2/25/98–Senator Lott Briefing"; "Talking Points," f. 077527, "3/5/98 Meeting with Vice-President Gore in preparation of GCC-10"; and f. 077541, "4/22/98–Meeting with Congressman Sensenbrenner," b. 179; f. 077569, "9/24/98–Meeting with Congressman Jerry Lewis"; Memorandum, Goldin to Sen. Christopher Bond, Sep. 29, 1998, in f. 077571 "10/7/98–Meeting with Sen. Barbara Mikulski"; Tab A, f. 077570, "10/5/98 Telecon with Yuri Koptev"; Talking Points, f. 077580, "January 28, 1999 Telecon with Yuri Koptev," b. 180; Goldin Recs.

18. US Congress, OTA, *U.S.—Russian Cooperation in Space,* OTA-ISS-618 (Washington, DC: GPO, Apr. 1995), 2, 7–16, 18, 22–23.

19. Krige, Callahan, and Maharaj, *NASA in the World,* 151.

20. LMSC, "A Concept for U.S. Based Station with Russian Enhancements," Sep. 21, 1993, f. 073656 "Lockheed Station Issues," b. 41, Goldin Recs.

21. Space Station Freedom Office, "Status of the Mir-2 Core Replacement," July 17, 1992, f. 073492, "Station 7/17/92," b. 38, Goldin Recs.; "Mir Space Station," *Shuttle-Mir: The U.S. and Russia Share History's Highest Stage,* https://history.nasa.gov/SP-4225/mir/mir.htm (accessed Dec. 17, 2020).

22. Doc. I-55, "Protocol to the Implementing Agreement," *Exploring the Unknown: Vol. II: External Relationships,* edited by Logsdon et al., 230–232; Nicogossian, "Status of OLMSA Programs," NASA-NIH Advis. Panel on Biomedical and Behavioral Res., Mtg. Mins. Jan. 9–10, 1994, 3, f. 073850 "NASA/NIH Advisory Committee meeting," b. 44, Goldin Recs.

23. "Implementing Agreement"; Lambright, "Leadership and Large-scale Technology: The Case of the International Space Station," 198. Lambright argues that Clinton proposed this to prevent the Russians from selling missile technology to India. (See, e.g., Margaret P. Finarelli to Goldin, Sep. 1, 1992 and US State Dept., "Missile Sanctions against ISRO and Glavkosmos," f. 073501, "Indian Visit 9/2/92," b. 38, Goldin Recs.) In May 1992 Bush had imposed a two-year ban on US sales to or imports from Glavkosmos and the Indian Space Research Organization (ISRO) when a proposed transfer of Russian missile engine technology to ISRO threatened to worsen global missile proliferation. To augment the Bush stick, Clinton would offer a carrot.

24. "Item 10, Russian Human Space Flight Issues and Concerns," f. 073659, b. 41, Goldin Recs. From a list of questions provided by the Russian representatives to the Space Station Transition Team to NASA: "Although they had a great deal of technical expertise in Space they felt that they did not have a similar level of expertise in the field of Economics. They indicated that [they] would welcome any guidance by NASA in the area of cost estimating and accounting."

25. By agreement, the United States retained ownership and intellectual property rights to "data produced by these instruments." See "Background and Talking Points regarding Utilization Plans (Science, Technology, and Payload Accommodation) for NASA-Mir Interim Program (Phase I) and Joint Station Program (Phases II and III), 1. Implementation of Mir-1 Activities," 2–3, f. 073659, b. 41, Goldin Recs. While it is tempting to credit such loans, some done at a midlevel or scientist-to-scientist basis, to a belief in science being above politics, John Krige challenges that view in "NASA as an Instrument of Foreign Policy," *Societal Impact of Spaceflight,* 208. NASA "served as a vector of U.S. foreign policy," he writes, deliberately using scientific collaboration as "a platform to consolidate the political and cultural solidarity of the free world."

26. f. 13.2.1.1.3, "Space Biomedical and Life Support Systems JWG (8/23–27/92)," sh. 1 / 2, cb. 1, Chambers Files; Roger D. Billica, June 17, 1998, interview by Wright, Carol Butler, and Davison, 10, "People: Oral Histories," *History of Shuttle-Mir;* Ken Souza, interview with the author, Jan. 15, 2008.

27. "Space Biomedical and Life Support Systems JWG," Chambers Files.

28. "Canadian Participation in Phases 1 and 2," f. 073673, "Briefing Book: Space Station Heads of Agencies Meeting & Bilaterals w/CSA & the Italian Space Agency, Nov. 6–8, 1993," b. 41; Schumacher to Goldin, May 5, 1994 visits by Heimann of Dornier, f. 073816, "Visit of Mr. Werner Heimann . . . Dornier 5/5/94," and Mennicken of DARA, f. 073817, "Visit by Dr. Jan-Baldem Mennicken German Space Agency 5/5/94," b. 44; Bryan O'Connor and [first name unspecified] Nguyen, "Shuttle-Mir Rendezvous Phase I," pres., Aug. 5, 1994, f. 073893, "Shuttle-Mir Rendezvous Phase I O'Connor-Nguyen 8/5/94," b. 45, Goldin Recs.

29. "Background and Talking Points . . . Phase III, 2. Free-flyer," f. 073659, b. 41; Robert W. Clarke and Schumacher to Goldin, May 4, 1994, f. 073816 & 073817, b. 44, Goldin Recs.; Nield and Vorobiev, eds., *Phase 1 Program Joint Report*, 2; Uri interview.

30. P. Maliga, "Support for the Science Community," f. 073766, "4/4/94 Meeting with RSA Officials," b. 43, Goldin Recs. For the $400 million figure, which in Oct. 1993 was to cover Phases 1 and 2, see "Strategy for the $400 Million Support to Russia," f. 073659, b. 41, Goldin Recs.

31. See, for example, Office of Technology Assessment, *U.S.-Russian Cooperation in Space*, OTA-ISS-618 (Washington, DC: GPO, 1995), iii, 20, 82–84 and "Talking Points," Mar. 4, 1998, f. 077527, b. 179, Goldin Recs.

32. Billy Goodman, "Noted Researchers Laud Donation to Russian Science," *The Scientist* (Jan. 11, 1993), https://www.the-scientist.com/news/noted-researchers-laud-donation-to-russian-science-59843 and Peter Gwynne, "George Soros Reduces Scope of his International Science Foundation," *The Scientist* (Jan. 8, 1996), https://www.the-scientist.com/news/george-soros-reduces-scope-of-his-international-science-foundation-58210 (accessed July 23, 2020); Talking Points, Tab B, Yakobashvili mtg., f. 074032, "6/07/95 Meeting with Deputy Minister of Science and Technology Policy of Russian Federation–Yakobashvili," b. 46, Goldin Recs.

33. L. Cline, "$400M Contract," Tab C; D. Jacobs and Cline, "Liaison Offices," Item E; handwritten notes, Tab 6, f. 073766, b. 43, Goldin Recs. See also Brian Dailey interview, Bethesda, MD, by Wright, Apr. 19, 1999, "People: Oral Histories," *History of Shuttle-Mir*. Dailey provided insight into NASA and Goldin's roles in identifying and aligning Koptev and the RSA as civil space counterparts in 1992.

34. A useful source to consult on people, events, politics, and opinions is the Shuttle-Mir section of the JSC Oral History Project, https://historycollection.jsc.nasa.gov/JSCHistoryPortal/history/oral_histories/shuttle-mir.htm, (accessed July 23, 2020).

35. "Mir Space Station," https://www.history.nasa.gov/SP-4225/mir/mir.htm (accessed July 23, 2020).

36. Talking Points I. B., "Approach to Joint Developments," 3 and I. C., "Proposed Schedule of Activities," 1, f. 073659, b. 41; Talking Point 4b, Meeting with Yuri Koptev, Sep. 27, 1994, f. 073920, "Meeting w/Yuri Koptev, RSA at NASA Headquarters, Wash DC 27 Sept 1994," b. 46, Goldin Recs.

37. The fire broke out early in Jerry Linenger's 1997 mission and the collision occurred during the stay of Michael Foale, in June of the same year.

38. Charles Stegemoeller, JSC, interview by Wright, Butler, and Tim Farrell, Aug. 6, 1998 and Richard W. Nygren, interview by Wright, Butler, and Summer Bergen, July 23, 1998, "People: Oral Histories," *History of Shuttle-Mir;* Goldin notes and Talking Points Tab 3c, mtg. with Koptev, Sep. 27, 1994, f. 073920, b. 46, Goldin Recs.

39. Stegemoeller interview. Per Stegemoeller, Spektr had been initiated as a top-secret military module during the Cold War, to study the spectra of missile rocket plumes.

40. Stegemoeller interview; Nield and Vorobiev, eds., *Phase 1 Program Joint Report,* 124–126.

41. Stegemoeller interview; Gary Kitmacher, JSC, interview by Wright, Butler, and Rollins, June 29, 1998, 22, "People: Oral Histories," *History of Shuttle-Mir.*

42. Stegemoeller interview; Nield and Vorobiev, eds.,*Phase 1 Program Joint Report,* 59.

43. Nield and Vorobiev, eds., *Phase 1 Program Joint Report,* 253–254, 277–280.

44. "Greenhouse—Integrated Plant Experiments on Mir: Greenhouse Hardware," *Science: Fundamental Biology,* last updated 07/16/1999, archived https://www.history.nasa.gov/SP-4225/science/science.htm#fb (accessed July 23, 2020).

45. F. B. Salisbury et al., "Plant Growth During the Greenhouse II Experiment on the Mir Orbital Station," *Advances in Space Research* 31, no. 1 (2003): 221–227. Problems during 1995 occurred in the data downlink, watering system, controllers, lights, fans, and camera.

46. The ups and downs of the collaboration have been detailed elsewhere, including James Oberg's *Star-Crossed Orbits: Inside the U.S.–Russian Space Alliance* (New York: McGraw-Hill, 2002); journalist Bryan Burrough's *Dragonfly: NASA and the Crisis Aboard Mir* (New York: HarperCollins, 1998); *Off the Planet: Surviving Five Perilous Months Aboard the Space Station Mir* (New York: McGraw-Hill, 2000) by astronaut-physician Jerry Linenger; and *Waystation to the Stars: The Story of Mir, Michael, and Me* (London: Headline Book Publishing, 1999), written by astronaut Michael Foale's father, Colin Foale. Other published sources include NASA's official web-based overview of the program, *History of Shuttle-Mir,* curated by Kim Dismukes and based on the NASA-published history, *Shuttle-Mir: The U.S. and Russia Share History's Highest Stage,* https://history.nasa.gov/SP-4225/ (accessed Dec. 27, 2020). The first three tended to focus on the negatives, the next two attempted to balance the terror with triumphs, and the last was a study in optimism. The Agency also compiled individual Shuttle-Mir program oral histories, and the Records of NASA Administrator Daniel S. Goldin at the National Archives in College Park, MD, contain detailed briefings that clearly express the frustrations of both sides.

47. Nield and Vorobiev, eds.,*Phase 1 Program Joint Report,* 250.

48. Many NASA employees interviewed for the Shuttle-Mir oral history project referred to these delays. None stated that there had been any bad intent on the part of the RSA. Rather, there had been a failure to communicate and understand Russian and Kazakh customs regulations and procedures, and some of the problems could have been avoided.

49. Panel Disc./Recs.–Gen., NASA-NIH Advis. Panel on Biomedical and Behavioral Res., 10. NASA was able to research provided it chose a Russian Co-Investigator. The RSA also required NASA astronauts to follow the same countermeasures as their cosmonauts.

50. Nield and Vorobiev, eds.,*Phase 1 Program Joint Report,* 170, 173.

51. Charles interview.

52. Billica interview, and Sam Pool interview Aug. 3, 1998, by Wright, Kevin Prusnak, and Franklin Tarazona, "People: Oral Histories," *History of Shuttle-Mir.*

53. Charles interview.

54. Charles interview.

55. Pool interview; and Bonnie Dunbar interview, June 16, 1998, by Wright, Butler, and Davison; C. Michael Foale interview, June 16, 1998, by Wright, Butler, and Davison; and Frank Culbertson, Mar. 24, 1998, interview by Davison, Wright, and Rollins, "People: Oral Histories," *History of Shuttle-Mir;* Nield and Vorobiev, eds.,*Phase 1 Program Joint Report,* 108, 148, 152–153, 157–158, 160, 240, 252; Roberta L. Gross to F. James Sensenbrenner, Aug. 29, 1997, Washington, DC, OIG, http://www.hq.nasa.gov/office/oig/hq/old/inspections_assessments/mir/Welcome.html (accessed Sep. 11, 2008).

56. Burrough, *Dragonfly,* 447–449, 455–456, 457–458, 472; Foale, *Waystation to the Stars,* 165, 171, 174, 187. The first commander of Foale's Mir mission, Valeriy Tsibliyev, developed a heart irregularity credited to the stress of command compounded by the Progress crash.

57. Burrough, *Dragonfly,* 508–511; Oberg, *Star-Crossed Orbits,* 136–146, 204–210; Linenger, *Off the Planet,* 248–251; Fax, James P. Connolly, Mar. 11, 1993 and Paula Cleggett-Haleim et al. to Sulzman and Chambers, Mar. 15, 1993; Draft, OSSA PA Q&A, n.d. [Mar. 1993]; Peter Judd, "Inhuman Sacrifice!" *Globe,* Mar. 9, 1993, n. p. For workplace safety in the late 1990s, see Office of Safety & Mission Assurance, NASA Procedural Requirements 8715.1, "NASA Occupational Safety and Health Programs," Aug. 9, 1999, NODIS Library, NASA-Goddard, Greenbelt, MD, http://nodis3.gsfc.nasa.gov/displayDir.cfm?Internal_ID=N_PR_8715_0001_&page_name=main (accessed Sep. 30, 2011).

58. Gross to Sensenbrenner, Aug. 29, 1997; Marcia S. Smith, Testimony before the US House of Reps. Comm. On Science, Sep. 18, 1997. Transcript, US House of Reps., http://commdocs.house.gov/committees/science/hsy126000.000/hsy126000_0.htm (accessed July 23, 2020).

59. Oberg, *Star-Crossed Orbits,* 278–281 and 342. Oberg quoted Merbold from an internal NASA memo summarizing his statements.

60. Ibid., 281–282 and 342.

61. Charles interview. Charles began as cardiovascular-discipline lead on the Thagard flight, then deputy project scientist, project scientist for all human life sciences investigation by the time of the Lucid mission, and finally, mission scientist for NASA-Mir, overseeing all the sciences for NASA 4 and NASA 5.

62. Charles interview. Defining success was an issue not confined to Shuttle-Mir. Charles and David Liskowsky of HQ had to calculate the success of *Columbia*'s last mission after it disintegrated during reentry. The SLS-2 team called the mission both 99.1 percent and 120 percent successful since the pilot and commander, less busy than a cockpit team on missions requiring rendezvous, docking, deployment, grappling, construction, or repair, had time to participate as test subjects. See Charles and

Liskowsky, "STS-107 Whole Payload % Science Gained," May 30, 2003, http://space-flight.nasa.gov/shuttle/Science_Gained_05–30–03.pdf (accessed Apr. 30, 2020) and OLMSA, Post Launch Mission Operation Report, No. U-420–68–93–58, Mar. 1994, 3–4, 10.

63. Nield and Vorobiev, eds., *Phase 1 Program Joint Report;* Harry Holloway, "OLMSA Overview" to NASA Alumni League annual mtg., June 6, 1994, f. 076046, "NASA Alumni League 1994 Annual Mtg. 6/6/94," b. 138, Goldin Recs. Spektr and Priroda both were to carry gear for biological, medical, and biotechnical experiments.

64. Culbertson, "Science Recovery" and "Preliminary Results/Phase 2/3 Benefits" in "Statement of Capt. Frank Culbertson (USN-Ret), Phase 1 Prog. Mgr., Before the Comm. on Science U.S. House of Representatives, Sep. 18, 1997," in "Hearing on Mir Safety, Sep. 18, 1997 before the Committee on Science, House of Representatives," https://www.history.nasa.gov/SP-4225/documentation/mir-safety/safety.htm (accessed July 23, 2020).

65. Nield and Vorobiev, eds., *Phase 1 Program Joint Report;* "Phase 1 Program Overview," Shuttle–Mir Science, https://www.history.nasa.gov/SP-4225/science/science.htm#research-overview (accessed July 23, 2020).

66. O'Connor and Nguyen, "Shuttle-Mir Rendezvous Phase 1," f. 073893, b. 45, Goldin Recs.

67. "Definitization Negotiations Contract NAS 15–10110," 80, f. 073851, "Briefing 6–16–94 Koptev Meeting," b. 44; "January 28, 1999—Telecon w/Yuri Koptev," Goldin Recs; "Agreement Among the Governments of . . . Concerning Cooperation on the Civil International Space Station," archived content, US Department of State, https://www.2009–2017.state.gov/documents/organization/190175.pdf (accessed Nov. 29, 2020). The ruble was worth one-fifth of its earlier value six months after devaluation, $9.95 million instead of $44.5 million, but the RSA asked for a new exchange rate that would have still doubled the value of the contract, possibly because inflation in Russia doubled as well. A helpful explanation of this crisis is Abbigail J. Chiodo and Michael T. Owyang, "A Case Study of a Currency Crisis: The Russian Default of 1998," Fed. Res. Bank of St. Louis *Review*, Nov./Dec. 2002, 7–18.

68. "Background and Talking Points," f. 077562 "8/4/98 Mtg. with Yuri Koptev," b. 180 Goldin Recs.

69. Schumacher to Goldin, Jan. 26, 1998, f. 077506 "1/27/98 Mtg. with Yuri Koptev," b. 179; Schumacher, Talking Points, f. 077527, b. 179; "Key Points," f. 077562, b. 180; Talking Points, Tab B, f. 077574, "10/16/98 Telecon with Yuri Koptev," b. 180; Talking Points, f. 077580, b. 180, Goldin Recs. A summer 1999 deorbit was to "signal that . . . 1st priority is joining partnership on Space Station" to ISS partners, Congress, and the American public. The Russians deorbited Mir, long a source of national pride, in Mar. 2001.

70. Notes, f. 076124, "7/15/97 telecon with Koptev," b. 139; Talking Points, f. 077550 "June 8, 9, 1998–Dinner and Meeting w/Dr. Bensoussan," b. 179 and f. 077580, b. 180, Goldin Recs.

71. Talking Points, f. 076124, b. 139 and Talking Points, f. 077580, b. 180, Goldin Recs.

72. Talking Points for Mar. 5, 1998 mtg. with V.P. Gore in preparation for the Gore-Chernomyrdin conf. 10, f. 077527, b. 179, Goldin Recs.

73. "Meeting with RSA General Director Koptev, August 4, 1998," f. 077562, b. 180, Goldin Recs.

74. "I. B. Approach to Joint Developments," Talking Points, p. 4, f. 073659, b. 41; Notes, "8/4/98 Mtg. with Yuri Koptev," f. 077562; Schumacher, Tab A, "10/5/98 Telecon with Yuri Koptev," f. 077570; "10/16/98 Telecon with Yuri Koptev," f. 077574, b. 180, Goldin Recs.; John H. Beall to Comptroller, Nov. 23, 1999 and attached "Verification of Payments to the Russian Space Agency" and "Protocol on NAS15–10110 ISS Funding Info. Exchange"; OIG, "Russian Involvement in the Intl. Space Station Program," JSC, Sep. 26, 1996, 1–2, 4, 6, obtained via FOIA requests 2008–37 & 2008–36; "International Space Station," https://www.nasa.gov/mission_pages/station/expeditions/expedition01/index.html (accessed Dec. 31, 2020). The issue of the Russian government not funding Energiya arose as early as 1995.

75. Braukus, pr. 97–8, Jan. 9, 1997, HQ, "Monkey Dies after Completing 14-Day Bion Mission."

76. "Spacelab Life Sciences (SLS) Payloads," Souza, Etheridge, and Callahan, eds., *Life into Space: Space Life Sciences Experiments. Ames Research Center, Kennedy Space Center, 1991–1998*. Spacelab 3 cancellation date from https://www.globalsecurity.org/space/library/report/gao/nsi95033.htm (accessed Sep. 16, 2020).

77. Colin Burgess and Chris Dubbs, *Animals in Space: From Research Rockets to the Space Shuttle* (Berlin: Springer-Praxis, 2007), 299; [Lawrence Chambers?], "Bion 12 White Paper Funding History," n.d. [May 16, 1997] and attached drafts, f. "Bion White Papers," sh. 3, cb. 1 Chambers Files. The manifest of experiments flown on Bion 11 can be found on the Life Sciences Data Archive website, https://lsda.jsc.nasa.gov/Mission/miss/135 (accessed July 23, 2020).

78. Fact Sheet: Bion 11 & 12–Joint US-Russian Biosatellite Missions, f. "Bion White Papers," sh. 3, cb. 1 and f. 13.2.1.19 "Bion-11 (General)," sh. 4, cb. 1, Chambers Files.

79. "Bion 11 Neuroscience Hardware," in *Life into Space, Space Life Sciences Experiments, Ames Research Center, 1998–2003*, edited by Souza, online only, https://history.arc.nasa.gov/bibliography.htm (accessed July 23, 2020).

80. Fact Sheet: Bion 11 & 12–Joint US-Russian Biosatellite Missions, f. "Bion White Papers," sh. 3, cb. 1 and "Bion 11, 12 Missions: US/French Experiments" and "Russian Experiments," attached to e-mail Mike Skidmore to Duncan Atchison, Sep. 8, 1994, f. 13.2.1.19, sh. 4, cb. 1, Chambers Files.

81. Atchison to multiple addressees Aug. 12, 1994, f. 13.2.1.19.6 "Bion IWG"; 2 memos Schneider to multiple addressees Sep. 10, 1994, f. 13.2.1.19.2.1 "Memo's [sic]," all sh. 4, cb. 1, Chambers Files.

82. Duane M. Rumbaugh to Souza, Aug. 11, 1995, f. 13.2.1.19.2.1, sh. 4, cb. 1, Chambers Files.

83. Ronald C. Merrell to Bradford W. Parkinson, Nov. 25, 1996; Nicogossian to Richard Dalbello, Dec. 6, 1996; "Bion Organization Chart for Primate Science Payload," attached to Souza to Vernikos, Mar. 27, 1996; letter of invitation, Bielitzki to prospective members of the Animal Care, Use and Bioethics Working Group, n.d., attached

to memorandum Souza to Chambers, Sep. 12, 1996, all in f. 13.2.1.19.2.1, sh. 4, cb. 1, Chambers Files.

84. "Bion Project: Science Selection," presentation, Apr. 10, 1996, 3, f. 13.2.1.19, sh. 4, cb. 1, Chambers Files.

85. Fact Sheet: Bion 11 & 12, draft, hand dated 9/96; "Animal Care at NASA ARC–Recent Concerns," f. "Bion White Papers," sh. 3, cb. 1, Chambers Files; Andrew Lawler, "Key NASA Lab Under Fire for Animal Care Practices," *Science*, vol. 268, no. 5218, 1995, p. 1692.

86. Tomko, Richard Grindeland, and Skidmore, "Bion 11/12 Integrated Science Proposal," 8, 12, 21, bound as 13.2.1.19.7, green spiral, sh. 4, cb. 1, Chambers Files. Bion 11/12 science stated that Ames PIs wanted to do the biopsies at R+ 12–24 hours (see p. 8); Cosmos 2044 postflight biopsies were taken at R + 24, and in Cosmos 2229 at 3–5 days; hoping for within 24 hours (p. 12); bone biopsies for R+ 24–36. Biopies of the iliac crest were done under general anesthesia and took 30–40 minutes (p. 21).

87. "Animal Care at NASA ARC–Recent Concerns," Chambers Files.

88. Greg Ganske and Tim Roemer to "Dear Colleague," n.d. [Jan. 1997], f. 13.2.1.19.10 "Bion Termination File"; Goldin to Christopher S. Bond, July 5, 1996, f. 13.2.1.19.2.1, all sh. 4, cb. 1, Chambers Files. The Ganske-Roemer memo, headed "No More Monkey Business," had a prominent photo of two sad-faced simians clinging to each other.

89. Patricia M. Riep-Dice to Mary Beth Sweetland, Aug. 16, 1995 attached to fax Sweetland to Michael Marlaire, July 25, 1995; Nicogossian to Sweetland, June 24, 1996; Koptev to Goldin Feb. 27, 1996 and Yuri S. Osipov to Goldin Feb. 16, 1996, f. 13.2.1.19.2.1; E. A. Ilyin, D. O. Meshkov, and V. I. Korolkov to The Chancellery of the Russian Federation President and draft, N. G. Khruschov to N. Sayenko, attached to Khruschov to V. E. Sokolov, Apr. 15, 1996, f. 13.2.1.19.13 "Protocol #24, IMBP Biomedical Ethics Committee Reports, sh. 4, cb. 1, Chambers Files.

90. Skidmore et al., "Incident Report," Jan. 9, 1997, attached to fax Skidmore in Moscow to Larry Chambers in Washington, DC, Jan. 9, 1997, f. 13.2.1.19.2.1, sh. 4, cb. 1, Chambers Files.

91. Vernikos to Antonio Güell, June 12, 1996, f. 13.2.1.19.2.1, sh. 4, cb. 1, Chambers Files.

92. Lawrence Chambers to "Distribution," Nov. 22, 1995, f. 13.2.1.19.1 "Trip Reports/Protocols," sh. 4, cb. 1, Chambers Files.

93. Tomko, Grindeland, and Skidmore, "Bion 11/12 Integrated Science Proposal," 8, 12, 21; Kozlovskaya et al., "Bion 11 Science Objectives and Results," *J Grav Physiol* 7, no. 1 (2000): S-19–S-25; Zérath et al., "Spaceflight affects bone formation in rhesus monkeys," *J Appl Physiol* (2002) 93: 1047–1056 The authors described the biopsy as 0.5–0.6 cm (0.2–0.25 in) wide and the procedure taking about 20 minutes. Both monkeys were reportedly "alert, active, without signs of distress" right after the flight.

94. Nicogossian to Merrell Jan. 17, 1997, f. 13.2.1.19.2.1; "Attachment #1: Veterinary Recovery Notes," Jan. 7, 1997 and Bielitzki, "Attachment #2: Veterinary Notes Surrounding the Death of 357," Jan. 8, 1997, attached to Skidmore et al., "Incident

Report," f. 13.2.1.19.2.1; Merrell to Nicogossian, Apr. 21, 1997, attached to Vernikos to multiple addressees, Apr. 23, 1997, f. 13.2.1.19.10 "Bion Termination File," all sh. 4, cb. 1, Chambers Files. Recktenwald et al. ("Effects of Spaceflight on Rhesus Quadrupedal Locomotion After Return to 1G") also noted lower limb shaking by both monkeys during short (10 steps) treadmill tests at R+1.

95. "Bion 11: Preliminary Results and Continued Research," n.d. [1997], f. 13.2.1.19 "Bion 11 (General)," sh. 4, cb. 1, Chambers Files.

96. Jeffrey M. Lupis to Victor I. Kozlov, Apr. 24, 1997, Joseph F. Kroener to Kozlov Mar. 31, 1998, and Kroener to Leonid Makridenko June 19, 1998; Grigoriev to Vernikos Jan. 16, 1998 and Vernikos to Grigoriev Jan. 28, 1998, all f. 13.2.1.19.2.1 "Memo's" [sic]; ARC, "Bion 12 Mission Replan," presentation May 7, 1997, f. 13.2.1.20.1 "Trip Reports/Protocols"; Skidmore, "Summary of Results, Bion 12 Tech. Rev. #1 & Russian Free-Flyer Wkg. Mtg.," Sep. 8–19, 1997, attached to e-mail Skidmore to multiple addressees, Sep. 21, 1997, f. 13.2.1.20.1 "Trip Reports/Protocols," all sh. 4, cb. 1, Chambers Files.

97. "NASA Sets a Bad Precedent," *Space News* May 19–25, 1997; "NASA Out of Space Monkey Business," *New York Times*, Apr. 22, 1997, n. p.

98. Jack Gibbons to Goldin, Feb. 9, 1996, f. 13.2.1.19.10, sh. 4, cb. 1, Chambers Files.

Chapter 9. More People, Less Science, Less NASA? International Participants, Centrifuge, and Nongovernmental Organizations

1. Notes, briefing book, Nov. 6–8, 1993, f. 073673 and Robert W. Clarke to Goldin, Dec. 7, 1993, f. 073687 "12/7/93 Dr. Jan-Baldem Mennicken, Director . . . (DARA)," b. 38; Clarke to Goldin, May 4, 1994, f. 073816, b. 44; Nicogossian, in "Minutes of SS & Center Directors Mtg. Nov. 21, 1994," 2, f. "Senior Staff Telecon," 073693, b. 42; Schumacher to Goldin and Tab L, Apr. 25, 1996, "4/25/96 Dinner at the French Embassy," f. 074239, b. 51; Schumacher to Goldin, Apr. 30, 1996, f. "5/1/96 Meeting with Dr. Mennicken of DARA," 074250, b. 52; Schumacher, talking points, Sep. 8, 1997, f. 076147, b. 139; Schumacher to Goldin, Nov. 5, 1997, f. 076168 "11/5/97 Meeting with Major General Ben-Eliahu . . . Israel Air Force," b. 140; J. Donald Miller et al., memo for record, Apr. 2, 1998, f. "Hungary," 077535; "Summary of Disc.," June 9, 1998, f. 077550; Schumacher to Goldin, June 10, 1998, f. "June 11, 1998 Meeting w/Korea's Minister of Sci & Tech," 077552, all b. 179; Schumacher to Goldin, July 16, 1998, f. "July 16, 1998 "Meeting w/Romanian Minister for R&T," 077557 and May 28, 1999, f. "June 1, 1999 Meeting w/Gerard Brachet," 077593, b. 180, Goldin Recs.; D.W.T.C., "Belgische laboratoria," "Dynamische Belgische onderzoekers," and "Welke rol voor de jongeren in het Internationale ruimtestation?" *Space Connection* 29 (Oct. 1999): 18–19, 20–22, 23.

2. NASA, Space Shuttle Mission STS-47 Press Kit, Sep. 1992, 29; "Israel Space Agency Investigation About Hornets (ISAIAH)," n.d. [May or June 1995], f. "Israeli hornet Expt," 074025, b. 48, Goldin Recs.

3. "Israel Space Agency," Wikipedia, https://en.wikipedia.org/wiki/Israel_Space_ Agency; "The Israel Space Weather and Cosmic Ray Center," https://www.iswc.space/;

"Research Field—Atmospheric Sciences," Porter School of the Environment and Earth Sciences, Dept. of Geophysics, Tel Aviv University, https://physics.tau.ac.il/earth/research_fields_atmospheric_science; "Space Weather Prediction Center," Natl. Weather Serv., https://www.swpc.noaa.gov (all accessed Sep. 27, 2020).

4. Arieh O'Sullivan, "IAF Pilot with the Right Stuff," *Jerusalem Post*, Apr. 30, 1977, 1; Danna Harman, "Engineer Chosen to Be First Israeli in Space," Associated Press Worldstream (Jerusalem), Apr. 30, 1997; "Israel Chooses Fighter Pilot as Candidate for First Space Flight," Agence France Presse (Jerusalem), Apr. 30, 1997; Schumacher to Goldin and Tabs C, E, F, Apr. 29, 1996 and Sep. 30 1996, f. 074247 "4/30/96 Meeting with Professor Ne'eman of Israel Space Agency," b. 51 and f. 074341 "Israel," b. 43; [Unknown] to Goldin, Nov. 5, 1997 and "Excerpt from Mr. Goldin's Interview–The First Israeli Astronaut: Not as Unreachable as the Sky,'" *Ma'ariv*, Apr. 28, 1997, f. 076168, b. 140, Goldin Recs.

5. "11/5/97 Mtg. with Maj. Gen. Ben-Eliahu, Commander IAF," f. 076168, b. 140, Goldin Recs. The Israeli AF put it out in the trade press that they had already selected "an Israeli Air Force Colonel who is an F-16 pilot and an electronics engineer" for the mission, identifying him only as 38-year-old "Col. A." Ilan Ramon was part of the delegation visiting HQ, and Israel could not formally announce a selection without approval from NASA.

6. "Freestar" was actually an acronym for a Small Shuttle Payload (SSP) on STS-107: Fast Reaction Experiments Enabling Science, Technology, Applications and Research. Developed by Omitron, an engineering firm in Beltsville, MD, the payload was six experiments, including MEIDEX.

7. Ilan Ramon, Astronaut Bios., Payload Spec. Astronauts, May 2004, JSC, https://www.nasa.gov/sites/default/files/atoms/files/ramon.pdf; "STS-107 Press Kit," 4, Jan. 25, 2003, *Shuttle Press Kit Online*, https://archive.org/details/nasa_techdoc_20030011376 (both accessed July 29, 2020); Charles and Liskowsky, "STS-107 Whole Payload % Science Gained"; "S*T*A*R*S Space Day Media Advisory," May 2, 2002, Archived News, *Astrotech Corporation*, http://www.spacehab.com/news-and-events/news/stars-space-day-media-advisory (accessed Mar. 23, 2011).

8. Darly Henriques da Silva, "Brazilian Participation in the ISS Program: Commitment or Bargain Struck?" *Space Policy* 21, iss. 1 (2005): 60.

9. "Add. 1: Statement of Work," *Agreement*, Board of Trustees, UA for UAB and Brazilian Commercial Space Services (BRAZSAT), n.d., for Apr. 1–July 1, 1997, www.uab.edu/images/finance/vpad/pdf/purchasing/feeforservice.PDF (accessed Sep. 24, 2010). The contract, which predated Brazil's 1997 invitation to construct six ISS components, called for Brazsat, the Center for Macromolecular Crystallography (CMC) at the U of AL Birmingham (UAB), Diversified Scientific, Inc., and Spacehab "to develop a complete business plan for the [protein crystal] X-ray Facility" for ISS. (No dedicated facility was ever built.) UAB also tasked Brazsat with helping raise the estimated $25 million to build it. In addition, "BRAZSAT will continue to support the CMC's commercial outreach efforts in Brazil and other countries . . . continue to foster and help coordinate ongoing collaborations with INPE, FEI University [Centro Universitário da Faculdade de Engenharia Industrial], the University of São Carlos [Universidade

Federal de São Carlos], and União Chemica, Inc. [União Química Farmacêutica Nacional S/A] . . . and continue to support and help coordinate the CMC's interaction with Brazilian companies and universities in regard to its PCG [Protein Crystal Growth] space flight program."

10. Michael L. Smith, "The Kiss of Death," Sep. 1996, updated Nov. 1, 1998, *Cocorí: Complete Costa Rica*, https://cocori.com/library/eco/chagas.htm and Ann Kellan, "Looking for Cures in Space for Parasite-Caused Disease," *CNN.com*, Nov. 5, 1998, www.cnn.com/TECH/science/9811/05/t_t/rainforest.to.shuttle/index.html (both accessed July 29, 2020); "Chile: Space and Future," Unispace III, Third UN Conf. on the Expl. and Peaceful Uses of Outer Space, Vienna, July 19–30, 1999, https://www.unispace3.co.cl/inicial5b.html (accessed Sep. 24, 2010) and "Hitos de la Historia Espacial Chilena: Investigación del Mal de Chagas," *Astronáutica Chile* V1.0, 2004, http://astronauticachile.cl/historia-espacial-chilena/historia-espacial-chilena/ (accessed Sep. 27, 2020).

11. STS-83 Press Kit, Apr. 1997, 16, 20; STS-94 Press Kit, July 1997, 17, 21.

12. Da Silva, "Brazilian Participation in the ISS Program," 58.

13. Ibid., 55, 58, 59; Geoffrey Allen Pigman, "The Win-Win Response: Promoting Exports and Democracy in the Clinton Administration 1993–1997," wkg. paper, Int'l. Studies Assoc., Mar. 1998, Mpls., MN, archived at Columbia International Affairs Online, http://www.ciaonet.org/ (accessed Apr. 1, 2008); Maria Helena Fonseca de Souza Rolim, "The USA-Brazil Implementing Arrangement on the ISS: Interpretation and Application," paper at the 44th Space Law Colloquium, 52nd Congress of the International Institute of Space Law, Toulouse, Oct. 1–5, 2001, repr. at *Revista Sociedade Brasileira de Direito Aeroespacial*, https://sbda.org.br/wp-content/uploads/2018/10/1754.htm (accessed Sep. 27, 2020).

14. Da Silva, "Brazilian Participation in the ISS Program," 57, 59–60.

15. Rolim, "The USA-Brazil Implementing Arrangement on the ISS," 4–5, 6, 9, 10, 11, 14, 15.

16. Alessandra Soares, "A Biotecnologia," *Revista Eletrônica de Ciências*, Instituto de Física de São Carlos, Universidade de São Paulo, No. 21, Aug./Sep. 2003, repr. of 1998 interview with Oliva, http://www.cdcc.sc.usp.br/ciencia/artigos/art_21/biotecnologia.html (accessed Mar. 30, 2011); STS-91 Press Kit, 53–54.

17. Da Silva, "Brazilian Participation in the ISS Program," 61.

18. "Marcos Pontes," Intl. Astronauts, Career Astronauts, Astronaut Bios., NASA-JSC, Nov. 2006, http://www11.jsc.nasa.gov/Bios/htmlbios/pontes.html (accessed Sep. 24, 2010).

19. Da Silva, "Brazilian Participation in the ISS Program," 61.

20. Frank Braun, "Brazil to Propose $10 Million Space Station Contribution," *Space News* 16, no. 1 (Jan. 10, 2005): 20.

21. Delphine Thouvenot, "Brazil's President in Moscow Signs Deal for Joint Space Mission," Agence France-Presse, Oct. 18, 2005, at *Space Daily*, https://www.spacedaily.com/news/iss-05zzzzp.html (accessed July 29, 2020).

22. Donald A. Thomas, Julie A. Robinson, Judy Tate, and Tracy Thumm, *Inspiring the Next Generation: Student Experiments and Educational Activities on the ISS,*

2000–2006, NASA/TP–2006–213721 (Houston: NASA-JSC, May 2006), 84; "Marcos Pontes," Astronaut Bios. SED was a gravitropism experiment in which students observed differences between the growth of *Phaseolus vulgaris* seeds in their classrooms and on ISS. The acronym may have stood for Science Experiment Demonstration.

23. Yuzhnoe Design Office, "Biomed Spacecraft for Biological and Medical-Biological Investigations," pres. Sep. 20, 1994 and L. Chambers, "Ukraine Evaluation/Comments, Flight Hardware/Access to Space," Sep. 21, 1994, f. 13.14 "Ukranian [sic] File," sh. 1 / 2, cb. 2, Chambers Files; Braukus, "Ukrainian Payload Specialists Selected for Shuttle Mission," pr. 97–99, HQ, May 16, 1997, http://www.nasa.gov/home/hqnews/1997/97–099.txt (accessed Dec. 12, 2012); Joint Summit Stmt., WH, Office of the Press Secretary, Nov. 22, 1994, in *Dispatch* 5, no. 49 (Dec. 5, 1994), Bur. Pub. Affairs, US Dept. of State, http://dosfan.lib.uic.edu/ERC/briefing/dispatch/1994/html/Dispatchv5no49.html (accessed Sep. 24, 2010).

24. Leonid K. Kadenyuk, Payload Spec. Astronauts, Astronaut Bios., JSC, Jan. 1998, https://www.nasa.gov/sites/default/files/atoms/files/kadenyuk.pdf (accessed July 29, 2020); Braukus, "Ukrainian Payload Specialists Selected for Shuttle Mission."

25. STS-87 press kit, NASA, Nov. 1997, 29, 40; Kenneth Souza, ed., "Collaborative Ukrainian Experiment/STS-87," *Life into Space, 1996–2003*, Internet Archive Wayback Machine, https://web.archive.org/web/20150213053302/http://lis.arc.nasa.gov/; "Mission/Study Information," STS-87, LSDA, https://lsda.jsc.nasa.gov/Mission/miss/101 (both accessed Sep. 25, 2020).

26. Sci., Aero., and Tech., FY 1998 Est., Budget Summ., OLMSA, Summ. of Resource Reqts., 20; Bridget Booher, "Astral Agriculture," *Duke Magazine*, May/June 1997, https://alumni.duke.edu/magazine/articles/duke-university-alumni-magazine-418 (accessed July 29, 2020).

27. Beth Schmid, "U.S./Ukrainian Students Collaborate on Shuttle Experiment," pr. H97–270, NASA HQ, Nov. 17, 1997, https://www.nasa.gov/centers/johnson/news/releases/1996_1998/h97–270.html (accessed July 29, 2020); OBPR, "Researchers Achieve Breakthrough by Growing Plants from 'Seed to Seed' in Space," SLS Research Highlights, Dec. 2000, http://spaceresearch.nasa.gov/common/docs/highlights/seed_to_seed_2000.pdf (accessed Mar. 23, 2011).

28. STS-87 press kit, 6; LSDA; "Fast Plants in Space," Teachers and Students Investigating Plants in Space—TSIPS, TEAMS Distance Learning, LA County Off. of Ed., n.d. [ca. 1997], https://teams.lacoe.edu/DOCUMENTATION/classrooms/gary/plants/projects/tsips/cue.html (accessed Sep. 25, 2010). Nine Ukrainian students attended the launch in Florida.

29. Louis Ostrach, interview with the author, June 13, 2006. "Ukraine," *CIA World Factbook*, Sep. 10, 2020, https://www.cia.gov/library/publications/the-world-factbook/geos/up.html (accessed Oct. 4, 2020).

30. "STScI," AURA, https://www.aura-astronomy.org/centers/space-telescope-science-institute/ (accessed Sep. 27, 2020).

31. Robert W. Clarke to Goldin, May 4, 1994, f. 073816, b. 44, Goldin Recs.

32. Potomac Institute for Policy Studies, *ISS Commercialization Study* (Arlington, VA: PIPS, Mar. 20, 1997), iii, 2, B-7, C-5, C-7.

33. See, for example, Joan Lisa Bromberg, *NASA and the Space Industry* (Baltimore: Johns Hopkins, 1999), with the examples of the McDonnell Douglas electrophoresis project and the Industrial Space Facility; Edward L. Hudgins, ed., *Space: The Free-Market Frontier* (Washington, DC: Cato Institute, 2002), addressing space privatization from a libertarian viewpoint; and Erik Seedhouse, *Tourists in Space: A Practical Guide* (Chichester: Springer Praxis, 2008), a physiologist's look at commercial space tourism.

34. PIPS, "International Space Station Commercialization Study," 5–6, 9–10, 11–12, 14, 17, 23, 24–29, 32, B-3, C-6–9, C-16. The report noted that "the proposed NASA Working Group [to assist with this study] did not materialize, so we lacked some approved Administration perspectives."

35. NASA, Commercial Development Plan for the ISS, final draft, Nov. 16, 1998, 2, 3, 4, Hist. Prog. Off., NASA HQ, Washington, DC, history.nasa.gov/31317.pdf (accessed Sep. 27, 2020).

36. Lawler, "Space Station Plans To Spark Battle in 2000," *Space.com*, Dec. 30, 1999, http://www.space.com/news/spacestation/station_institute_991230.html (accessed Sep. 24, 2010).

37. NRC, *Institutional Arrangements for Space Station Research* (Washington, DC: National Academies Press, 1999), 3–7, https://www.nap.edu/catalog/9757/institutional-arrangements-for-space-station-research (accessed Sep. 27, 2020).

38. Leonard David, "Space Station Declared Open-For-Business," *Space.com*, http://www.space.com/news/spacestation/commercial_iss_000203.html (accessed Sep. 24, 2010). See Chap. 1 for the rise and fall of MDAC's electrophoresis project.

39. Lawler, "Space Station Plans to Spark Battle in 2000."

40. OLMSA, LMSAAC, "Mtg. Rept.," NASA HQ, Oct. 19–20, 2000, 3, 7–8, 10, App. D, 1–2, www.nasa.gov/pdf/189494main_LMSAAC_minutes_102000.pdf (accessed Jan. 6, 2021).

41. OLMSA, LMSAAC, "Mtg. Rept.," NASA HQ, Feb. 15–16, 2001, 2, 8, www.nasa.gov/pdf/189493main_LMSAAC_minutes_022001.pdf (accessed Jan. 6, 2021).

42. US House of Representatives, 106th Congress, 2d sess., H. Rept. 106–843 (Conference Report), 15, https://www.congress.gov/bill/106th-congress/house-bill/1654 (accessed Dec. 31, 2020); OLMSA, BPRAC, "Mtg. Rept.," NASA HQ, Oct. 25–26, 2001, 8, www.nasa.gov/pdf/189488main_BPRAC_minutes_102001.pdf (accessed Jan. 6, 2021). The LMSAAC changed its name to the Biological and Physical Research Advisory Committee (BPRAC) to reflect the change from OLMSA to OBPR.

43. "Statement of A. Thomas Young, Chairman, ISS Management and Cost Evaluation Task Force Before the House Science Committee," released Nov. 7, 2001, SpaceRef. com, www.spaceref.com/news/viewsr.html?pid=4021 (accessed Jan. 6, 2021).

44. Lawler, "Space Station Plans To Spark Battle in 2000"; OLMSA, BPRAC, "Mtg. Rept.," NASA HQ, Feb. 19–20, 2002, 9, http://www.nasa.gov/pdf/189500main_BPRAC_ minutes_022002.pdf (accessed Sep. 24, 2010).

45. "Testimony by Sean O'Keefe on NASA's FY 2003 Budget before the House Sci. Comm. (Part 2)," SpaceRef.com, http://www.spaceref.com/news/viewsr. html?pid=4802 (accessed Jan. 24, 2022).

46. OLMSA, BPRAC, "Mtg. Rept.," NASA HQ, Aug. 29–30, 2002, 4–5, 9, http://

www.nasa.gov/pdf/189487main_BPRAC_minutes _082002.pdf (accessed Sep. 24, 2010). FTE stood for Full Time Equivalent, i.e., personnel, whether contractor or civil servant.

47. OLMSA, BPRAC, "Mtg. Rept.," NASA HQ, Feb. 13–14, 2003, 6, 22, 23–24, http://www.nasa.gov/pdf/189499main_BPRAC_ minutes_022003.pdf (accessed Sep. 24, 2010).

48. OLMSA, BPRAC, "Mtg. Rept.," NASA HQ, Oct. 23–24, 2003, 5, 8–9, http://www.nasa.gov/pdf/189489main_BPRAC_minutes_102003.pdf (accessed Sep. 24, 2010).

49. OLMSA, BPRAC, "Mtg. Rept.," NASA HQ, Feb. 13–14, 2003, 24; OLMSA, BPRAC, "Mtg. Rept.," Feb. 12–13, 2004, 3, http://www.nasa.gov/pdf/189485main_BPRAC_%20minutes_022004.pdf (accessed Aug. 14, 2009).

50. OLMSA, BPRAC, "Mtg. Rept.," NASA HQ, Oct. 23–24, 2003, 8–9, http://www.nasa.gov/pdf/189489main_BPRAC_ minutes_102003.pdf (accessed Sep. 24, 2010).

51. "Consensus on Critical Issues on which OBPR Needs Feedback," in OLMSA, BPRAC, "Mtg. Rept.," NASA HQ, Oct. 23–24, 2003, 17, http://www.nasa.gov/pdf/189489main_BPRAC_ minutes_102003.pdf (accessed Sep. 24, 2010).

52. Berger, *Sp. News Bus. Rept.*, June 3, 2003, archived at Space News, http://www.spacenews.com/ (accessed Mar. 15, 2008).

53. Dolores Beasley, "ISS Research Institute On Hold," HQ pr. 04–029, Jan. 22, 2004, https://www.nasa.gov/home/hqnews/2004/jan/HQ_04_029_iss_research_institute.html (accessed Sep. 25, 2020); OLMSA, BPRAC, "Mtg. Rept.," Feb. 12–13, 2004, 3–4.

54. NASA OIG, "NASA's Efforts to Maximize Research on the International Space Station," Report No. IG-13-019, July 8, 2013, i and "NASA's Management of the Center for the Advancement of Science in Space," Report No. IG-18-010, Jan. 11, 2018, 3–4; ProOrbis, "ISS U.S. National Laboratory, Vol. 1: ProOrbis's Role in the Genesis of the Center for the Advancement of Science in Space (CASIS)," Apr. 12, 2012 (http://proorbis.com/thought-leadership/iss-u-s-national-laboratory-series/) accessed Sep. 25, 2020.

55. Charles A. Fuller, "Acute Physiological Responses of Squirrel Monkeys Exposed to Hyperdynamic Environments," ASEM (1984) 55(3): 226–230; Fuller, "Effects of Centrifuge Diameter & Operation on Rodent Adaptation to Chronic Centrifugation," Final Report, NAG2–795 (Washington, DC: NASA, 1992), 5, 7, 8, 11–14, 47.

56. Timothy Castellano, "An Investigation of Acoustic Noise Requirements for the Space Station Centrifuge Facility," TM-108811 (Moffett Field, CA: NASA-ARC, Feb. 1994), 1–2, 12, 16, 19–22. This study also used data from 1991–93.

57. Robert H. Benson, "Life Sciences Centrifuge Facility Assessment—Final Report," Rept. No. MSFC-1, NASA-CR-196848, Sep. 15, 1994, 1–4.

58. Benson, "Life Sciences Centrifuge Facility Assessment—Final Report," 4.

59. Benson, "Life Sciences Centrifuge Facility Assessment—Final Report," 4–6.

60. Laurence R. Young, "Life Sciences Centrifuge Facility Review, Final Report," CR-196849, Sep. 13, 1994, MSFC, 1–5. His final note was that better verification would

lessen the "risk" of "antivivisection community" objections, "real or perceived," limiting "the performance of key animal experiments."

61. Code U [OBPR] Res. Facilities Assessment Team, "Host Syst. Reqs. Status," "Habitat Syst. Reqs. Status," "Support Equip. Reqs. Status," and "Areas of Concern," Feb. 14–16, 1995, white binder 56.14.31, sh. 2, cb. 3, Chambers Files.

62. John Givens, "Summary" and Frank Rubarth, "Lab. Support Equip. Reqs. and Status," Code U Res. Facilities Assessment Team, Space Station Biological Res. Proj., Feb. 14–16, 1995, white bdr. 56.14.31, sh. 2, cb. 3, Chambers Files.

63. Souza, memorandum "SSBRP Costing Options," Feb. 16, 1995, white bdr. 56.14.31, sh. 2, cb. 3, Chambers Files.

64. Schumacher to Goldin, Nov. 3, 1995, f. 074121 "11/04/95–phone call to Mac Evans re RADARSAT launch," b. 50, Goldin Recs.

65. E-mails, Bob Soltess to Earl Ferguson et al., "2.1 B Freeze on New Procurements," Mar. 20, 1996; Roger Sachse to Soltess, reply, Mar. 21, 1996, f. 56.14.45, "Centrifuge Termination CAM U.S. Procurement," sh. 2, cb. 3, Chambers Files.

66. Trafton to AA for Procurement, and "Reasons for Extension of 90-Day Hold on Res. Facilities," f. 56.14.45, sh. 2, cb. 3, Chambers Files. The June memo made no mention of the hold being for new procurements only.

67. "Centrifuge/Glovebox Procurement: Telecon with ISS/HQ Mgmt.," June 13, 1996; Chambers telecon notes; fax, "MDA Position on Extension of the Centrifuge and Glovebox Procurement," June 12, 1996, 56.14.45, sh. 2, cb. 3, Chambers Files. Chambers's notes specify management to be "George," presumably Abbey.

68. "Centrifuge Termination CAM U.S. Procurement," sh. 2, cb. 3, Chambers Files.

69. "Reasons for Extention [sic]," Chambers notes, "Glovebox Development by Japanese," June 13, 1996, and B. Ostroumov to R. H. Brinkley, Moscow, Feb. 14, 1996, f. 56.14.45, sh. 2, cb. 3, Chambers Files; Tab D3, "Japanese offsets for JEM Launch Costs (Unlikely to be raised)," f. 074310 "7/24/96 Meeting with Chairman Inoue . . . Japanese Diet," b. 52, Goldin Recs.

70. Tab D3, "Japanese Offsets," f. 074310 "7/24/96 Meeting with Chairman Inoue," b. 52, Goldin Recs.

71. Takeshi Fujita and Nobuyoshi Fujimoto, "Potential Projects for Offset (Centrifuge)," May 29–31, 1996; Kazuo Suzuki to Brinkley, June 10, 1996; e-mail, Joellen Jarvi to ARC list, forwarded by Souza to Vernikos and others at HQ, both June 19, 1996, f. 56.14.45, sh. 2, cb. 3, Chambers Files.

72. R. Barbera to Brinkley, Paris, June 7, 1996, 56.14.45, sh. 2, cb. 3, Chambers Files.

73. E-mail Jarvi to ARC list and forwarded, f. 56.14.45, sh. 2, cb. 3, Chambers Files.

74. Kathie Olsen and Ann Carlson, interview with the author, Apr. 19, 2006.

75. [ARC?], "Impacts of SSBRP Responsibilities to NASDA," ca. Aug. 12, 1996, f. 56.14.45, sh. 2, cb. 3, Chambers Files.

76. "Agenda," "Agreements and Actions," and "JEM Launch Slip," NASA-STA Prog. Coord. Comm., Sep. 6, 1996, Washington, DC; Suzuki to Brinkley, Sep. 20, 1996; and "Summ. of Oct. 28–31 Mtgs.," f. 56.14.45, sh. 2, cb. 3, Chambers Files.

77. Kuniaki Shiraki, "Centrifuge Project," Our Missions, JAXA, Oct. 1, 2003, https://global.jaxa.jp/article/special/pm_messages2004/006_e.html (accessed Dec. 31, 2020).

78. David Liskowsky, interview with the author, June 14, 2006.

79. Shiraki, "Centrifuge Project."

80. William E. Berry, interview with the author, May 16, 2006.

81. Shiraki, "Centrifuge Project."

82. W. Michael Hawes, "ISS Status," PPT to the BPRAC, June 14, 2001, repr. on NASAWatch, http://www.nasawatch.com/iss/06.14.01.hawes.bprac.ppt (accessed Sep. 25, 2010).

83. "NASA/NASDA Negotiations on Japanese Provision of Centrifuge Rotor, LSG, and CAM, Summ. of Oct. 28–31 Mtgs.," f. 56.14.45, sh. 2, cb. 3, Chambers Files and "NASA Adv. Council Mtg. Min. Sep. 10–11, 2002–Pt. 2" rel. 9–25–02, repr. on SpaceRef, http://www.spaceref.com/news/viewsr.html?pid=6652 (accessed July 29, 2020).

84. For a very helpful explanation of the framing concept of social sciences and media studies as it applies to NASA, see Valerie Neal, "Framing the Meaning of Spaceflight in the Shuttle Era," in *Societal Impact of Spaceflight*, edited by Dick and Launius, 67–87.

85. See, for example, George Sarver, "Monthly Review: Centrifuge Status Summary," Space Station Biological Research Project, PPT, June 17, 2003, 21, 22, 24, 26 and Sep. 15, 2005, 12, 14, 23, 27, 28, 30, 32, 35, in f. 27.2 "SSBRP Monthlys," sh. 2/3, cb. 2, Chambers Files. By June 2003 "severe budget problems" at NASDA signaled an "impact" on its ability to deliver the centrifuge and life sciences glovebox. NASDA petitioned the US space agency to take over responsibility for part of the projects, equal to an estimated $200 million.

86. Steve Berner, "Japan's Space Program: A Fork in the Road?" (Santa Monica, CA: RAND Corp., 2005), 2, 9–11, 12–14, 20–22, 25, 26–27; Mtg. Rept., BPRAC, Oct. 23–24, 2003, 15.

87. McCarthy, notes, Vest Committee mtg. June 7–8, 1993, 18, f. 075980, b. 137; "Bkgd. and Talking Points . . . Phase III, 3. Centrifuge," 21, f. 073659, "Briefing Book . . . Moscow, Oct. 2–6, 1993," b. 41, Goldin Recs.

88. FY 2002 budget, HSF 1–26, HSF 1–37.

89. Mtg. Rept., BPRAC, Oct. 25–26, 2001, 8, App. D 1–2.

90. Mtg. Rept., BPRAC, Nov. 29, 2001, n. p. (5–6). Picosatellites are very small orbiters, generally with a "wet" mass of 0.1–1 kg (0.22–2.2 pounds), i.e., including fuel.

91. Mtg. Rept., BPRAC, Feb. 19–20, 2002, Aug. 3–4, 6, and 29–30, 2002, App. D, 4.

92. Olsen and Carlson interview; Terri Lomax, interview with the author, Apr. 20, 2006.

93. *Rept. by the NASA Bio. and Phys. Res. Research Maximization and Prioritization (ReMAP) Task Force to the NASA Advisory Council Aug. 2002,* http://www.spaceresearch.nasa.gov/docs/remap/remap_final_sum.pdf (accessed Sep. 25, 2010).

94. Office of Bioastronautics, *Bioastronautics Roadmap*, A-2, A 31–34.

95. Mtg. Rept., BPRAC, Oct. 23–24, 2003, 15.

96. Mtg. Rept., BPRAC, Feb. 12–13, 2004, 8.

Chapter 10. The Vision for Space Exploration

1. George W. Bush, "A Renewed Spirit of Discovery: The President's Vision for U.S. Space Exploration," *The Vision for Space Exploration* (Washington, DC: NASA, Feb. 2004).

2. Courtney Stadd and Jeff Bingham, "The US Civil Space Sector: Alternate Futures," *Space Policy* 20 (2004): 241–252.

3. Craig Cornelius, "Science in the National Vision for Space Exploration: Objectives and Constituencies of the 'Discovery-Driven' Paradigm," *Space Policy* 21 (2005): 41–48. For the downside to this philosophy, see Jeffrey F. Bell, "The Bush Space Initiative: Fiscal Nightmare or . . . Fiscal Nightmare?" *SpaceDaily*, Mar. 17, 2004, http://www.spacedaily.com/news/spacetravel-04j.html (accessed May 28, 2020).

4. "Biography of President George W. Bush," (ca. 2002), White House, http://www.whitehouse.gov/president/biography.html (accessed May 2, 2008).

5. Bush, "A Renewed Spirit of Discovery."

6. Dennis Wingo, "Establishing the Vision for Space Exploration," Apr. 22, 2008, Spaceref.com, www.spaceref.com/news/viewnews.html?id=1285 (accessed May 18, 2020). Credit must go to Wingo's "Comments" piece for introducing economics to this discussion as Bush's frame of reference.

7. "Biography of President George W. Bush."

8. Bush, "A Renewed Spirit of Discovery."

9. Draft Fact Sheet: A Renewed Spirit of Discovery, Jan. 12, 2004, f. 606310–57510, box 2, sub series OS001 (Space Research and Development), subject files OS (Outer Space), George W. Bush Presidential Library, https://catalog.archives.gov/id/32200199 (accessed May 11, 2020); "Summary of FY 2005 Budget Request," 1–2, 1–5, 1–7, 1–27, NASA's FY 2005 Budget and Planning Documents, https://www.nasa.gov/about/budget/FY05_budget.html (accessed May 11, 2020).

10. Sean O'Keefe, "Message from the NASA Administrator," *The Vision for Space Exploration*, NP-2004-01-334-HQ (Washington, DC: NASA, Feb. 2004), I; "Sean O'Keefe (Dec. 21, 2001–Feb. 11, 2005)," official biography, NASA History Division, http://history.nasa.gov/okeefe.html (accessed May 12, 2020).

11. Bush, "A Renewed Spirit of Discovery."

12. Bush, *The Vision for Space Exploration*, 15.

13. Glenn Mahone/Bob Jacobs, "NASA Begins New Exploration Journey With FY 2005 Budget," pr. 04–047, Feb. 3, 2004, http://www.nasa.gov/home/hqnews/2004/feb/HQ_04047_FY05_budget.html (accessed May 12, 2020); FY 2005 Budget Summary, "Summary of FY 2005 Budget Request," SUM 1–2 and Biological Sciences Research, "Theme: Biological Sciences Research," ESA 12.5, 12.6, 12.12 both at https://www.nasa.gov/about/budget/FY05_budget.html (accessed May 7, 2020).

14. FY 2006 began Oct. 1, 2005 and the NASA budget request was submitted Feb. 7, 2005.

15. Mahone/Jacobs, "NASA Administrator Sean O'Keefe Resigns," pr. 04–400, Dec. 13, 2004, https://www.nasa.gov/home/hqnews/2004/dec/HQ_04400_okeefe_resigns.html; Mahone/Sarah Keegan, "NASA's Budget Enables New Age of Exploration,"

pr. 05–039, Feb. 7, 2005, http://www.nasa.gov/home/hqnews/2005/feb/HQ_05039_ok_budget.html; "Sean O'Keefe (Dec. 21, 2001–Feb. 11, 2005)," NASA History Division, last updated Aug. 31, 2006, https://www.history.nasa.gov/okeefe.html (all accessed May 18, 2020).

16. "NASA Administrator Michael Griffin," Oct. 2007, "People," About NASA, http://www.nasa.gov/about/highlights/griffin_bio.html (accessed May 25, 2020); Michael D. Griffin and James R. French, Space Vehicle Design (Reston, VA: AIAA, 1991 and 2004).

17. NASA, "National Aeronautics and Space Administration, President's FY 2006 Budget Request," https://www.nasa.gov/about/budget/FY_2006/index.html (accessed Dec. 17, 2020) as a chart inside the front cover; Marcia S. Smith and Daniel Morgan, The National Aeronautics and Space Administration's FY2006 Budget Request: Description, Analysis, and Issues for Congress, Nov. 17, 2005, Congressional Research Service, CRS–1–2, 7, 19, 21–22, 24–25, 35.

18. "House Committee on Science Holds a Hearing on the Future of NASA," June 28, 2005, 9, 18, 28, NASA, https://www.nasa.gov/pdf/119619main_Griffin_Hil_testimony_062805.pdf (accessed May 18, 2020).

19. Smith and Morgan, The National Aeronautics and Space Administration's FY2006 Budget Request, C–20.

20. Ronald L. Schaefer and the AIAA Life Sciences Technical Committee, "Life Sciences," Aerospace America, Dec. 2005, 80–81.

21. "Frequently Asked Questions," ELMS website, n.d. (2006), http://www.elmscoalition.org/about/about_faq.php (accessed May 9, 2008). Other professional groups included the Aerospace Medical Association and the American Society of Plant Biologists. The ASGSB became the American Society for Gravitational and Space Research (ASGSR) in 2012, expanding beyond the biological sciences.

22. Statement by the ASGSB Governing Board, in ASGSB Newsletter 21, no. 1 (Winter 2005): 3.

23. "ELMS provides key legislators an overview of NASA's life sciences programs and the impacts of anticipated NASA budget reductions," Successes, (2005), ELMS, http://www.elmscoalition.org/successes/ (accessed May 9, 2008).

24. Makoto Asashima, Ichiro Asukata, and Christopher S. Brown to Michael Griffin, Sep. 6, 2005, "Issues," and Daniel Beysens, Gerhard Haerendel, H.-C. Gunga, and Brown to Griffin, Oct. 19, 2005, "Successes," ELMS, (2006), http://www.elmscoalition.org (accessed May 9, 2008).

25. Craig Covault, "Moon Stuck" and "2025 Asteroid Landing Eyed," AWST, Jan. 21, 2008, 24–27.

26. Michael Griffin, speech at GSFC, Sep. 12, 2006, repr. in Ad Astra, National Space Society, Winter 2006; John Marburger, Keynote Address, Mar. 16, 2006, repr. National Space Society, https://space.nss.org/2006-goddard-memorial-symposium-speech-by-john-marburger/ (accessed Jan. 6, 2021).

27. Marburger, Keynote Address.

28. Ibid.

29. Ibid., emphasis added.

30. Wendell Mendell, "Space Activism as Epiphanic Belief System," *Societal Impact of Spaceflight*, edited by Dick and Launius, 581–582.

31. Jake W. Spidle Jr., *The Lovelace Medical Center: Pioneer in American Health Care* (Albuquerque: University of New Mexico Press, 1987), 53–54 and Charles A. Dempsey, *Air Force Aerospace Medical Research Laboratory: 50 Years of Research on Man in Flight* (Wright Patterson Air Force Base, OH: USAF, 1985), xxix, 6–7, 24. Three Mayo Aero Medical Unit researchers, with US Army Captain Harry Armstrong, M.D., founder of the Physiological Research Laboratory at Wright Field, won the 1940 Collier trophy for developing the first high-altitude oxygen mask in regular use.

32. R. W. Krauss, "NASA–A New Course? A Biologist's View of the Augustine Report," *The FASEB Journal* 5 (1991): 251, cited in Francis J. Haddy, "NASA—Has Its Biological Groundwork for a Trip to Mars Improved?" *The FASEB Journal*, Vol. 21 (Mar. 2007): 643–646.

33. Haddy, "NASA—Has Its Biological Groundwork for a Trip to Mars Improved?" 644–646.

Parting Thoughts

1. Sylvia K. Kraemer, "Organizing for Exploration"; Memorandum, George M. Low to Administrator, "NASA as a Technology Agency," May 25, 1971, in *Exploring the Unknown: Selected Documents in the History of the U.S. Civil Space Program*, Vol. I: *Organizing for Exploration* (Washington, DC: NASA, 1995), edited by Logsdon et al., 623, 685–686.

2. James Hansen, *Engineer in Charge: A History of the Langley Aeronautical Laboratory, 1917–1958* SP-4305 (Washington, DC: NASA, 1987), 3, 7–9, 11, 16, 21. At first a purely nonpaid, federal advisory committee with a few military detailees doing studies, NACA was not granted congressional funds to build a lab (Langley) until 1916. It was not completed nor any staff engineers hired until 1920.

3. E. C. "Pete" Aldridge Jr. et al., *Report of the President's Commission on Implementation of United States Space Exploration Policy, A Journey to Inspire, Innovate, and Discover*, June 2004 (Washington, DC: NASA), 22–24.

4. Jill M. Hruby et al., "The Evolution of Federally Funded Research & Development Centers," *Public Interest Report*, Spring 2011, 24–30.

5. Aldridge Jr. et al., *Report of the President's Commission*, 22–24; "The NASA Astrobiology Institute Concludes its 20-year Tenure," Dec. 20, 2019, NASA Astrobiology Institute, https://nai.nasa.gov/articles/2019/12/20/the-nasa-astrobiology-institute-to-end-its-20-year-tenure/ (accessed June 2, 2020). The NAI became entirely virtual Jan. 1, 2020, as a Research Coordination Network (RCN), a concept originated by the National Science Foundation in 2000.

6. Kevin R. Kosar, *The Quasi Government: Hybrid Organizations with Both Government and Private Sector Legal Characteristics*, RL30533 (Washington, DC: Congressional Research Service, Feb. 13, 2007), CRS-15-6.

7. "ISS National Lab Project Pipeline Map," Projects, ISS National Laboratory, Center for the Advancement of Science in Space," https://www.issnationallab.org/projects/ (accessed June 29, 2021).

8. National Space Council, *A New Era for Deep Space Exploration and Development* (Washington, DC: White House, July 23, 2020), 1–2, 7.

9. Aerospace Safety Advisory Panel, *Annual Report for 2020*, 3, 7, 39–40;

10. Michael Johnson, ed., "Commercial and Marketing Pricing Policy," Apr. 29, 2021, NASA, https://www.nasa.gov/leo-economy/commercial-use/pricing-policy (accessed June 29, 2021).

11. Roger Launius also reflected on this strategy and concluded that "NASA could . . . [pursue] transformational technologies while private firms operate space systems. Turning low-Earth orbit over to commercial entities empowers NASA to focus on deep space exploration, perhaps eventually visiting Mars." See Launius, *Historical Analogs for the Stimulation of Space Commerce*, SP-2014-4554 (Washington, DC: NASA, 2014), 94.

Selected Bibliography

Bowles, Mark D., and Robert S. Arrighi. *NASA's Nuclear Frontier: The Plum Brook Reactor Facility.* Washington, DC: NASA, August 2004.

Bromberg, Joan Lisa. *NASA and the Space Industry.* Baltimore: Johns Hopkins University Press, 1999.

Brooks, Courtney G., James M. Grimwood, and Loyd S. Swenson Jr. *Chariots for Apollo: A History of Manned Lunar Spacecraft.* SP-4205. Washington, DC: NASA, 1979.

Buckey, Jay C., Jr., and Jerry L. Homick. *The Neurolab Spacelab Mission: Neuroscience Research in Space. Results from the STS-90, Neurolab Spacelab Mission.* SP-2003-535. Houston: NASA-JSC, 2003.

Bugos, Glenn E. *Atmosphere of Freedom: Sixty Years at the NASA Ames Research Center.* SP-2000-4314. Washington, DC: NASA, 2000.

Bungo, Michael W., Tandi M. Bagian, Mark A. Bowman, and Barry M. Levitan, eds. *Results of the Life Sciences DSOs Conducted Aboard the Space Shuttle 1981–1986.* TM-58280. Houston: NASA-JSC, 1987.

Burgess, Colin, and Chris Dubbs. *Animals in Space: From Research Rockets to the Space Shuttle.* Berlin: Springer-Praxis, 2007.

Burrough, Bryan. *Dragonfly: NASA and the Crisis Aboard Mir.* New York: HarperCollins, 1998.

Chaikin, Andrew. *A Man on the Moon: The Voyages of the Apollo Astronauts.* New York: Penguin, 1994.

Clark, Lenwood G., William H. Kinar, David J. Carter Jr., and James L. Jones Jr. *The Long Duration Exposure Facility (LDEF) Mission 1 Experiments.* SP-473. Washington, DC: NASA-LaRC, 1984.

Clément, Gilles, and Angie Bukley, eds. *Artificial Gravity.* Hawthorne, CA and New York: Microcosm Press and Springer, 2007.

Clément, Gilles, and Klaus Slenzka, eds. *Fundamentals of Space Biology: Research on Cells, Animals, and Plants in Space.* El Segundo, CA and New York: Microcosm Press and Springer, 2006.

Compton, W. David, and Charles D. Benson. *Living and Working in Space: A History of Skylab.* SP-4208. Washington, DC: NASA, 1983.

Corn, Joseph. *The Winged Gospel: America's Romance with Aviation, 1900–1950*. New York: Oxford University Press, 1983.

Daelemans, Gerard, and Frances L. Mosier, eds. *1999 Shuttle Small Payloads Symposium*. NASA Conference Publication 1999-209476. Greenbelt, MD: NASA-GSFC, 1999.

Dempsey, Charles A. *Air Force Aerospace Medical Research Laboratory: 50 Years of Research on Man in Flight*. Wright-Patterson Air Force Base, OH: USAF, 1985.

Dethloff, Henry C. *Suddenly Tomorrow Came: A History of the Johnson Space Center*. SP-4307. Houston: NASA-JSC, 1993.

Dick, Steven J., and Roger D. Launius, eds. *Critical Issues in the History of Spaceflight*. SP-2006-4702. Washington, DC: NASA, 2006.

Dick, Steven J., and Roger D. Launius, eds. *Societal Impact of Spaceflight*. NASA SP-2007-4801. Washington, DC: NASA, 2007.

Dick, Steven J., and James E. Strick. *The Living Universe: NASA and the Development of Astrobiology*. New Brunswick, NJ: Rutgers University Press, 2004.

Dunar, Andrew J., and Stephen P. Waring. *Power to Explore: A History of Marshall Space Flight Center, 1960–1990*. SP-4313. Washington, DC: NASA, 1999.

Emond, John, ed., N. Bennett, D. McCauley, and K. Murphy, compilers. *The Spacelab Accomplishments Forum*. Proceedings, Washington, DC, March 10–11, 1999. NASA/CP-2000–210332. CD. Huntsville, AL: NASA-MSFC, September 2000.

Evans, Charles H., Jr., and Suzanne T. Ildstad, eds. *Small Clinical Trials: Issues and Challenges*. Washington, DC: National Academy Press, 2001.

Ezell, Linda Neuman. *NASA Historical Data Book*, Vol. II, *Programs and Projects 1958–1968*, SP-4012. Washington, DC: NASA, 1988.

Ezell, Linda Neuman. *NASA Historical Data Book*, Vol. III, *Programs and Projects 1969–1978*, SP-4012. Washington, DC: NASA, 1988.

Foale, Colin. *Waystation to the Stars: The Story of Mir, Michael, and Me*. London: Headline Book Publishing, 1999.

Fragola, Joseph R., Gaspare Maggiore, Michael V. Frank, Luis Gerez, Richard Mcfadden, Erin P. Collins, Jorge Ballesio, Peter L. Appignani, and James A. Karns. *Probabilistic Risk Assessment of the Space Shuttle: A Study of the Potential of Losing the Vehicle during Nominal Operations*. CR-197808–CR-197811. New York: SAIC and Washington, DC: NASA, 1995.

General Accounting Office. *Space Station: Impact of the Expanded Russian Role on Funding and Research*. GAO/NSIAD-94–220. Washington, DC: US GAO, June 1994.

Glenn, John, with Nick Taylor. *John Glenn: A Memoir*. New York: Bantam, 1999.

Goldsmith, Frann, and Frances L. Mosier, eds. *1995 Shuttle Small Payloads Symposium*. CP-3310. Greenbelt, MD: NASA-GSFC, 1995.

Habitability and Human Factors Office, Space Life Sciences Directorate. *Anthropometry and Biomechanics, Man-Systems Integration Standards*, NASA STD-3000, Rev. B. Houston: NASA-JSC, 1995.

Hacker, Barton C., and James M. Grimwood. *On the Shoulders of Titans: A History of Project Gemini*. NASA SP-4203. Washington, DC: NASA, 1977.

Hale, Wayne, Helen Lane, Gail Chapline, and Kamlesh Lulla, eds. *Wings in Orbit: Sci-*

entific and Engineering Legacies of the Space Shuttle. NASA SP-2010-3409. Washington, DC: NASA, 2010.

Hall, Rex D., and David J. Shayler. *Soyuz: A Universal Spacecraft.* London: Springer-Praxis, 2003.

Halstead, Thora W., and Patricia A. Dufour. *Biological and Medical Experiments on the Space Shuttle 1981–1985.* TM-108025. Washington, DC: NASA-OSSA, 1986.

Hansen, James R. *Engineer in Charge: A History of the Langley Aeronautical Laboratory, 1917–1958.* SP-4305. Washington, DC: NASA, 1987.

Hansen, James R. *Spaceflight Revolution: NASA Langley Research Center from Sputnik to Apollo.* SP-4308. Washington, DC: NASA, 1995.

Hogan, Thor. *Mars Wars: The Rise and Fall of the Space Exploration Initiative.* SP 2007-4410. Washington, DC: NASA, 2007.

Hudgins, Edward L, ed. *Space: The Free-Market Frontier.* Washington, DC: Cato Institute, 2003.

Johnston, Richard S., and Lawrence F. Dietlein, eds. *Biomedical Results from Skylab.* SP-377. Washington, DC: NASA, 1977.

Johnston, Richard S., Lawrence F. Dietlein, and Charles A. Berry, eds. *Biomedical Results of Apollo.* SP-368. Washington, DC: NASA, 1975.

Karash, Yuri Y. *The Superpower Odyssey: A Russian Perspective on Space Cooperation.* Reston, VA: American Institute of Aeronautics and Astronautics, 1999.

Krige, John, Angelina Long Callahan, and Ashok Maharaj. *NASA in the World: Fifty Years of International Collaboration in Space.* New York: Palgrave Macmillan, 2013.

Launius, Roger D., and Howard E. McCurdy, eds. *Spaceflight and the Myth of Presidential Leadership.* Urbana: University of Illinois Press, 1997.

Light, Michael. *Full Moon.* New York: Knopf, 1999.

Linenger, Jerry M. *Off the Planet: Surviving Five Perilous Months Aboard the Space Station Mir.* New York: McGraw-Hill, 2000.

Logsdon, John M. *Together in Orbit: The Origins of International Participation in the Space Station.* Monograph in Aerospace History, No. 11. Washington, DC: NASA, 1998.

Logsdon, John M., Dwayne A. Day, and Roger D. Launius, eds. *Exploring the Unknown: Selected Documents in the History of the U.S. Civil Space Program.* Volume II: *External Relationships.* SP-4407. Washington, DC: NASA, 1996.

Logsdon, John M., Stephen J. Garber, Roger D. Launius, and Ray A. Williamson, eds. *Exploring the Unknown: Selected Documents in the History of the U.S. Civil Space Program.* Volume VI: *Space and Earth Science.* SP-4407. Washington, DC: NASA, 2004.

Logsdon, John M., Roger D. Launius, David H. Onkst, and Stephen J. Garber, eds. *Exploring the Unknown: Selected Documents in the History of the U.S. Civil Space Program.* Volume III: *Using Space.* SP-4407. Washington, DC: NASA, 1998.

Logsdon, John M., Linda J. Lear, Jannelle Warren-Findley, Ray A. Williamson, and Dwayne A. Day, eds. *Exploring the Unknown: Selected Documents in the History of the U.S. Civil Space Program.* Volume I: *Organizing for Exploration.* SP-4407. Washington, DC: NASA, 1995.

Logsdon, John M., Ray A. Williamson, Roger D. Launius, Russell J. Acker, Stephen J. Garber, Roger D. Launius, and Jonathan L. Friedman, eds. *Exploring the Unknown: Selected Documents in the History of the U.S. Civil Space Program.* Volume IV: *Accessing Space.* SP-4407. Washington, DC: NASA, 1994.

Longnecker, David E., and Ricardo A. Molins, eds. *A Risk Reduction Strategy for Human Exploration of Space: A Review of NASA's Bioastronautics Roadmap.* Washington, DC: National Academies Press, 2006.

Mackowski, Maura Phillips. *Testing the Limits: Aviation Medicine and the Origins of Manned Space Flight.* College Station: Texas A&M University Press, 2006.

Mains, Richard C., and Edward W. Gomersall. *Final Reports of U.S. Monkey and Rat Experiments Flown on the Soviet Satellite Cosmos 1514.* TM-88223. Moffett Field, CA: NASA-ARC, 1986.

Morey-Holton, Emily, P. D. Sebesta, Alan M. Ladwig, J. T. Jackson, and W. M. Knott III. *NASA Newsletters for the Weber Student Shuttle Involvement Project.* TM-101001. Moffett Field, CA: NASA-ARC, 1988.

Morgan, Clay. *Shuttle-Mir: The U.S. and Russia Share History's Highest Stage.* SP-2001-4225. Houston: NASA-JSC, 2001.

Mudgway, Donald J. *Uplink-Downlink: A History of the Deep Space Network.* SP-2001-4227. Washington, DC: NASA, 2001.

National Aeronautics and Space Administration. *Bioastronautics Roadmap: A Risk Reduction Strategy for Human Space Exploration.* SP-2004-6113. Washington, DC: NASA, February 2005.

National Aeronautics and Space Administration. *NASA Report to Congress Regarding a Plan for the International Space Station National Laboratory.* May 2007.

National Aeronautics and Space Administration. *National Aeronautics and Space Administration Biological and Physical Research Enterprise Strategy.* NP-2003-10-298-HQ. Washington, DC: NASA, 2003.

National Aeronautics and Space Administration. *Results of Life Sciences DSOs Conducted Aboard the Space Shuttle 1988–1990.* Houston: NASA-JSC, January 1991.

National Aeronautics and Space Administration. *Results of Life Sciences DSOs Conducted Aboard the Shuttle 1991–1993.* Houston: NASA-JSC, July 1994.

National Aeronautics and Space Administration. *Science in Orbit: The Shuttle & Spacelab Experience: 1981–1986.* NP-119. Huntsville, AL: NASA-MSFC, 1988.

National Aeronautics and Space Administration. Task Force on Countermeasures, Final Report, May 1997. Space Life Sciences Division. Available online at "NASA Documents Online," NASA Headquarters Library, http://www.hq.nasa.gov/office/hqlibrary/find/nasadoc.htm.

National Commission on Space. *Pioneering the Space Frontier.* New York: Bantam Books, 1986.

National Research Council, Committee on Advanced Space Technology. *Space Technology to Meet Future Needs.* Washington, DC: National Academies Press, 1987.

National Research Council, Committee on Advanced Technology for Human Support in Space. *Advanced Technology for Human Support in Space.* Washington, DC: National Academies Press, 1997.

National Research Council, Committee on Human Exploration of Space. *Human Exploration of Space: A Review of NASA's 90-Day Study and Alternatives.* Washington, DC: National Academy Press, 1990.

National Research Council, Task Group on Life Sciences. *Space Science in the Twenty-First Century: Imperatives for the Decades 1995 to 2015.* Washington, DC: National Academies Press, 1988.

Nicogossian, Arnauld E., comp. *The Apollo-Soyuz Test Project Medical Report.* SP-411. Washington, DC: NASA, 1977.

Nicogossian, Arnauld E., Stanley R. Mohler, Oleg G. Gazenko, and Anatoliy I. Grigoryev, eds. *Space Biology and Medicine,* Vol. III: *Humans in Spaceflight, Book 1.* Reston, VA: American Institute of Aeronautics and Astronautics, 1996.

Nicogossian, Arnauld E., Stanley R. Mohler, Oleg G. Gazenko, and Anatoliy I. Grigoryev, eds. *Space Biology and Medicine,* Vol. III: *Humans in Spaceflight, Book 2.* Reston, VA: American Institute of Aeronautics and Astronautics, 1996.

Nield, George C., and Pavel Vorobiev, eds. *Phase 1 Program Joint Report.* SP-1999-6108. Houston: NASA-JSC, 1999.

Newkirk, Roland, and Ivan D. Ertel with Courtney G. Brooks. *Skylab: A Chronology.* SP-4011. Washington, DC: NASA, 1977.

Oberg, James. *Star-Crossed Orbits: Inside the U.S.–Russian Space Alliance.* New York: McGraw-Hill, 2002.

Pitts, John A. *The Human Factor: Biomedicine in the Manned Space Program to 1980.* SP-4213. Washington, DC: NASA, 1985.

Portree, David S. F. *Mir Hardware Heritage.* JSC 26770. Houston: NASA-JSC, October 1994.

Presidential Commission on the Space Shuttle Challenger Accident. *Report to the President.* Washington, DC: Presidential Commission on the Space Shuttle Challenger Accident, 1986.

Quayle, Dan. *Standing Firm: A Vice-Presidential Memoir.* New York: HarperCollins, 1994.

Rajulu, Sudhakar L., and Glenn K. Klute. *Anthropometric Survey of the Astronaut Applicants and Astronauts from 1985 to 1991.* RP-1304. Houston: NASA-JSC, 1993.

Report by the NASA Biological and Physical Research Research Maximization and Prioritization (ReMAP) Task Force to the NASA Advisory Council August 2002. Washington, DC: NASA-OBPR, 2002.

Rumerman, Judy A., comp. *NASA Historical Data Book,* Vol. V: *NASA Launch Systems, Space Transportation, Human Spaceflight, and Space Science 1979–1988.* SP-4012. Washington, DC: NASA, 1999.

Rumerman, Judy A., comp. *NASA Historical Data Book,* Vol. VI: *NASA Space Applications, Aeronautics and Space Research and Technology, Tracking and Data Acquisition/Support Operations, Commercial Programs, and Resources 1979–1988.* SP-4012. Washington, DC: NASA, 2000.

Rumerman, Judy A., comp. *NASA Historical Data Book,* Vol. VII: *NASA Launch Systems, Space Transportation, Human Spaceflight, and Space Science, 1989–1998.* SP-4012. Washington, DC: NASA, 2009.

Sadeh, Eligar, ed. *Space Politics and Policy, An Evolutionary Perspective.* Dordrecht: Kluwer Academic Publishers, 2002.

Sahm, P. R., M. H. Keller, and B. Schiewe. *Scientific Results of the German Spacelab Mission D-2.* Bonn: Deutsche Agentur für Raumfahrtangelegenheiten and Göttingen: DLR, 1996.

Sawin, Charles F., Gerald R. Taylor, and Wanda L. Smith, eds. *Extended Duration Orbiter Medical Project: Final Report 1989–1995.* SP-1999-534. Houston: NASA-JSC, 1999.

Seedhouse, Erik. *Tourists in Space: A Practical Guide.* Chichester, UK: Praxis Publishing, 2008.

Sietzen, Frank, Jr., and Keith L. Cowing. *New Moon Rising: The Making of America's New Space Vision and the Remaking of NASA.* Burlington, ON: Apogee Books, 2004.

Snyder, Robert S., Percy H. Rhodes, and Teresa Y. Miller. *Continuous Flow Electrophoresis System Experiments on Shuttle Flights STS-6 and STS-7.* TP-2778. Huntsville, AL: NASA-MSFC, 1987.

Sonnenfeld, Gerald, ed. *Experimentation with Animal Models in Space.* Amsterdam: Elsevier, 2005.

Souza, Kenneth, Guy Etheridge, and Paul X. Callahan, eds. *Life Into Space: Space Life Sciences Experiments, Ames Research Center, Kennedy Space Center 1991–1998: Including profiles of 1996–1998 Experiments.* SP-2000-534. Moffett Field, CA: NASA-ARC, 2000.

Souza, Kenneth, Robert Hogan, and Rodney Ballard, eds. *Life Into Space: Space Life Sciences Experiments, NASA Ames Research Center 1965–1990.* RP-1372. Moffett Field, CA: NASA-ARC, 1995.

Space Science Division Annual Report 2002–2003. Moffett Field, CA: NASA-ARC.

Space Station Redesign Team. *Final Report to the Advisory Committee on the Redesign of the Space Station.* TM-109241. Washington, DC: NASA, 1993.

Spidle, Jake W., Jr. *The Lovelace Medical Center: Pioneer in American Health Care.* Albuquerque: University of New Mexico Press, 1987.

Sulzman, Frank M., and A. M. Genin, eds. *Space Biology and Medicine* Vol. II: *Life Support and Habitability.* Reston, VA: American Institute of Aeronautics and Astronautics and Moscow: Nauka Press, 1994.

Sutton, Jeffrey P., Bobby R. Alford, Jeffrey R. Davis, and Jefferson D. Howell Jr. *National Space Biomedical Research Institute Revised Strategic Plan, July 17, 2003.* www.nsbri.org/About/StrategicPlan.pdf.

Swenson, Loyd S., Jr., James M. Grimwood, and Charles C. Alexander, *This New Ocean: A History of Project Mercury.* SP-4201. Washington, DC: NASA, 1966.

Technical & Administrative Services Corporation. *Space Station Freedom Media Handbook.* Washington, DC: NASA-OSSD, 1992.

Thomas, Kenneth S., and Harold J. McMann. *U.S. Spacesuits.* Berlin: Springer Praxis, 2005.

US Congress, Office of Technology Assessment. *Civilian Space Stations and the U.S. Future in Space.* OTA-STI-241. Washington, DC: GPO, November 1984.

US Department of Defense. *Military Handbook, Anthropometry of U. S. Military Personnel (Metric).* DOD-HDBK-743A. Washington, DC: Department of Defense, 1991.

Wolverton, B. C., and John D. Wolverton. *Growing Clean Water: Nature's Solution to Water Pollution*. Picayune, MS: Wolverton Environmental Systems, Inc., 2001.

Weitekamp, Margaret A. *Right Stuff, Wrong Sex: America's First Women in Space Program*. Baltimore: Johns Hopkins University Press, 2004.

Index

Page numbers in *italics* indicate illustrations and tables.

Abbey, George, 107, 166, 313n90, 345n67
Abrahamson, James, 45–46, 66. *See also* Space stations: speculative
Advisory Committee on Redesign of Space Station, 103–4, 202, 203, *205*, 206, 326n42, 327n51, 329n76
Advisory Committee on the Future of the U.S. Space Program (AC-FUSSP), 40, 181–82, 318n42, 325–26n32. *See also* Space Exploration Initiative
Aerospace Medicine Advisory Committee, 183–85, 189. *See also* Radiation research
Agenzia Spaziale Italiana. *See* Italian Space Agency
Aging, 49–52, 94, 132, 277n18. *See also* Glenn, John H.; Neurolab
Agriculture, Department of, U.S., 146–47, 198. *See also* Animal rights: protective legislation
Air Force, U.S., 46, 81, 82, 83, 105, 145, 193, 302n34; animal tests and, 175–76, 186, 188, 235, 316n20, 320n62; CAM and, 176, 316n20; radiation studies by, 175–76, 178–79, 186, 188, 314n7, 316n20, 320n62; space stations and, 193, 207; USAFSAM, 48, 80, 210, 289n66, 320n62, 328n64, 329n74
Alford, Bobby R., 163–64, 166, 188–89. *See also* National Space Biomedical Research Institute

Allen, Joseph P., 30, 40, 284n87, 291n1. *See also* Faget, Maxime; Industrial Space Facility; Space Industries, Inc.
American Institute of Aeronautics and Astronautics, 265–66. *See also* Space Exploration Initiative
American Institute of Biological Sciences, 112, 208, 231, 268. *See also* Peer review
American Society for Gravitational and Space Biology, 85–86, 266, 348n21. *See also* LifeSat
Ames Research Center (ARC), 68, 150, 154, 200, 253, 255, 286–87n32; animal rights groups and, 146, 231–32; animal studies, 9–11, 136, 147, 174, 211; Astrobiology Institute, 155, 272, 310n50, 349n5; bone loss studies, 9, 49; CELSS research and, 58, 59, 61, 62, 289n65; centrifuges, 88, 138–39, *139*, 208, 211–12, 249, *250*; circadian rhythm studies, 48, 51; Cosmos/Bion program, 215, 216, 231, 234, 338n86; ISF and, 27, 40; LifeSat and, 85–86, 87, 88; Neutral Buoyancy Test Facility, 53, *54*; organization of, *69*, 154, 155, 157; radiation studies, 87, 150, 174, 187; Russia, collaboration with, 187, 220, 231; Soviet Union, collaboration with, 174, 215, 216; space suit design, 52–53, *54*; SSIP and, 9–12
Analog studies, 92, *93*, 94, 99, 296n63. *See also* Operational Medicine
Animal Care and Use Committees, 147, 231, 234. *See also* Animal rights

Universities Space Research
Association
NASA Life Sciences Advisory Committee,
108, 109, 138, 208, 258
NASA Specialized Center of Research
and Training (NSCORT), 65, 178, 182,
183, 317n30. *See also* NASA: university
collaboration
National Academies of Science, 37, 113,
148; Space Studies Board, 66, 68
National Advisory Committee for Aeronautics (NACA), 168, 170–71, 271, 272,
349n2
National Commission on Space, 212, 260,
329n77. *See also* Paine, Thomas
National Institutes of Health, 3, 12–13,
50–52, 111, 128, 146, 151–52, *151*, 164;
animal tests and, 231, 235; potential
space station user, as, 198
National Oceanographic and Atmospheric
Administration, 183, 187, 198, 238. *See
also* Radiation research
National Research Council, 138, 167, 177,
199; NASA funding and, 208–11; NGO
management of ISS, 244–47; privatization of space and, 37, 40; Space Studies
Board of, 128, 208
National Science Foundation, 3, 111, 128,
164, 349n5
National Space Biomedical Research
Institute (NSBRI), 151–66, *161*; Alford,
Bobby R. and, 163–64, 166, 188–89;
funding, 152–54, 156–60, 163–64,
165–66; Goldin, Daniel S. and, 152,
154, 155, 158, 163, 166; organization of, 160–63; radiation research
and, 188–89; space privatization and,
153, 159, 164; ZBR and, 154–55, 158,
309n43, 309–10n45
National Space Development Agency of
Japan. *See* Japan and space
Navy, U.S., 78, 80–82, 90, 92, *93*, 100,
176, 286n26, 296n63. *See also* Analog
studies; Orbiters, shuttle: safety of
Neurolab, 118, 127–36, *133*, *135*, 149, 157;
animal experiments, 128, 131–35, *131*,
136, 235

Neurological/neurovestibular research,
132–34, 147
Nicogossian, Arnauld, 50, 66, 189, 215,
234, 244, 318–19n47. *See also* NASA:
Office of Life and Microgravity Sciences and Applications
Noise, 11, 96–97, *125*, *126*, 135, 199,
225, 249, 325n25. *See also* Contamination, environmental
Nongovernmental Organizations
(NGOs), 148, 243–48. *See also* Federally Funded Research and Development Centers
North American Rockwell, 87; CERV
and, 102, 103; shuttle orbiter safety
studies, 75, 80, 81
Nuclear Regulatory Commission. *See*
Atomic Energy Commission

Oak Ridge Institute of Nuclear Studies
(ORINS), 143, 144, 145, 272. *See also*
Test subjects, human
Odor, control of in space, 11, 57,
121, 198. *See also* Contamination,
environmental
O'Keefe, Sean, 240, 247; VSE and, 14,
262–63
Operational Medicine, 111, 182, 183,
211; planning for, 90–92, 94–97, 99,
182, 183. *See also* Medical complica-
tions of spaceflight
Orbiters, shuttle: design of, 42, 44–45,
120; escape systems, 45, 75–76,
77, 78, 79, 80, 81, 82, 84, 291n1;
expected flight rate, 284n108;
extended duration missions, *119*, *120*,
301n25; follow-ons to, 81, 98; safety
of, 75, 76, 78, 80, 81, 82, 83; shuttle
stack, destruction of, 82, 83, 84
Ortho Pharmaceuticals, 17, 18, 19, 23,
24, 26, 281n79. *See also* Johnson &
Johnson
Orthostatic Intolerance (OI), 48–49, 51,
114, 115, 116, 134. *See also* Countermeasures; Extended Duration
Orbiter Medical Project; Gender
differences, astronaut

funding, 74, *137*, 159, 199, 200, 201, 204, 206; cancellation threatened, 74; centrifuge on, 138–42, *139*, *140*, 190, 208, 212; contamination, environmental, 90, 95–97; crew size, 103, 204; emergency rescue vehicles and, 97–107; free flyers and, 27–37, 40; international participation and, 202–4, 206; medical/health sciences and, 90–91; National Laboratory, 113; other options, comparison with, *32*, *33*, *34*; power for, 201, 204, 206; promotion of, 26–27, 29, *36*; redesign of, 201, 202–4, *205*, 206; safety of, 68, 99–100, 238

Space stations: AG, 207; CELSS and, 62; crew size, 46, 201; design of, 193, 197–99, 201, 207; medical facilities of, 199; political purpose of, 74; predictions about, 46, 58–59; Skylab, 196; Soviet, 194–97; speculative, 62, 193–95, 207, 326n33; users, potential, 198

Space suits, 75, 80; AX series, 52, *54*; Biomedical Instrumentation Port, 80; designers of, 53, 57, 58; design issues, 42, 44, 52–53; gloves, 55, *56*; hard suits, 52–53; LES, 80; maintainability of, 53; Mark III series, *54*; prebreathing and, 53; radiation resistance of, 173; Russian, 57, 58; testing of, 53–55

Space Telescope Science Institute, 155, 272. *See also* National Space Biomedical Research Institute; Nongovernmental Organizations

Stennis Space Center, John C., 58, 65. *See also* Closed Environmental Life Support Systems

Stofan, Andrew, 31, 99, 144. *See also* International Space Station

Student programs, 238–39; experiments, 110, 149, 238–39, 242–43, 277n20; outcomes, 277n19. *See also* Get Away Special program; Shuttle Student Involvement Program; Universities Space Research Association

Success, defining, 7, 191; CFES, 19–20; Columbia, STS-107, 335–36n62; NSBRI, 163–65; Shuttle-Mir, 224, 226–28

Teacher in Space program, 11, 13, 25, 276n9. *See also Challenger* accident

Techshot, Inc., 13, 278n21. *See also* Shuttle Student Involvement Program; Vellinger, John

Test subjects, animal, 95, 145, 211, 215, 249; alternatives to, 146–47; challenges to using, 9–11, 95–96, 132, 256, 258–59, 304n62; Cosmos/Bion program, 214–17, 230–35; LifeSat, 86–87; military, 175–76; NASA funding for, 147–48; Neurolab, 128, 131–36; public reaction to, 225, 231, 235, 306–7n14, 344–45n60; purpose of, 145; student experiments with, 9–13, 277–78n20. *See also* Centrifuge, space; Cosmos/Bion program; Fish, as test subjects; Insects, as test subjects; Shuttle Student Involvement Program; U.S. Air Force School of Aerospace Medicine

Test subjects, human, 42, 80, 90–92, 95, 113–14, 123, 128, 140, 143–45. *See also* Centrifuge, space; Cobb, Jerrie; Extended Duration Orbiter Medical Project; Glenn, John H.; Human subjects research; Lovelace, Randolph, II; Medical ethics; Mercury Thirteen; Shuttle-Mir program; U.S. Air Force School of Aerospace Medicine

Tethers, 141, 212–13. *See also* Artificial Gravity; Countermeasures, exercise

Thagard, Norman, 187, 220, 221. *See also* Shuttle-Mir program

Translational Research Institute. *See* National Space Biomedical Research Institute

Transport codes (models), 176–79, 186, 191. *See also* Radiation research

Truly, Richard, 81, 101, 136, 185, 202, 291n87, 318n42, 325–26n32. *See also* Crew Emergency Rescue Vehicle; Space Exploration Initiative

Ukraine and space, 241–43. *See also* NASA: international collaboration

United Kingdom and space, 150, 168, 203, 222, 291n1

Maura Phillips Mackowski is a research historian based in Arizona. Before earning her MA in historical studies from the University of Maryland–Baltimore County and PhD in US history from Arizona State University she was a freelance writer specializing in science, technology, and high-tech business. Maura also worked in the personnel department of a St. Louis–area steel service center, earning an MA in management from Webster University there. She is the author of *Testing the Limits: Aviation Medicine and the Origins of Manned Space Flight*.

Printed in the United States
by Baker & Taylor Publisher Services